Spectrum Sharing in Cognitive Radio Networks

Shweta Pandit · Ghanshyam Singh

Spectrum Sharing in Cognitive Radio Networks

Medium Access Control Protocol Based Approach

Springer

Shweta Pandit
Jaypee University of Information Technology
Solan, Himachal Pradesh
India

Ghanshyam Singh
Department of Electronics and Communication Engineering
Jaypee University of Information Technology
Solan, Himachal Pradesh
India

ISBN 978-3-319-53146-5 ISBN 978-3-319-53147-2 (eBook)
DOI 10.1007/978-3-319-53147-2

Library of Congress Control Number: 2017930956

Printed on acid-free paper

This Springer imprint is published by Springer Nature
The registered company is Springer International Publishing AG
The registered company address is: Gewerbestrasse 11, 6330 Cham, Switzerland

Preface

Presently, we rely on the wireless devices and systems to not only enable on-demand, pervasive communications for a large proportion of the population, but also other critical application areas such as scientific and medical research, industrial control and automation, and public safety. As the wireless communication systems and its applications continue to flourish, the demand for precious spectrum resources will continue to grow. In the foreseeable future, we expect that the demand for spectrum will continue to increases as new wireless technologies and applications with high data throughput requirements continue to emerge. This voracious enthusiasm for additional spectrum resources cannot be met by simply allocating new spectrum. The usable capacity of spectrum must be expanded with innovative technologies, regulatory reforms, and removal of market barriers. The cognitive radio is one of the innovative technologies that have the potential to effectively address the spectrum shortage problem and radically change the way we utilize spectrum. Due to its potential impact, various stakeholders—including regulatory policymakers, wireless device manufacturers, telecommunication operators, and academic researchers—have shown strong interest in it, especially with respect to research and development.

Therefore, the cognitive radio has emerged as a prone candidate for exploiting the increasingly flexible licensing of the wireless communication system. The regulatory bodies have come to realize that most of the time, a large portion of certain licensed frequency band remain empty/unused. To remedy this, new regulations would allow for devices which are able to sense and adapt to their spectral environment, such as cognitive radio to become secondary user and such users are wireless devices that opportunistically employ the spectrum already licensed to the primary users. The primary users generally associated with the primary spectral licensed holder and thus have higher priority right to the spectrum. The intuitive objective behind secondary spectrum licensing is to improve the spectral efficiency of the network, whereas depending on the type of licensing and not affecting higher priority users.

In the cognitive radio network, the medium access control (MAC) protocols play an important role to exploit the spectrum opportunities, manage the interference to the primary users and coordinate. The dynamic leasing, in which some wireless devices opportunistically employ the spectrum rather than choose for a long-term sub-lease. In order to exploit the spectrum, we require a device which is able to sense the communication opportunity and then take actions based on the sensed environment. The cognitive radio offers a novel way of solving spectrum under-utilization problems. The emergence of FCC's secondary market initiative, it has been brought on by both the obvious desire for spectral efficiency, as well as empirical measurements showing that most of the time certain licensed frequency remain unused. The goal of secondary market initiative is to remove unnecessary regulatory barriers to new secondary market oriented policies such as

(1) Spectrum leasing, which allow non-licensed users to lease any part, or the entire spectrum from the licensed user.
(2) Dynamic spectrum leasing, which is a temporary and opportunistic usage of spectrum rather than a large term sub-lease.
(3) Private commons, which a licensee could allow non-licensed user access to his/her spectrum without a contract, optional with access fee.
(4) Interruptible spectrum leasing, which would be suitable for a lesser that wants a high level of assurance that any temporally in use, or leased, to an incumbent cognitive radio could be efficiently reclaimed if needed. A prime example would be the leasing of the generally unoccupied spectrum allocated to the government or local enforcement agencies, which in time of emergency could be quickly reclaimed.

This book puts together a rich set of research articles featuring recent advances in theory, design, and analysis of cognitive radio wireless communication networks. The book consists of 10 chapters, which cover a wide range of topics related to the cognitive radio technology, in particular, the topics covered in this book include fundamental challenges and issues in designing cognitive radio systems, information-theoretic analysis of cognitive radio systems, spectrum sensing and co-existence issues, adaptive physical layer protocols and link adaptation techniques for cognitive radio, different techniques for spectrum access by distributed cognitive radio, cognitive medium access control (C-MAC) protocols. The book is organized as follows. Chapter 1 provides a comprehensive survey with state of the art of the various spectrum sharing techniques and the fundamental issues related to cognitive radio design and the major research challenges mostly from a signal processing and communication-theoretic perspective. The potential advantages, limiting factors, and characteristic features of the existing cognitive radio spectrum sharing domains are thoroughly discussed. As the complexities of wireless technologies increase, novel multidisciplinary approaches for the spectrum sharing/management are required with inputs from the technology, economics and regulations. To identify the available spectrum resource, decision on the optimal sensing and transmission time with proper coordination among the users for spectrum access are the important characteristics of spectrum sharing methods.

The spectrum sensing is the key requirement and one of the most challenging issues of the cognitive radio system. In this context, Chap. 2 presents a survey of the physical layer spectrum sensing techniques for cognitive radios. The major challenges in spectrum sensing are outlined and several techniques for improving spectrum sensing performance are discussed. Further, a hybrid model for non-cooperative spectrum sensing has been presented, with this terminology the proper channelization of the three techniques has been performed with relevant discussion. The presented approach helps in detecting the idle spectrum opportunistically with better utilization of the spectrum under non-cooperative sensing with enhanced spectrum efficiency. We have also explored the sensing under cooperative environment. The presented approach helps in detecting the idle spectrum bands (spectrum holes that is the underutilized sub-bands of the radio spectrum) opportunistically with better utilization of the spectrum under non cooperative sensing with increase in the overall spectrum efficiency.

In Chap. 3, we have proposed a novel multichannel cooperative MAC protocol for the distributed cognitive radio network which has the back-off algorithm for contention solving among the competing cognitive users. The back-off algorithm for resolving collision among the competing users has allowed the collided cognitive users to become successful by selecting another contention slot from the increased contention window. The increased number of successful users has enhanced the throughput of the cognitive radio network by transmitting their data over the detected idle licensed channels. Moreover, the optimum number of contention slots have been achieved which has maximized the number of successful cognitive users as well as throughput.

In Chap. 4, the cognitive radio MAC protocol in practical scenario is considered and the perfect and imperfect sensing effect on the performance of throughput and energy efficiency of the cognitive radio network is presented. The imperfect sensing resulted due to false alarm has affected the system performance of cognitive radio network by missing the opportunities of spectrum use in comparison to the perfect sensing, as demonstrated in the simulation results. In addition to this, the optimum number of contention slots has been obtained for the proposed MAC protocol which has avoided contention slots throughput tradeoff problem. Moreover, the performance of MAC protocol for different licensed channels utilization probability has been simulated. The simulation results have illustrated that throughput and energy efficiency of the MAC protocol for imperfectly sensed environment is less as compared to that of the perfect sensing scenario and the interference to the primary user is less in the proposed protocol for lower values of miss detection probability.

In Chap. 5, the scheme for maximizing the bandwidth efficiency by util.ng the wasted bandwidth of the licensed channels in the distributed cognitive radio MAC protocol has been proposed. In addition to this, the contention resolving algorithm has been also applied in this proposed bandwidth maximization scheme as discussed in Chap. 3. Further, the bandwidth wastage in the cooperative distributed MAC protocol has been minimized by transmitting data of the cognitive users over the idle licensed channels, which are unutilized in the sensing-sharing and contention interval. The proposed technique has significantly enhanced the throughput of the cooperative distributed network. Moreover, the comparison of the proposed scheme in this chapter has been performed with the SMC-MAC protocol.

Chapter 6 concerns the energy efficiency of cognitive radio terminal and have obtained the optimum transmit power for the cognitive terminal at which the energy efficiency is maximum. It is further shown that the complexity of proposed algorithm for computing the optimum transmit power is very less. We have considered different scenario of channel conditions at different channel gain and have maximized the energy efficiency of the cognitive radio terminal.

In Chap. 7, a technique to eliminate the sensing-throughput trade-off of the conventional approach by increasing throughput of cognitive radio user and simultaneously reducing interference with the primary users has been explored. This presented technique is also reducing the data loss rate by decreasing collision of frames of primary and secondary users. Finally, the simulation results have been provided which is compared with conventional and proposed approach. From these simulation results, it is demonstrated that the throughput is more for proposed approach as compared to that of the conventional approach.

In Chap. 8, we have explored an optimal power allocation scheme for the spectrum sharing with imperfect channel state information between the cogni-tive/secondary user (CU) and licensed/primary user (PU) over Rayleigh fading environment. We have analyzed the ergodic capacity of CU link under the combi-nation of peak transmit power and peak/average interference power constraints with or without primary user interference. In addition to this, the outage capacity with multiple primary user interference is also analyzed with the error variance under the joint peak transmit power and peak interference power constraint as well as individual peak interference power constraint. Moreover, the power expenditure is also investigated to achieve the lower limit of ergodic and outage capacity. The minimum mean square channel estimation technique is used for the channel estimation between CU and PU. However, the convex optimization method is used for the optimal power allocation.

In Chap. 9, we have considered two adaptation policies for spectrum sharing in cognitive radio such as power adaption policy and rate and power adaptation for multilevel quadrature amplitude modulation (M-QAM) format. We have obtained the channel capacity for both these policies under Rayleigh and Rician fading environment. The rate and power of secondary transmitter is varied based upon the channel state information (CSI) of the secondary link and sensing information, which shows the activity of the primary user. We also considered the channel fading in between the secondary user and primary user and obtained the secondary transmitter power adaption policy for Rayleigh and Rician fading environment under peak transmit power and interference power constraint.

Chapter 10 presents a cross-layer optimized design framework for cognitive radios in a dynamic spectrum access environment. In general speaking, layered architectures like Open Systems Interconnection (OSI) and Transmission Control Protocol (TCP) models forbids direct communication between the non-adjacent layers and communication between the adjacent layers is also limited in such a way that higher layer protocol only makes use of the services at the lower layers and is not concerned about the details of how the service is being provided. This in turn becomes bottleneck for new emerging wireless services. Therefore, cross-layer optimization work related to wireless and cognitive radio network has been reviewed in this chapter. In addition, MAC layer parameters optimization with the help of cross-layer interaction has been emphasized and various challenges in this interaction have been presented.

In summary, the book provides a unified view of the state of the art of cognitive wireless communications and networking technology, which should be accessible to a readership with basic knowledge about wireless communications and telecommunications networking. The readership may find the rich set of references in each of the chapters very useful. The authors have performed a good job by providing a concise summary of all the chapters at the preface of the book. I would strongly recommend the book to graduate students and researchers and engineers working or intending to work in the area of cognitive radio.

Although numerous journal/conference publications, tutorials, and books on cognitive radio have been published in the last few years, the vast majority of them focus on the various physical-layer attributes of the technology. More importantly, these technical publications discuss the cognitive radio in isolation, essentially as a standalone system or network, with little regard for how it may interact with legacy wireless systems or how heterogeneous cognitive radio systems may collaborate with each other. Although this book's main theme is cognitive radio, its specific focus areas are quite different from the existing literature. The prime intent of this book is to provide a comprehensive discussion on how cognitive radio technologies can be employed to enable efficient wireless communication system. In other words, the discussions in this book revolve around how cognitive radio technologies can be used to enable various wireless networks to coexist and efficiently share spectrum. The intended readership of this book includes wireless communications industry researchers and practitioners as well as researchers in academia. The readership is assumed to have background knowledge in wireless communications and networking, although they may have no in-depth knowledge of cognitive radio technologies. The intention of this book is to introduce communication generalists to the technical challenges of the various coexistence techniques and mechanisms as well as solution approaches which are enabled by cognitive radios.

This book distinguishes itself from the existing prosperous literature of cognitive radio networks. The existing literature presents a self-contained introduction of the emerging cognitive radio networking paradigm outlining the theoretical fundamentals and requirements for enabling such a technology. The emphasis of such books is on the theoretical design, optimization, and performance evaluation of opportunistic spectrum access in cognitive radio networks.

The main challenge of existing distributed opportunistic spectrum management schemes is that they do not consider the unavoidable practical limitations of today's cognitive radio networks such as the inability to measure the interference at the primary receivers. Consequently, optimizing the constrained cognitive radio network performance based only on the local interference measurements at the cognitive radio senders does not lead to truly optimal performance due to the existence of hidden or exposed primary senders. More specifically, the existing schemes have a cognitive radio sender decide its transmission strategy based on its local interference measurement—while such decisions should have been made based on the interference measurement at the nearby primary receivers to be interfered with its transmission. However, there does not exist a practical mechanism that enables a cognitive radio to determine the interference at nearby primary receivers. Furthermore, the existing transceiver technologies and spectrum measurement techniques are incapable of accurately assessing the spectrum usage over a wide frequency range due to the limitations imposed by the transceiver hardware.

This book is an extension of the Ph.D. dissertation of Dr. Shweta Pandit submitted to the Jaypee University of Information Technology Waknaghat, Solan under the supervision of Dr. Ghanshyam Singh. This book targets a wide range of readers including but not limited to researchers, industry experts, and senior undergraduate as well as graduate students. On the one hand, the readers with theoretical interests will experience an unprecedented treatment of the conventional cognitive radio network performance optimization problem that takes into account the practical limitations of recent technologies. Further, the readers interested in real-life distributed cognitive radio network realization will be exposed to a first-of-its-kind clean-slate implementation approach that demonstrates the significant multi-faced performance improvement. This book offers the reader a range of interesting topics portraying the current state of the art in cognitive radio technologies. In simple terms, while several existing opportunis-tic spectrum access approaches have been developed and theoretically optimized, they are challenged by the inherent constraints of practical implementation technologies. Analyzing these constraints and proposing an attractive and practical solution to counter these limitations are the basic aims of this book.

The authors are indebted to a numerous colleagues for the valuable suggestions during the entire period of manuscript preparation and especially thankful to the Professor B.N. Basu, and Professor S.K. Kak, IIT (BHU), India, for the motivation. We would like to acknowledge and thank a number of colleagues as well as M. Tech. and Ph.D. students who have made this book possible. In particular, we would like to acknowledge Ms. Shipra Kapoor and Ms. Bindu Bharti who have provided invaluable inputs that helped shape the contents of this book. The authors would also like to thank publishers at Springer, in particular Brian P. Halm, Charles B. Glaser, and Nicole Lowary for their helpful guidance and encouragement during the creation of this book. The authors would not justify their work without showing the gratitude to their family members who have always been the source of strength to tirelessly work and to accomplish the assignment.

Special Acknowledgements

We sincerely thank the authorities of Jaypee University of Information Technology, Wakanaghat, India for their kind support to come up with this book.

Solan, India

Shweta Pandit
Ghanshyam Singh

Contents

About the Authors

Shweta Pandit received a B.Tech. (Honors) degree from Himachal Pradesh University, Shimla, India, in 2010. She also received the M.Tech. and Ph.D. degrees in electronics and communication engineering from the Department of Electronics and Communication Engineering, Jaypee University of Information Technology, Waknaghat, Solan, India, in 2012 and 2015, respectively. She is currently working as an assistant professor in the Department of Electronics and Communication Engineering, Jaypee University of Information Technology. Her area of research interest is next-generation communication systems, cognitive radio, wireless networks, and capacity enhancement and interference reduction in wireless networks.

Prof. Ghanshyam Singh received a Ph.D. degree in electronics engineering from the Indian Institute of Technology, Banaras Hindu University, Varanasi, India, in 2000. He has served as research scientist with the Central Electronics Engineering Research Institute, Pilani, and the Institute for Plasma Research, Gandhinagar, India. He has also worked as assistant professor at the Electronics and Communication Engineering Department, Nirma University of Science and Technology, Ahmedabad, India, and visiting researcher at Seoul National University, Seoul, South Korea. He is presently a professor with the Department of Electronics and Communication Engineering, Jaypee University of Information Technology, Waknaghat, Solan, India. He is an author/co-author of more than 200 scientific papers in refereed journals and international conferences. His research and teaching interests include RF/microwave engineering, millimeter/terahertz-wave antennas and their applications in communications and imaging, next-generation communication systems (OFDM and cognitive radio), and nanophotonics. He has more than 18 years of teaching and research experience in the areas of electromagnetic/microwave engineering, wireless communication, and nanophotonics. He has supervised number of Ph.D. and M.Tech. theses and has worked as a reviewer for several reputed journals and conferences. He is the author of two books, Terahertz Planar Antennas for Next Generation Communication and MOSFET Technologies for Double-Pole Four-Throw Radio-Frequency Switch, published by Springer.

List of Figures

List of Tables

Chapter 1
Cognitive Radio Communication System: Spectrum Sharing Techniques

1.1 Introduction

Spectrum resource demand has greatly increased in recent years with the emergence of new wireless services and products, posing numerous challenges for service providers and regulators. Frequency allocation by regulatory bodies of different countries has allotted spectrum for various services through a static/fixed allocation scheme. This process aims to avoid frequency interference among users, which suggests that most of the frequency bands have already been assigned [1, 2]. However, actual spectrum usage measurements obtained by the Federal Communications Commission (FCC) Spectrum Policy Task Force [1] tell a different story. At any given time and location, much of the licensed spectrum lies idle. For example, utilization of frequency bands below 3 GHz ranges from 15 to 85%, and the frequency range above 3 GHz is even less utilized [3]. This inconsistency indicates that the spectrum shortage results from spectrum management policy rather than physical scarcity of usable frequencies. Therefore, spectrum is not scarce, but allocated spectrum is underutilized due to the fixed spectrum allocation scheme. This issue has sparked a flurry of exciting research activity within the engineering, economics, and regulatory communities in search of better spectrum management policies and techniques. One proposed solution for the next generation of wireless networks is a spectrum regulatory framework based on secondary spectrum access. In such a framework, secondary (unlicensed) systems coexist with primary (licensed) user systems and access spectrum on an opportunistic/sharing basis. This unutilized or underutilized spectrum of certain service providers/licensed users is known as “spectrum hole”. Figure 1.1 shows the utilization of spectrum between 9 kHz and 1 GHz [4].

Due to spectrum scarcity created by fixed spectrum allocation, new services trying to enter the communications arena may not be able to obtain enough spectrum to be commercially viable. The limitations of a fixed spectrum allocation-based scheme have been discussed in detail in [5]. Efforts to mitigate this

S. Pandit and G. Singh, *Spectrum Sharing in Cognitive Radio Networks*,
DOI 10.1007/978-3-319-53147-2_1

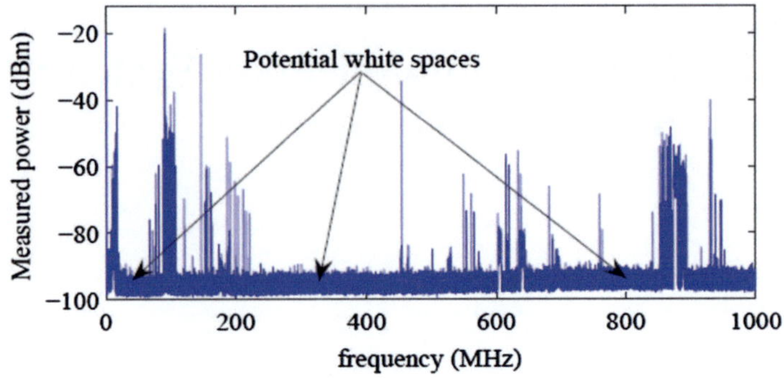

Fig. 1.1 Spectrum utilization measurements in 9 kHz–1 GHz band [4]

challenging issue have given rise to the evolution of the cognitive radio, which uses dynamic spectrum access (DSA) [6] and opportunistic spectrum access (OSA) [7] schemes rather than fixed spectrum allocation. DSA and OSA offer flexible methods of assigning spectrum to users by defining a set of techniques and models to support dynamic management of the spectrum. More broadly, these methods benefit society by enabling growth in wireless applications and services. Cognitive radio thus represents a promising wireless communication technology geared towards solving the spectrum scarcity problem by opportunistically identifying the unused portions of the spectrum, achieving optimal frequency band usage, and establishing communication through observation, learning, optimization, and intelligent adaptation, while ensuring that the licensed or primary users of the spectrum are not affected [5]. It is able to operate in multiple frequency bands, maximizes the utilization of limited radio spectrum, and accommodates an increasing number of services and applications in wireless communication systems. The driving force behind this cognitive radio technology is the new spectrum licensing methods initiated by the FCC. These methods are more flexible, allowing the unlicensed (or secondary/cognitive) users to access the spectrum provided that they do not interfere with licensed (primary) users.

In general, cognitive radio refers to a radio device with the ability to sense its radio frequency (RF) environment and to modify spectrum usage based on what it detects. In other words, the cognitive radio device senses the RF environment, analyzes resource availability, considers changes to its operation parameters, and then adapts to the changes it made. To increase its ubiquity, regulators and standard bodies have been putting policies and standards concerning cognitive radio and coexistence of secondary users (SU) with primary users (PU). Figure 1.2 illustrates the cognitive radio cycle and describes the functions of cognitive radio. Cognitive radio observes the RF environment and determines the device control parameters such as transmit power, carrier frequency, and modulation. Based on this decision, cognitive radio reconfigures itself for data transmission. Various research communities use different definitions of cognitive radio, and each community has a unique view on its defining features. According to communication theorists, cognitive radio is primarily about dynamic spectrum sharing, whereas networking/information technology researchers interpret cognitive radio as a device capable of cross-layer optimization. Computer scientists perceive it as a device capable of learning and adapting with assumed capabilities, and the hardware/RF community often views it as an evolutionary step from software-defined radio (SDR) [8–12]. The fundamental concept of cognitive radio has largely been adopted from SDR, which can operate on multiple frequency bands without any hardware modification. However, the selection of frequency band and operating parameters are manually controlled by the user through the software. In contrast to cognitive radio, in SDR the artificial intelligence component for learning and decision making is not available. Cognitive radio is SDR combined with the capability for sensing its environment and making decisions about parameters such as the modulation scheme and transmission power, without human intervention. A primary network is not aware of the cognitive network behavior, and it does not need any specific functionality to coexist with it. When a primary user transmission is detected, the secondary users should immediately react by changing their RF power, rate, codebook, and channel so that their transmissions do not degrade the primary user's quality of service (QoS).

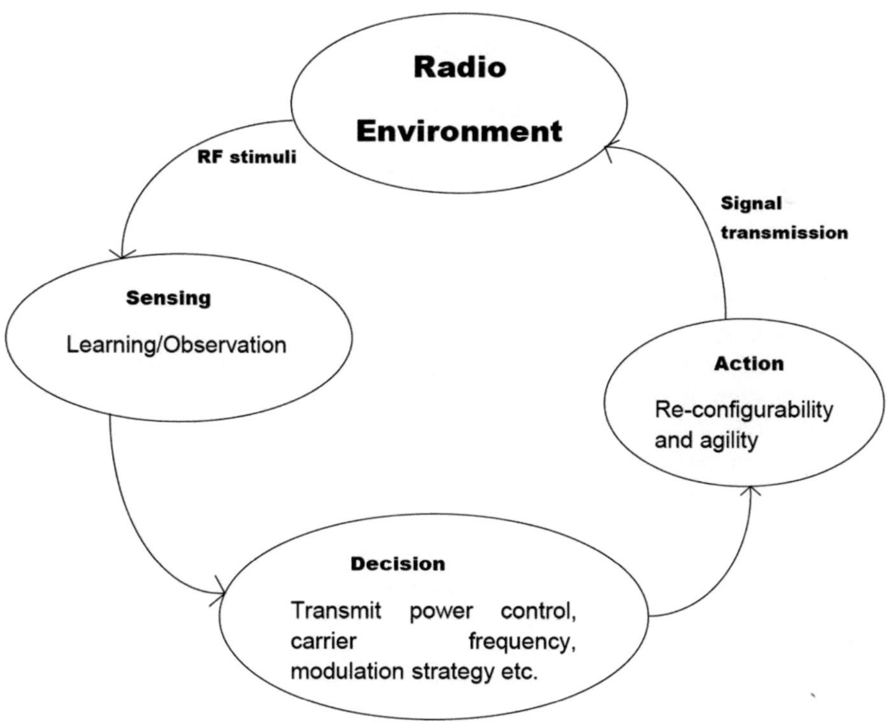

Fig. 1.2 The cognitive radio cycle

1.2 Functions of Cognitive Radio

The main functions of cognitive radio are centered on maintaining intelligent and efficient DSA, and are classified as follows [13].

1.2.1 Spectrum Sensing

One of the key requirements of cognitive radio networks is the ability to scan the entire frequency band for the presence of primary users. This function is known as spectrum sensing, and it is performed either locally by a cognitive user or collectively by a group of cognitive users. Spectrum sensing techniques play a very important role in cognitive radio networks, enabling them to be aware of their surroundings. Cognitive users utilize unused spectrum bands adaptively by enabling spectrum sensing and recording the activity of the primary and secondary users. Energy detection, matched filter detection and cyclostationary feature detection [14–21] are used to detect the activity of the primary and cognitive users. However, there is always a trade-off between accuracy and complexity when selecting the spectrum sensing technique for cognitive radio networks. The details of these techniques are discussed in Chap. 2.

In addition, available frequency bands are analyzed to determine their appropriateness for communication. Potential characteristics such as signal-to-noise ratio (SNR), link error rate, delays, interference, and holding time can be used to determine the most appropriate frequency band. After the selection of a frequency band, the cognitive user transmission in that band takes place. If a cognitive user/network detects a primary user transmission, it must vacate the corresponding frequency band and search for an unused frequency band, a process known as spectrum handoff [21] which is discussed in detail in Sect. 1.2.4. The delay associated with spectrum handoff makes cognitive radio networks vulnerable to various types of attacks. A cognitive radio transceiver detects unused spectrum, or a spectrum hole, and determines the method of accessing it without interfering with the transmission of a primary user. Spectrum sensing can be either centralized or distributed [20, 22]. In centralized spectrum sensing, a sensing controller (e.g., access point or base station) senses the target frequency band, and the information thus obtained is shared with other nodes in the system. This type of sensing can reduce the complexity of user terminals, because all sensing functions are performed at the sensing controller. However, centralized spectrum sensing suffers from location diversity. For example, the sensing controller may not be able to detect an unlicensed user at the edge of the cell. In distributed spectrum sensing, unlicensed users perform spectrum sensing independently, and the spectrum sensing results can be either used by individual cognitive radios (i.e., non-cooperative sensing) or shared with other users (i.e., cooperative sensing) [23–33]. Although cooperative sensing incurs communication and processing overhead, the accuracy of spectrum sensing is higher than that of non-cooperative sensing.

1.2.2 Spectrum Analysis

The information obtained from spectrum sensing is used to schedule and plan spectrum access by unlicensed users. In this case, the communication requirements of unlicensed users are also used to optimize the transmission parameters. The major components of spectrum management are spectrum analysis and spectrum access optimization [21]. In spectrum analysis, information from spectrum sensing is analyzed to gain knowledge about spectrum holes (e.g., interference estimation, duration of availability, and probability of collision with a licensed user due to sensing error). A decision to access the spectrum (e.g., frequency, bandwidth, modulation mode, transmit power, location, and time duration) is then made by optimizing the system performance, given the desired objective (e.g., maximizing the throughput of the unlicensed users) and constraints (e.g., maintaining the interference caused to licensed users below the target threshold).

1.2.3 Spectrum Sharing/Management

After a decision on spectrum access is made based on spectrum analysis, spectrum holes are accessed by the unlicensed users. Spectrum access is performed based on a cognitive medium access control (MAC) protocol, which is designed to avoid collision with the licensed users as well as with other unlicensed users [20, 21]. The cognitive radio transmitter must also negotiate with the cognitive radio receiver to synchronize the transmission to ensure that the transmitted data is received. A cognitive MAC protocol could be based on a fixed allocation MAC (e.g., frequency division multiple access [FDMA], time division multiple access [TDMA], code division multiple access [CDMA]) or a random access MAC (e.g., ALOHA [Additive Links On-line Hawaii Area], CSMA/CA) [21]. The DSA method significantly improves the utilization of frequency bands and enhances the performance of communication systems. A key component of DSA in cognitive radio technology is spectrum sharing, which allows for efficient and fair spectrum allocation or scheduling solutions among licensed and cognitive users. In the spectrum sharing model, the radio spectrum can be shared between a primary and cognitive user network simultaneously. In addition, unlicensed or cognitive users can opportunistically access the radio spectrum if it is unoccupied or not fully utilized by the primary users. Spectrum access by an unlicensed user is also allowed as long as it maintains an interference level at the primary receiver below a defined tolerable threshold so as not to interrupt the primary user's ongoing communication. Such sharing occurs without the primary user's awareness of the cognitive user, as the transmission of the cognitive user has a negligible impact on the operating conditions for which the primary user devices are designed. This spectrum sharing model is an attractive alternative, as it increases spectrum access and utilization while ensuring coexistence with existing legacy systems.

1.2.4 Spectrum Mobility

Spectrum mobility is a function related to a cognitive radio user's ability to change operating frequency bands. When a licensed user initially accesses a radio channel occupied by an unlicensed user, the unlicensed user can switch to a spectrum band that is idle. During this spectrum handoff, the protocol parameters at the different layers in the protocol stacks must be adjusted to match the new operating frequency band. The spectrum handoff must ensure that data transmission by the unlicensed user can continue in the new spectrum band. Spectrum mobility allows cognitive radio users to switch to unutilized frequency bands if the primary user appears during ongoing cognitive radio communication. However, this primary and secondary user mobility adds complexity to the cognitive radio network spectrum design. The presence or absence of a licensed channel for a stationary or pedestrian cognitive user in a particular location will be ambiguous when the licensed user is moving very rapidly. In addition, the sensing decision of a particular channel in a scenario may not be accurate for the fast-moving cognitive user, because the channel availability status at the current location of the cognitive user may be different. Therefore, frequent spectrum sensing is recommended for fast cognitive users to minimize false alarms and to increase the probability of detecting a licensed channel. In addition, efficient spectrum handoff techniques should be developed for cognitive users so that they will anticipate the channel when the primary user transmission on that channel begins. A Markov process has been utilized to predict the behavior of primary users based on past behavior, so that the cognitive user can vacate the channel before the primary user resumes its transmission, thus avoiding forced termination of the cognitive user's transmission [34]. This method of vacating the licensed channel is called proactive handoff. However, in reactive spectrum handoff, the spectrum is immediately vacated without the cognitive user's prior knowledge. Cognitive users should also reserve some of the idle channels to minimize disruption to their ongoing communication. This reallocation of the spectrum band to cognitive users can be done by either the central coordinator or the control channel in the distributed MAC protocol. In [35], inter-cell and intra-cell spectrum handoff techniques for cognitive users are proposed. In [36], the authors propose a protocol for an inbuilt spectrum mobility feature with the help of Poisson distribution.

1.3 Cognitive Radio Network Architecture

Cognitive radio networks based on the dynamic spectrum access technique may adopt either a cooperative or non-cooperative network architecture. Each cognitive node in the non-cooperative cognitive radio network architecture is responsible for its own decision. Therefore, it has minimal communication requirements (less overhead). However, spectrum utilization may be low. Further, within the cooperative, centralized cognitive radio network architecture, an integrated server retains a database of the spectrum availability and access information received from cognitive users. Therefore, spectrum management is simpler and enables efficient spectrum sharing. The dynamic spectrum access technique within the cooperative but distributed cognitive radio network architecture relies on coordinated local actions to achieve performance close to global optimal performance. The cognitive radios in this architecture periodically exchange information among themselves on their local environment, communication requirements, and performance, and use their local information as well as information received from their peers to determine their communication parameters. However, network performance may suffer from the hidden node problem and large control overhead. Additionally, the primary user in both the centralized and distributed scenarios may or may not cooperate. The communication protocols at the different layers need to perform such that the utilization of the radio spectrum is maximized while satisfying policy constraints. The important functions of the lower layers of the protocol stack in cognitive radio networks are outlined below, and are discussed in detail in Chap. 10.

Physical layer Spectrum sensing is the most important task of the physical layer, and includes functionalities such as detecting spectrum opportunities over a wide frequency band, estimating or predicting opportunity, and estimating interference at the primary receiver. Spectrum sensing involves the detection of spectrum holes across multiple dimensions, including time, frequency, space, and code. Further, in the physical layer, the RF-front end is implemented on the basis of SDR, and requires a high sampling rate, high-resolution analog-to-digital converters with a large dynamic range, multiple analog front-end circuitries, and high-speed signal processors [37].

MAC layer The MAC layer schedules the spectrum sensing activity and makes decisions regarding spectrum access on a channel. Transmission decisions are made by considering that the spectrum sensing could be imperfect, and decisions regarding the modulation and power level that should be used are again made by the MAC layer. Other important MAC layer tasks include synchronizing transmission parameters between transmitter and receiver, facilitating negotiation between primary and cognitive users, and facilitating spectrum trading functions [37].

Network layer The network layer primary tasks include topology construction, addressing, and routing. Under topology construction, spectrum detection, neighbor discovery, and topology management (e.g., spectrum mobility) are considered. Addressing can be static (e.g., an extension of physical and MAC address) or dynamic (e.g., using a dynamic host configuration protocol). In addition, in multi-hop cognitive radio networks, routing decisions need to be made on the basis of topology, MAC congestion, and link quality [37].

Transport layer The transport layer is responsible for flow and congestion control, which is affected by MAC protocol performance and spectrum mobility. The throughput performance of the traditional transmission control protocol (TCP) is a function of round-trip time (RTT) and packet loss probability. These depend on the spectrum management/sharing protocols, bandwidth of spectrum opportunities, transmission power, and interference level. During rerouting, RTT and packet loss rate change, and spectrum handoff latency (due to spectrum mobility) may increase the RTT. Therefore, minimizing packet delay and loss during spectrum handoffs (e.g., through efficient queue management) can improve the performance of TCP in a cognitive radio network [37].

The adaptive protocols in the MAC, network, transport, and application layers should take into account the variations in the cognitive radio environment, such as the traffic activity of primary users, transmission requirements of the secondary users, and variations in channel quality. For this, MAC layer beacons can be sensed for information regarding cognitive radio sleep and wake-up schedules and node connectivity. To link all the protocol modules in the stack and enable cross-layer interaction, a cognitive radio control is used to establish the interfaces among the SDR transceiver, adaptive protocols, and wireless applications and services. The cognitive radio uses intelligent algorithms to process the measured signal from the physical layer and to receive information on transmission requirements from the applications to control the protocol parameters in the different layers. A programmable radio architecture and related protocol stack with cross-layer interaction, along with the wideband radio front ends, will be the key enabling technologies for cognitive radio networks.

The spectrum sharing system model based on network architecture, as shown in Fig. 1.3, is classified as follows: (1) centralized architecture—an infrastructure-based network whose secondary users are managed by secondary base stations, which are in turn connected by a wired backbone, and (2) distributed/decentralized architecture—the secondary users communicate with each other in an ad hoc manner. The spectrum sensing operation in a decentralized architecture is usually performed collaboratively. This type of architecture also encompasses the coexistence of two or more wireless networks operating in unlicensed bands. The architecture of a cognitive radio network is an important feature for sharing the licensed spectrum with multiple cognitive users. There are two main types of cognitive radio network architecture, described in detail as follows [5]:

(a) *Centralized cognitive radio network*

In a centralized cognitive radio network, the control of spectrum allocation and access to a particular regime of the spectrum by cognitive users is maintained by a central controller, for example, a base station [20, 21]. In addition, all cognitive user communication is followed through this central controller, and spectrum access

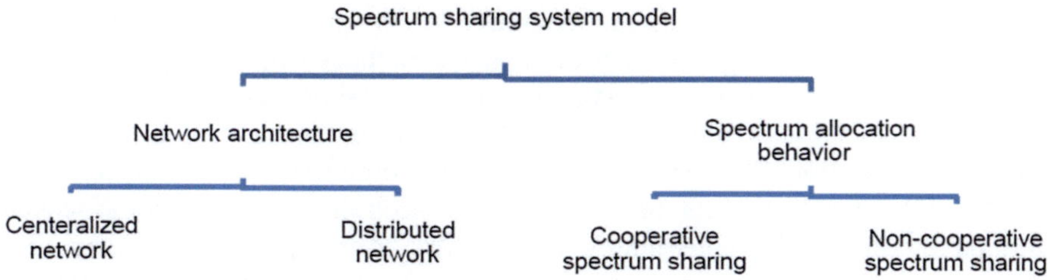

Fig. 1.3 The spectrum sharing system model for a cognitive radio communication system

decisions, such as the duration of spectrum allocation and transmit power by the cognitive users, are also controlled through this central base station. Consequently, the central controller must collect information on both the spectrum usage of the licensed users and the spectrum requirements of the cognitive users. Based on this information, an optimal solution can be obtained that maximizes the total network throughput, provides high QoS, and reduces latency. Central controller decisions are broadcast to all cognitive users in the network. However, the collection and exchange of information between the central controller and cognitive users incurs considerable overhead [5].

(b) *Distributed cognitive radio network*

In a distributed cognitive radio network, unlicensed users communicate with each other directly, in a peer-to-peer manner, without the need for any base station or central controller [5, 21]. In addition, decisions by unlicensed users regarding spectrum access can be made independently and autonomously. Each unlicensed user must collect information about the ambient radio environment and make its decision locally. Thus, the cognitive radio transceiver of each unlicensed user requires greater computational resources than that in the centralized network. However, the communication overhead in this case would be smaller. In multi-hop communication, unlicensed users may sometimes assume the role of relay stations [5].

1.4 Spectrum Allocation Performance

The spectrum sharing system modeled on spectrum allocation behavior is grouped into two classes: cooperative spectrum sharing and non-cooperative spectrum sharing.

(a) *Cooperative spectrum sharing*

In the cooperative spectrum sharing scheme [28], all the cognitive users cooperate with one another either through a centralized base station or through a common control channel in the centralized or distributed cognitive radio networks, respectively. Cooperation between cognitive users enables the sharing of the spectrum with maximum efficiency through the exchange of sensing information [29], thus reducing sensing time while improving sensing accuracy, resulting in a good degree of fairness, and higher complexity and overhead with an increase in energy consumption [30]. However, to reduce communication overhead, complexity, and power consumption in cooperative spectrum sensing, only the sensing information useful for determining the primary user's presence is used [31]. Communication overhead is further minimized in the cognitive radio spectrum sharing system through clustering [32], in which the spectrum sensing results are combined and processed locally by a cluster head. The heads of each cluster report the result to a central controller to make a final decision regarding channel access. Other techniques have been proposed for spectrum sharing that involve combining the spectrum sensing results of different unlicensed users and making spectrum sharing decisions based on cooperative sensing. The simplest such technique involves the use of an OR operation among the received sensing results [33] and weighted data-based fusion [38]. Sensing and combining techniques based on maximum ratio combining (MRC) and equal gain combining (EGC), with the help of multiple antennas and under different fading channels, are investigated in [39], which demonstrates that this method improves the probability of detecting primary users.

(b) *Non-cooperative spectrum sharing*

Unlike cooperative spectrum sharing, in non-cooperative spectrum sharing, cognitive users do not exchange any kind of information with one another [5]. This method of sharing is advantageous for networks with few users and incurs less communication overhead. In large user networks, however, it will cause severe degradation of spectrum efficiency because of the selfish nature of each cognitive user. Since the spectrum sensing information of a single user is utilized for decisions regarding sharing of the primary licensed channel, the odds of a false alarm are significantly higher in non-cooperative versus cooperative spectrum sharing, which results in performance degradation for either the primary or/and secondary user.

1.5 Spectrum Access Techniques

In a shared-use model, spectrum can be accessed by the unlicensed user or cognitive user in three different modes [5], namely, spectrum interweave/opportunistic spectrum access, spectrum underlay, and spectrum overlay, each of which is discussed in detail below.

(a) *Spectrum interweave/opportunistic spectrum access (OSA)*

At a particular time, frequency, or space, if the spectrum is unutilized/underutilized by the primary user, it can be opportunistically accessed by cognitive users with the help of the spectrum interweave access method [40], as shown in Fig. 1.4a. Therefore, a cognitive user attempting to access a spectrum band using the spectrum interweave technique must perform spectrum sensing to detect the activity of a primary user in that regime of the spectrum. If a spectrum white space that is an inactive primary user is detected, the cognitive user may access that unutilized spectrum, as clearly seen in Fig. 1.4a. Once the primary user resumes its transmission, the cognitive user must vacate the spectrum. The spectrum interweave method can be used by cognitive radio in FDMA, TDMA, or orthogonal frequency division multiplexing (OFDM) wireless systems.

(b) *Spectrum underlay*

In the spectrum underlay access method, the cognitive user can transmit concurrently with a primary user, as shown in Fig. 1.4b. However, the transmit power of a

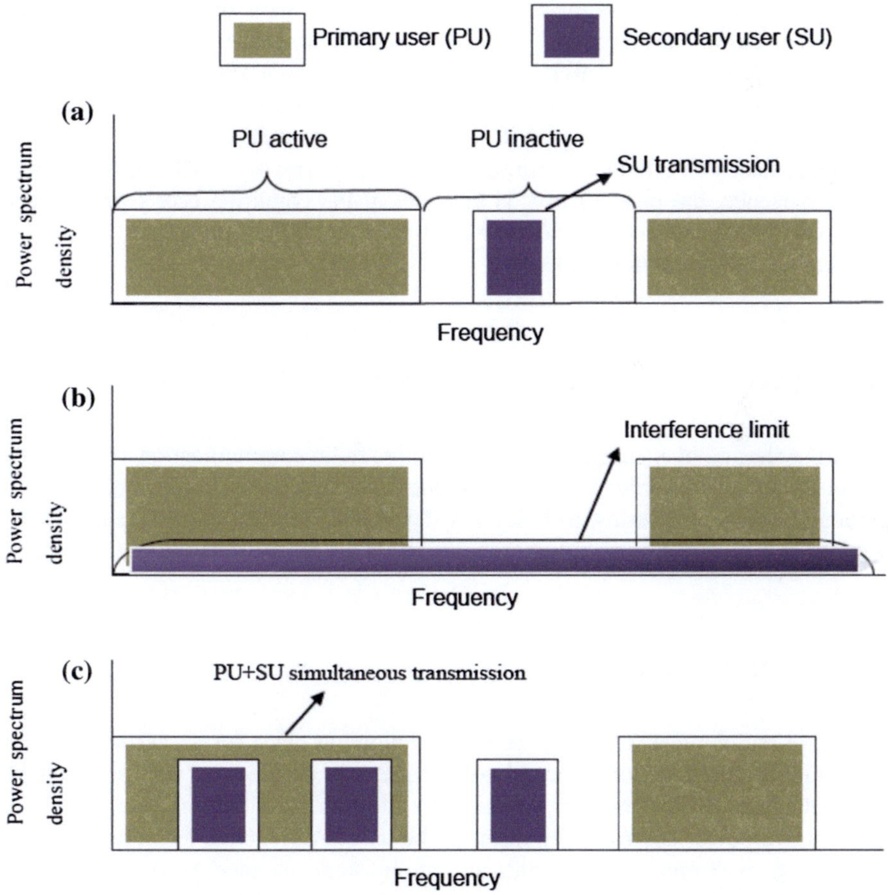

Fig. 1.4 Spectrum access techniques **a** interweave **b** underlay, and **c** overlay

cognitive user should be limited such that the interference caused by the cognitive user to the primary user remains below the interference temperature [40], defined as the interference limit set at the primary user's receiver up to which it can tolerate interference with no effect on its operation. The spectrum underlay can be used for cognitive radio systems using CDMA or ultra-wideband (UWB) technology [5]. Therefore, in the spectrum underlay access technique, spectrum sensing used to detect the spectrum hole for cognitive user transmission is not needed. However, the cognitive user's transmission should not cross the threshold for interference avoidance.

(c) *Spectrum overlay*

In the spectrum overlay mode, concurrent primary and cognitive user transmission is allowed, as shown in Fig. 1.4c. However, interference at the cognitive and primary receiver is mitigated by advanced coding and interference cancellation techniques [41–43]. Although spectrum overlay is a promising spectrum sharing technique, it requires a high degree of cooperation with the primary user and knowledge of the primary user's message signal. Moreover, the cognitive user helps to relay the primary user's information by utilizing some of its power, and uses its remaining power for transmitting its own data [44, 45]. Therefore, this increase in the primary user's SNR is offset by a decrease in its SNR due to the secondary user's interference, resulting in no net difference in SNR at the primary receiver. Hence, the primary user is unaware of the cognitive user's presence. In addition, dirty paper coding [46] is used by the cognitive transmitter to mitigate interference at the cognitive receiver.

1.6 Spectrum Sharing Techniques and Related Work

Spectrum sharing plays a major role in cognitive radio communication systems, and it can be achieved using various techniques. However, the implementation of a particular spectrum sharing technique is dependent on QoS requirements. In this section, several spectrum sharing techniques are presented [47].

1.6.1 Power Control

Cognitive radios must follow the rules/restrictions in place for accessing spectrum [5], which necessitates an appropriate management protocol as well as a reliable and scalable access mechanism. In dealing with protocol violations, both proactive and reactive power control techniques can be used. A proactive technique includes a rule (e.g., maximum power limit) and an enforcement mechanism (e.g., power allocation), and is applied prior to potential misbehavior on the part of the cognitive radio user before a violation of spectrum access rules can occur. A reactive technique, on the other hand, is required to punish misbehavior of cognitive radio users. Because cognitive users coexist with primary users in an operating spectrum, simply consideration of transmission power limits on a channel may be insufficient [48]. The presence of primary users in adjacent channels forces a reduction in the demand for signal power transmission on an available channel to ensure minimal adjacent channel interference. Hence, the occupancy of the neighboring channels is also a critical parameter for improved spectrum sharing in transmit power mode. Furthermore, in the opportunistic spectrum access transmission model, the cognitive user can transmit only when it detects the spectrum white space, which is the period during which the primary user is not transmitting over the band. In [49], the authors propose a new spectrum sharing transmission model in which the secondary user can transmit at any time without detecting the primary user, whether active or not, but with restricted transmission power to avoid harmful interference at the primary user. This is an important consideration in the case when the perfect channel state information (CSI) is not available, and it operates similarly to that of the ultra-wideband (UWB). However, the transmit power restriction affects the transmission range of cognitive radio user data and is unable to take full advantage of unutilized licensed spectrum in which it can transmit with maximum power. Therefore, the authors in [50] have proposed that sensing be performed to vary the transmission power of the secondary user, so that when the primary user is active, the secondary user transmits with low power to avoid interference with the primary user, and vice versa. Incorrect channel information also results in the degradation of cognitive radio system performance [51]. The variations in transmission power and rate according to the fading conditions are discussed in [52, 53]. Kang [54] determined the optimal power allocation to cognitive users under the Rayleigh fading environment based on the assumption that CSI is available to cognitive users, and calculated the ergodic and outage capacity closed-form expressions.

The OFDM-centric cognitive radio network has also been exploited by researchers, and several authors have described different methods for allocating optimal power to the subcarriers of cognitive radio users resulting from side-by-side coexistence of cognitive and primary users. Initially, the power loading method [55] developed for the OFDM cognitive radio network allocates optimal power to the subcarriers. This is achieved by keeping the interference constraint satisfied and using the location information of the secondary users with respect to the primary users. A comparison of various power allocation methods in OFDM-based cognitive radio networks is illustrated in Table 1.1. Fairness is an important parameter for network performance. Wang et al. [56] considered cognitive user fairness in the OFDM-based cognitive radio network, proposing optimal power and a simple power distribution algorithm with complexities of $O(L^2N)$ and $O(L + N)$, respectively (where N is the number of subcarriers of cognitive user and L is the number of primary user transmitter receiver pair). The cognitive user capacity optimization problem was solved in [56] with interference, fairness, and total power constraints. In [57], the joint rate and power optimization problem was considered in the max-min and proportional fairness scenario.

Table 1.1 Comparison of power allocation methods in OFDM-based cognitive radio networks

Power allocation method	Description	Complexity	No. of iterations
Gradient-based approach [61]	Allocates power to the cognitive users in time-varying channel with adaptive step size, while transmitting in only unutilized licensed frequency band and considering adjacent channel interference. Multiple primary and cognitive users are considered	$O(N)$	3
Power loading scheme [62]	Considers both co-channel and adjacent channel interference due to transmission by cognitive users in active and inactive licensed frequency bands; power allocation is performed in time-varying channel	$O(NlogN) + O(LM)$	$L + 1$
Geometrical programming approach [63]	Considers coexistence of a primary user and multiple cognitive users in the same frequency band and allocates power to cognitive users with the aim of saving power	Depends on the number of iterations	fixed
Iteration minimum algorithm [64]	Considers single cognitive user pairs and multiple primary users, and cognitive users transmit only in unutilized licensed channels	$O(Tf_s logN + N)$	2

A new power domain spectrum sharing method, non-orthogonal multiple access (NOMA) [58], has recently been explored. NOMA offers various advantages including higher throughput due to the wide bandwidth, exploitation of channel gain for optimal power allocation, and demonstrated ability to outperform the OFDM scheme [59], and it is beneficial for spectrum sharing in cognitive radio networks. All users of NOMA utilize whole available bandwidth, in contrast to the orthogonal frequency division multiple access (OFDMA), where available bandwidth is divided into subcarriers. This results in enhanced throughput [60], and power is allocated to cognitive users in consideration of channel conditions, with more transmission power allocated to the user with good channel conditions compared to the user with a more severe environment. However, since the same frequency is utilized for all user transmission in NOMA, the receiver must have the capacity to carefully decode its own signal and should minimize co-channel interference. Therefore, this system is more complex than OFDM in terms of its receiver decoding scheme. NOMA is an efficient spectrum sharing scheme in cognitive radio, because it avoids competition among cognitive users for specific channels, and requires power control only according to the environment. The base station, or central coordinator, controls the power allocation to different users, although the NOMA concept remains open for research in the distributed environment.

1.6.2 Game Theory

Game theory in a cognitive radio network is developed primarily for spectrum sharing through trading and fairness rules, with the main objective of fulfilling cognitive radio network demand while maximizing the revenue of the primary network. Therefore, game theory can effectively guarantee fairness and rationality in spectrum management within a cognitive radio network [65, 66]. In [65], the authors proposed the OODA (orient, observe, decide, and act) method for sharing the primary user network's spectrum among multiple heterogeneous cognitive radio networks with different QoS requirements, while taking into account the behavior modeling of the cognitive users. The authors in [66] considered the varying bandwidth subcarriers of multicarrier communication networks allocated to cognitive users and a utility function, with the aim of maximizing the data rate of cognitive users with constraints on resources such as defined power, spectrum, and bandwidth. The main contribution of this work lies in the definition of utility function based on proportional fairness, harmonic mean fairness, and max-min fairness with allocation problems. In [67, 68], the authors considered that a node cannot transmit and receive on the same channel simultaneously, and allocated resources to competing users in the ad hoc network. The convex optimization problem is solved in [67], and in [68], resource allocation is performed by the connectivity graph coloring method. The advantage of the former over the latter is that it requires fewer iterations and achieves significantly higher throughput. However, in both schemes [67, 68], there is only one homogeneous primary user network, which is utilized by cognitive users without considering the heterogeneity of the primary user system. The authors in [69] maximized cognitive radio link capacities using an incremental sub-gradient optimization approach both with and without fairness constraints, assuming that each cognitive radio user is half-duplex. In the aforementioned works [65, 67–69], the entire available spectrum from the spectrum pool is divided into orthogonal subcarriers for the OFDM access scheme, thus minimizing interference and enhancing spectrum efficiency. Utilizing game theoretical spectrum sharing using the OFDM access scheme in an ad hoc cognitive network, Niyato and Hossain [70] performed licensed spectrum sharing using the TDMA mode in a centralized cognitive network where all available bandwidth is accessed by the multiple cognitive users at different times. This technique is simpler than that of multicarrier communication, but it degrades throughput in comparison to the multicarrier OFDM access scheme. Since, spectrum trading is the process needed to share idle channels detected during spectrum sensing, the authors emphasize the various factors of spectrum trading between the primary and secondary users [70].

Three kinds of trading markets are defined in the literature: monopoly, oligopoly, and exchange market [71]. In [72], Nie and Comaniciu investigated the design of channel sharing etiquette in cognitive radio networks for both cooperative and non-cooperative scenarios. The performance of different components of a game theoretical framework for radio resource management—network-level bandwidth allocation, connection-level bandwidth allocation, capacity reservation, and admission control—have been analyzed in detail in [73].

In [74], the authors explore a method of spectrum sharing between several primary and secondary users based on the cost and amount of required bandwidth. In [75, 76], the authors discuss the most common application of game theory, auction theory, in cognitive radio spectrum sharing through interaction between cognitive users and primary users. An optimality solution for obtaining equilibrium in demand and supply of auctioned spectrum is discussed in [77]. In addition, the Nash equilibrium [78] is used in non-cooperative game theory for allocating spectrum to multiple cognitive users and the Nash bargaining solution in cooperative game theory among cognitive and primary users [77]. A static game spectrum sharing method employed for spectrum allocation in [78] was found to reduce the efficiency of the wireless network as a result of inefficient Nash equilibrium outcomes due to user selfishness, deriving individual benefit at the expense of overall and fair spectrum sharing. Spectrum sharing through cooperative game theory gives all cognitive users a single objective function and provides an optimal solution by considering each user's interest: the linear proportional fairness method of spectrum sharing. In a competitive environment with multiple cognitive users, the most common auction schemes are sequential auctions and Vickrey auctions [79], and the time-definite assignment of the spectrum [75] in the Vickrey auction makes it more advantageous than sequential auction for cognitive radios. Single and double auction methods are also defined in the classification of auction methods [80–82]. In the single auction trading method, there is one seller and many buyers, and the buyer with the highest bid wins the item. In the case of a large number of sellers and buyers, however, the double auction is an efficient method for spectrum trading. In the double auction [80, 83], the sellers/buyers submit their selling/buying prices to the auctioneer (spectrum broker), who allocates spectrum to a specific buyer at a price higher than the seller's asking price, thus generating a profit for itself [84]. In [85], the authors discuss the double auction in primary and cognitive radio networks, with the primary and secondary users being the bidders for the available channels. In addition, they considered that a broker would allocate a single channel to only one primary user network and to single/multiple cognitive user networks, with a primary network receiving higher priority than cognitive user networks. The benefit a primary user network will receives after trading spectrum to a cognitive network is dependent upon the amount of spectrum and the amount of time necessary to perform the allocation. However, to achieve greater benefit, the primary network must not degrade its own users' services. Chang and Chen [75] thus considered the QoS for the primary user through its blocking rate in order to ensure proper allocation. In [75], the benefits of primary users, cognitive users, regulatory systems, and service providers are considered, and a super-frame structure for competition among cognitive users is explored. A Vickrey auction scheme based on the signal-to-interference-plus noise ratio (SINR) and power is discussed in [75], and the min-max fair SINR allocation is performed for cognitive radio game spectrum allocation. In contrast to pricing and auction theory, revenue-based sharing is proposed in [86], in which the revenue shared by the primary user network is determined by the resources allocated among the primary and cognitive users.

1.6.3 Multiple Antennas

The concept of multiple antennas has also been exploited as a potential method for spectrum sharing in cognitive radio communication systems, due to throughput enhancement and interference cancellation. A system model for a cognitive radio network where the multiple antennas are implemented at a cognitive user transmitter is presented in [87], which provides significantly enhanced channel capacity compared to that of the single antenna at the cognitive user transmitter. It is also able to transmit on the same spectrum currently being used by the primary user, due to multiple-antenna beamforming [88]. Multiple antennas are also used to allocate transmit dimensions in space, providing the cognitive transmitter more degrees of freedom in space, in addition to the time and frequency to balance between maximizing its own transmit rate and minimizing the interference powers at the primary receivers. Two algorithms, direct-channel singular value decomposition (D-SVD) and projected-channel SVD (P-SVD), which enhance cognitive radio user capacity and avoid interference at the primary receiver by projecting null to the primary receiver through beamforming, are proposed in [87]. Bakr et al. [89] used the antenna weights to place nulls at primary receivers whereas secondary radio receivers use adaptive techniques to decode in the presence of interference from primary users. To obtain the antenna weights, channel estimation is performed through feedback from the primary receivers, and these estimates are used to compute the appropriate weights. The antenna weights are then adapted by the cognitive radio transmitter antennas to form the radiation pattern which nullifies the interference at the primary receiver and provides efficient communication to its respective cognitive radio receiver.

In [90, 91], the authors discuss characteristic function and its application in computing channel capacity under a fading environment with multiple antennas. Moment-generating function (MGF) and characteristic function (CF) are used to compute the error rate and channel capacity in [92]. Fading channel capacity using the MGF approach [93–96] in the multiple-antenna scenario with a different correlation coefficient in fading environments is formulated in [97]. In [98], the authors consider a cognitive radio spectrum sharing scenario without conventional constraints on the cognitive user transmit power and primary user received interference power, achieving results without degradation of the cognitive or primary services, in the multiple-antenna spatial domain. The authors also consider the imperfect CSI effect on system performance; however, the proposed method is not suitable for cognitive users sharing full-duplex primary user spectrum. In [99], a single cognitive user system capacity is computed by considering the interference constraint at the primary receiver, and hence the need to limit its transmission power. In addition, multiple antennas are considered for both cognitive and primary users. However, the pre-whitening versus post-whitening multi-antenna spectrum sharing technique is considered for cognitive users, because the amount of interference at the primary receiver is lower compared to that of the post-whitening scheme. An underlay multicast method of spectrum sharing in cognitive radio communication has been proposed [100] using multiple antennas only at the cognitive access point, then broadcasting the same information to all cognitive receivers with beam-steering and limiting the side-lobe power to the primary receiver. However, the perfect CSI is needed; otherwise, coexistence of the cognitive and primary users in the same spectrum may degrade the performance of both primary and cognitive user. Sridharan and Vishwanath [101] derived multiple-input, multiple-output (MIMO) cognitive channel (MCC) capacity with CSI knowledge at the cognitive user. However, transmit power limits exist at both the primary and secondary user transmitter, and MCC capacity is maximized by considering these two transmit power constraints at both transmitters with the help of Lagrange's optimization. Since the cognitive user system does not want to change the primary user network and should not impose any restriction on the primary network, the primary user transmit power constraint [101] is not a feasible solution for enhancing cognitive radio system performance. Adian and Aghaeinia [102] jointly considered transmission time and power allocation to heterogeneous cognitive users in centralized and distributed cognitive networks. In addition, the authors considered the advantage of multiple antennas with constrained by resource allocation fairness in the heterogeneous cognitive radio network. In [103], a new multiple-antenna channel model, the cognitive interference channel, was considered in place of the classical interference channel, where the cognitive transmitter is provided with the knowledge of the primary user data. This additional information at the cognitive transmitter facilitates knowledge of neighboring nodes.

MIMO systems have great potential to enhance throughput in the framework of wireless cellular networks [104, 105]. Multiple antennas can achieve many desirable goals for wireless communications, such as increased capacity without bandwidth expansion, transmission reliability enhancement via space–time coding, and co-channel interference suppression for multiuser transmission. By using multiple antennas in cognitive radio, one can allocate transmit dimensions in space, and hence can obtain many design benefits for the MIMO cognitive radio network. In particular, we can obtain high spatial multiplexing gain by sending independent information streams simultaneously over any transmit-receive antenna pair to enhance the system throughput of the cognitive radio network [106]. In addition, multiuser interference can be suppressed by applying transmit beamforming [107]. However, while multiple antennas can typically be deployed at the base station, they cannot be easily used at the mobile terminals due to size and cost constraints. This may limit the capacity of the system when a limited number of antennas at the receivers are considered. The problem can be addressed by serving multiple users with single antennas simultaneously, effectively creating a virtual MIMO system. In a cognitive radio network, spectrum sharing can also be considered to further improve spectrum utilization efficiency. However, the primary user will always have higher priority than the secondary users in spectrum resource utilizations.

Hence, the fundamental challenge of spectrum sharing is to ensure the QoS of the primary user by limiting interference to it. Therefore, it is crucial in the design of cognitive radio systems to take into consideration two main conflicting objectives, namely, maximizing the throughput of the cognitive radio system and minimizing the interference at the primary receiver. In [108, 109], the authors designed a capacity-achieving transmit spatial spectrum for a single secondary link in a cognitive radio network under both its own transmit-power constraint and interference-power constraint at the primary receivers. The proposed problem was formulated as a convex optimization problem. In [110], the problem of joint power control and beamforming in the downlink of a cognitive radio network was studied for a limited number of users. Hamdi et al. [111] presented spectrum sharing between a large number of cognitive radio users and a licensed user in order to enhance spectrum efficiency with the deployment of a number of antennas at the cognitive base station; an opportunistic spectrum sharing approach was proposed to maximize the downlink throughput of the cognitive radio system and limit the interference to the primary user. In the proposed approach, cognitive users whose channels are nearly orthogonal to the primary user channel are pre-selected in order to minimize the interference with the primary user, and a lower bound of the proposed cognitive system capacity is derived. The simulation results show that the proposed approach is able to achieve high sum-rate throughput, with affordable complexity, when either single or multiple antennas at the cognitive radio mobile terminals are considered. Moreover, simulation results have shown that when the cognitive user is equipped with multiple antennas, the proposed method combined with receiver antenna selection can further reduce selection complexity, with little loss in the sum-rate throughput. However, these results are based on an assumption of perfect CSI at the transmitter, which may not be practical. Therefore, this interesting topic remains open for future consideration, and should be investigated, with emphasis on the robustness of the proposed broadcast scheduling algorithm with respect to channel estimation errors. Manna et al. [112] presented a cognitive radio network consisting of a primary transmitter-primary receiver pair and a secondary base station-secondary receiver pair. To improve the performance of both the primary and secondary pairs, an overlay spectrum sharing scheme is employed wherein the primary user (PU) leases half of its time slots to the secondary user (SU) in exchange for the SU cooperatively relaying the PU's data, using an amplify-and-forward scheme. The proposed scheme also involves the design of antenna weights and power allocation to meet a certain error or rate design criterion for both PU and SU. New closed-form expressions are derived for the rate and bit error rate for arbitrary SNR to evaluate the performance of the proposed scheme, and an asymptotic analysis is performed in the high-SNR regime to obtain the diversity order. Consequently, numerical analysis of these expressions reveals that the proposed cooperative overlay scheme can achieve significant performance gains for both PU and SU compared to the conventional non-cooperative underlay scheme, which provides incentive to both users to cooperate. The proposed design of a spectrum sharing scheme provides a higher rate and low error performance for both PU and SU networks compared to that of the conventional underlay system where no cooperation occurs. Maham and Popovski [113] considered a downlink primary multiple-input, single-output (MISO) system operating under a controlled interference from the downlink MISO cognitive radio. This is a secondary system deriving exact expressions for the outage probability of the primary user under Rayleigh fading, when the primary system is exposed to interference from a secondary base station (BS). Further, the authors considered three different operating modes for the primary BS, namely, the space–time coding, antenna selection, and beamforming, each having different channel information requirements. The outage probability is analyzed when the primary BS uses a fixed rate. In a high-SNR scenario, a closed-form asymptotic formula for the outage probability is derived. In addition, optimum transmit power in the secondary system is investigated for maximizing the ergodic capacity when an outage constraint exists at the primary system. An adaptive-rate antenna-selection primary system for increasing throughput was also proposed. Sboui et al. [114] investigated the spectral efficiency gain of an uplink cognitive radio MIMO system in which the secondary user is allowed to share the spectrum with the primary user by utilizing a specific precoding scheme to communicate with a common receiver. At the common receiver, the authors adopted a successive interference cancellation (SIC) technique to eliminate the effect of the detected primary signal transmitted through the exploited eigenmodes. They also analyzed SIC operation inaccuracy and CSI estimation imperfection on PU and SU throughputs. The numerical results show that the proposed scheme significantly enhances the cognitive achievable rate. Lastly, they investigated the behavior of PU and SU rates by studying the rate achievable region. The secondary user exploits the unused eigenmodes of the primary user and shares those that are used by respecting both the total power and interference temperature constraints. The authors also showed that the secondary achievable rate is significantly increased when the secondary user exploits the free eigenmodes and shares the used eigenmodes. The impact of an imperfect CSI estimation on the primary and secondary rates is also highlighted. Lastly, the authors investigated the rate region of the system and characterized the trade-off between the primary and secondary rates.

1.6.4 Medium Access Control (MAC) Protocol

In spectrum sharing, users traditionally obtain access to the channel through a MAC protocol. The difference in MAC protocols between traditional wireless communication and cognitive radio systems is that multiple channels must be shared by multiple cognitive users versus single channel sharing by multiple users in conventional MAC protocols. In addition, cognitive users must differentiate between the primary user and cognitive user transmission in order to stop transmission and protect the primary user or to retransmit if cognitive user interference occurs. The available licensed channels for communication vary with time and location; thus each cognitive user does not have a fixed number of channels for transmission. All functions must be incorporated into the MAC protocol of the cognitive radio communication system. Because the cognitive user has intelligent capability and is able to switch among multiple channels, sensing and switching features must be incorporated into cognitive radio MAC protocol spectrum sharing. In addition, there may be multiple cognitive radio users trying to access the spectrum, and so access must be coordinated in order to prevent collisions among multiple users in overlapping portions of the spectrum. Cross-layer design and optimization strategies [115, 116] have been developed for cognitive radio to address the layered protocol and structure limitations. The physical layer deals directly with the physical environment/channel that is followed by the MAC layer, which needs attention in the design of the communication system and various layer parameters such as frame type, frame size, data rate, channel/time slot allocation, scheduling scheme, and retransmission probability. All these MAC layer parameters are part of the MAC protocol and are responsible for spectrum sensing and spectrum access decisions [117]. The major objectives of the MAC protocol design in cognitive radio networks are as follows:

(a) Optimizing spectrum sensing and spectrum access decisions
(b) Controlling multiuser access in a multichannel network
(c) Allocating radio spectrum and scheduling traffic transmission.

Various cognitive radio MAC protocols are presented in Chap. 3, along with detailed discussion.

1.7 Potential Challenges

A major challenge in spectrum sharing is achieving a significant improvement in spectrum efficiency without losing the advantages associated with static spectrum allocation. The spectrum policy domain should develop strategies for spectrum sharing that lead to efficient spectrum use, protect the rights of license holders, and maintain the QoS. There are also significant economic considerations. Policies must protect the interests of primary users, who have made significant infrastructure investments. However, in all spectrum decisions and sharing techniques, the channel is considered a spectrum unit, and the development of a protocol/set of rules is a crucial issue. In general, the common control channel facilitates several spectrum sharing functionalities. However, it must be vacated when a primary user returns, and so implementation of a fixed common control channel is not feasible. Moreover, in cognitive radio networks, a channel common to all users is highly dependent on the topology, and varies over time. Therefore, a solution to this issue is also critical in cognitive radio communication systems. In addition, spectrum sharing in the cognitive radio network is highly dependent on the number of users in the system. Cognitive users increase competition and may degrade performance; therefore, the spectrum sharing system must be highly scalable. The need for an energy-efficient cognitive radio terminal is another challenge in cognitive communication networks. Further, because cognitive radio works on unutilized licensed channels and receives lower priority than licensed users, the risk of blocking of communication is significant, creating a severe problem in particular for the real-time cognitive radio user's traffic. Therefore, spectrum sharing methods in cognitive radio networks must be carefully designed to meet the QoS requirements of cognitive users.

The potential problem addressed in this book is how to efficiently share the spectrum of licensed users with cognitive users. Since the range of methods of spectrum sharing is very broad, the scope of this book is limited to the second layer of the Open Systems Interconnection (OSI) model for spectrum sharing, and more specifically, on the MAC protocol for the multichannel distributed cognitive radio network. Stevenson et al. [118] presented a standardized cognitive MAC protocol (IEEE 802.22) for centralized cognitive radio networks. However, the MAC protocol for the distributed cognitive network has not yet been standardized. The primary objective is the design of a suitable frame structure for the cognitive radio network in the primary user interference environment, and computation of the key performance indicators for the system, such as throughput and energy efficiency. Since sensing errors have a significant adverse effect on the performance of cognitive radio and the primary user's communication system, it is another important parameter to consider in designing the cognitive radio MAC protocol. Cognitive users are unlicensed, and should not cause disruption to the licensed users. Therefore, the transmit power control algorithm should be in place to prevent the degradation of primary user network performance and also to enhance the energy efficiency of the cognitive users. In addition, a fading phenomenon is present in the channel, and therefore spectrum sharing in the fading environment of a cognitive radio network is also an important issue for discussion.

Further, a new security threat has arisen with the development of cognitive radios that have not been studied previously. Adversaries can exploit several vulnerabilities of this new technology and cause severe performance degradation. Testing of the technology through large-scale experimentation is essential for ensuring that it is robust, secure, and efficient for users of the spectrum, and that it will not harm legacy systems. The development of advanced and adaptable test beds using advances in hardware, software and policy, proof-of-concept demonstrations, and standardization of current/future test beds is imperative for assessing the performance of new technologies.

Beyond the technical issues, there are also policy-domain challenges in dynamic spectrum sharing. The future of spectrum sharing systems may employ dynamic spectrum markets in which the primary licensees can sell spectrum access to SUs on a temporary basis [119]. Further, the security threats are mainly related to two fundamental characteristics of cognitive radios, cognitive capability and reconfigurability. The threats related to cognitive capability include attacks launched by adversaries that mimic primary transmitters, and the transmission of false observations related to spectrum sensing. Reconfiguration can be exploited by attackers through the use of malicious code installed in cognitive radios. In addition, cognitive radio networks face all the classic threats present in conventional wireless networks. In general, due to their open nature, wireless networks are susceptible to several attacks targeting the physical or medium access control (MAC) layers. The attacks targeting the physical layer through RF jamming can severely disrupt a network's operation [120, 121]. Attacks at the MAC layer include MAC address spoofing and transmission of spurious MAC frames [122], as well as greedy behavior by cheating on backoff rules [123, 124]. Moreover, the unique cognitive characteristics of these networks create new security threats and challenges. The basic operation of cognitive radios is spectrum sensing, and whenever a cognitive radio detects a primary user signal, it must vacate the specific spectrum band. Malicious users can mimic incumbent transmitters to force cognitive radios to vacate a specific band, which is known as a primary user emulation attack (PUEA). Another type of attack involves collaborative spectrum sensing, a technique used to improve spectrum sensing in fading environments where multiple cognitive radios collaborate. Here, a malicious cognitive radio can purposely provide false observations, known as a spectrum sensing data falsification (SSDF) attack. PUEA and SSDF attacks are targeted towards the physical layer of a cognitive radio network. MAC threats and specific threats for IEEE 802.22 cognitive radio networks target the MAC layer. However, adversaries can launch so-called cross-layer attacks targeting multiple layers that can affect the entire cognitive cycle (Fig. 1.2), as attacks at all layers have now become feasible. In [125], the authors introduced two types of attacks against a cognitive radio network: reporting false sensing data (RFSD) attack and small backoff window (SBW) attack. SBW is a very common attack in wireless networks, where malicious users choose a very small value for the minimum contention window (CW_{min}) (e.g., see [124, 126, 127]), aiming to monopolize bandwidth. SBW attacks are feasible against cognitive radios with MAC layers using a carrier sense multiple access with collision avoidance (CSMA/CA) type of access. Several MAC protocols designed for cognitive radio networks are of this type [128].

As cognitive radios adopt the layered architecture of the conventional networks, several cross-layer attacks are possible. These can include a combination of an SSDF attack and an SBW, and the so-called lion attack [129]. Cognitive radio networks are usually based on software-defined radio (SDR) systems, devices with radio functionalities implemented in software, which are vulnerable to a number of software- and hardware-related threats. Initially, PUEAs affect the RF environment by polluting it with fake incumbent signals. An immediate effect of RF pollution is a cascading phenomenon affecting spectrum sensing, analysis, and decision. Energy detection is the most widely used method because of its simplicity and low computational overhead [38, 130–132]. Nevertheless, this is the method most vulnerable to PUEAs, because it does not perform well in low-SNR environments. Furthermore, PUEAs can be launched against cognitive radio networks that detect energy from non-sophisticated adversaries because energy levels generated using an incumbent carrier frequency is a trivial task. A PUEA can be more effective on learning cognitive radios [133], as these radios establish long-term behavior based on their observations from the environment. Avoiding interference with PUs is of paramount importance in cognitive radio networks, and to this end, the MAC layer strictly collaborates with the physical layer and the hardware components. Using simulations, the authors in [134] show how denial-of-service (DoS) attacks using spurious MAC frames affect the performance of a multi-hop cognitive radio network.

Cognitive radio capabilities also breed new demand for access to spectrum, and the steady stream of technical, market, and policy innovations is continuously creating potential cognitive users seeking access to spectrum. Various issues remain in both the technological and policy domains. Some of these are briefly summarized in Fig. 1.5. Effective management of spectrum resources is a key challenge in dynamic spectrum sharing, which requires advancements in allocation and assignment mechanisms that not only facilitate spectrum sharing, but also support measurement and dynamic assessment of the costs and benefits of sharing. Research is also needed for developing and refining the ability to quantify spectrum efficiency, harmful interference, spectrum value, and fair access to spectrum. Advanced spectrum sensing techniques must be able to promptly and precisely identify transmission opportunities over a very wide spectrum band that may host multiple different wireless services. However, designing a framework that jointly enables database-driven and sensing-driven spectrum sharing approaches remains a challenge. Facilitating harmonious coexistence among heterogeneous wireless technologies is another challenge in dynamic spectrum sharing. Specific metrics must be established for assessing how well devices are coexisting. In addition, modulation schemes must be developed that adapt in concert with other system components to mitigate/prevent interference. Realistic propagation models, including inferential models, for frequencies being considered for new applications help regulators and policymakers anticipate the merits of coexistence in both a technological and non-technological context. Improving spectral efficiency and radio configurability for hardware- and software-defined radios is crucial for enabling the commercialization of appropriate spectrum sharing consumer and network equipment. This requires advancements in smart radio architectures that support a high dynamic range for wideband operation. Advances in the areas of radio hardware, software, signal processing, protocols, and access theory must be developed such that they will work in concert, flexibly and over time, to support diverse wireless technology needs. The fundamental limits in these areas also need to be explored. The successful deployment of new spectrum access technologies, such as the cognitive radio, and the realization of their benefits will depend in part on the placement of essential security mechanisms in sufficiently robust form to resist misuse of the technologies.

A cognitive radio system is characterized by a time-dependent structure of primary user traffic patterns and imperfect spectrum sensing. Therefore, developing a method for secondary users to estimate channel parameters without any help from the primary user is crucial for the design of a practical cognitive radio system. Practical methods for spectrum sharing among cognitive radios need to be developed given the limited availability of control channels for coordination of spectrum access in multiuser, multichannel, and multi-radio scenarios. Both horizontal and

Fig. 1.5 Various open issues in cognitive radio networks [37, 119]

vertical spectrum sharing scenarios need to be considered. In this context, distributed adaptation and resource management techniques that rely only on local information, with low control overhead and implementation complexity, are most advantageous. Further, the design and evaluation of higher-layer protocols, such as transport layer protocols for cognitive radio networks in a wired-cum-wireless scenario, have not been adequately addressed in the literature. Cross-layer adaptation and optimization are key components in the design of cognitive radio protocols for adaptive, flexible, and stable dynamic spectrum access networks. Such optimization will need to consider spectrum availability as well as dynamic channel and network selection, in addition to parameters such as QoS requirements, mobility, traffic load, node density, and channel parameters currently considered for traditional cross-layer optimization in wireless networks. Moreover, spectrum policy research is needed to define spectrum etiquette for cognitive radios, to analyze their impact on technology and business strategies and their economic and social benefits, to develop mechanisms to enforce etiquette protocols derived from spectrum policy, and to define the role of different stakeholders [135]. The development of policy and legal frameworks for cognitive radios for the different spectrum sharing models (or market models) will be a significant driver of the direction of future research [136].

Because of the complex nature of cognitive radio networks with many degrees of freedom, evaluating their system-wide performance in real-world environments and studying their emergent behavior is critical for assessing the viability of their practical implementation. The deployment of small cells has been identified as a potential approach for increasing cellular wireless network capacity. However, small cells would be underlain in a macrocell. If the small cells and the macrocell use the same spectrum, interference would occur, and the performance of users in both cells would be degraded. Cognitive radio techniques can be adopted to mitigate this problem. The introduction of device-to-device communications technology has enabled offloading of data traffic from a base station by allowing them to communicate with each other directly. Similar to the cognitive radio small cell networks, device-to-device communications can reuse the spectrum allocated to the cellular users. Energy is an important resource for mobile devices. Therefore, the techniques available in cognitive radios (e.g., spectrum sensing, DSA, and sharing) can also be optimized to maximize energy efficiency for data transmission [37].

1.8 Summary

Cognitive radio technology has been proposed as an intelligent technique for wireless networks to mitigate the problem of spectrum scarcity and to significantly enhance spectrum efficiency. However, many open issues and challenges remain that must be resolved prior to implementation of the technology. In this chapter, we have provided a comprehensive survey of the research activities in cognitive radio communication, particularly with regard to spectrum sharing techniques. However, random data channel selection in cognitive radio ad hoc networks reduces the potential for successful communication among cognitive users due to interference and/or the frequent return of the primary user, resulting in increased energy consumption. This will be more critical when cognitive users have real-time traffic such as voice and disaster information. The frequent return of primary users may require a restart of the entire process, including spectrum sensing, channel selection, and communication over the control and data channels, which consumes additional time and has a direct impact on throughput and energy efficiency. We have also presented potential issues in the design of cognitive radio communication networks and reviewed various approaches to spectrum sharing in cognitive radio. This chapter also addressed potential challenges and issues related to the regulatory authorities and wireless service providers as well as vendors for spectrum sharing of the licensed spectrum bands. Finally, we discussed future trends and directions in research and outlined open research issues.

References

1. NTIA, U.S. frequency allocations. [online] http://www.ntia.doc.gov/files/ntia/publications/2003-allochrt.pdf
2. Federal Communications Commission, Spectrum policy task force report. Technical report 02–135, Nov 2002
3. D. Cabric, I.-D. O'Donnell, M.S.-W. Chen, R.-W. Brodersen, Spectrum sharing radios. IEEE Circ. Syst. Mag. **6**(2), 30–45 (2006)
4. D. Datla, Spectrum surveying for dynamic spectrum access networks. MSc Thesis, University of Madras, 2004
5. E. Hoossain, D. Niyato, Z. Han, *Dynamic Spectrum Access and Management in Cognitive Radio Networks* (Cambridge University Press, New York, 2009)
6. Qing Zhao, B.M. Sadler, A survey of dynamic spectrum access: Signal processing, networking, and regulatory policy. IEEE Sig. Process. Mag. **24**(3), 79–89 (2007)
7. C. Santivanez, R. Ramanathan, C. Partridge, R. Krishnan, M. Condell, S. Polit, Opportunistic spectrum access: challenges, architecture, protocols, in *Proceeding of 2nd Annual International Workshop on Wireless Internet* (WICON), New York, NY, USA, 2006, pp. 1–9
8. J. Mitola III Cognitive radio: An integrated agent architecture for software defined radio. Ph. D. Thesis, Royal Institute of Technology (KTH) Sweden, May 2000
9. F.K. Jondral, Software-defined radio: basics and evolution to cognitive radio. EURASIP J. Wireless Commun. Network. **5**(3), 275–283 (2005)
10. U. Ramacher, Software-defined radio prospects for multi-standard mobile phones. IEEE J. Comput. **40**(10), 62–69 (2007)
11. R. Bagheri, A. Mirzaei, M.-E. Heidari, S. Chehrazi, M. Lee, M. Mikhemar, W.K. Tang, A.A. Abidi, Software-defined radio receiver: Dream to reality. IEEE Commun. Mag. **44**(8), 111–118 (2006)
12. H. Arslan, S. Yarkan, *Cognitive Radio, Software Defined Radio, and Adaptive Wireless Systems* (Springer, Netherlands, 2007)
13. I.-F. Akyildiz, W.-Y. Lee, M.-C. Vuran, S. Mohanty, NeXt generation/dynamic spectrum access/cognitive radio wireless networks: a survey. Comput. Netw. **50**(13), 2127–2159 (2006)

14. X. Liu, S. Shankar, Sensing-based opportunistic channel access. Mobile Netw. Appl. **11**(4), 577–591 (2006)
15. T. Yucek, H. Arslan, A survey of spectrum sensing algorithms for cognitive radio applications. IEEE Commun. Surv. Tutorials **11**(1), 116–130 (2009)
16. J. Lehtomaki, Analysis of energy based signal detection. Ph.D. Thesis, University of Oulu, Finland, Dec 2005
17. D. Cabric, S.M. Mishra, R.W. Brodersen, Implementation issues in spectrum sensing for cognitive radios, in *Proceedings of Asilomar Conference on Signals, Systems and Computers*, Pacific Grove, California, USA, Nov 2004, pp. 772–776
18. K. Kim, I.-A. Akbar, K.-K. Bae, J.-S. Um, C.-M. Spooner, J.-H. Reed, Cyclostationary approaches to signal detection and classification in cognitive radio, in *Proceeding of IEEE International Symposium on New Frontiers in Dynamic Spectrum Access Networks*, Dublin, Ireland, April 2007, pp. 212–215
19. A. Ghasemi, E.-S. Sousa, Spectrum sensing in cognitive radio networks: requirements, challenges and design trade-offs. IEEE Commun. Mag. **46**(4), 32–39 (2008)
20. I.-F. Akyildiz, W.-Y. Lee, M.-C. Vuran, S. Mohanty, A survey on spectrum management in cognitive radio networks. IEEE Commun. Mag. **46**(4), 40–48 (2008)
21. B. Wang, K.-J.-R. Liu, Advances in cognitive radio networks: a survey. IEEE Journal of Selected Topics in Signal Processing **5**(1), 5–23 (2011)
22. N. Noorshams, M. Malboubi, A. Bahai, Centralized and decentralized cooperative spectrum sensing in cognitive radio networks: a novel approach, in *Proceedings of IEEE International Workshop on Signal Processing Advances in Wireless Communications* (SPAWC), Marrakech, Morocco, 20–23 June 2010, pp. 1–5
23. A. Ghasemi, E.S. Sousa, Collaborative spectrum sensing for opportunistic access in fading environment, in *Proceedings of IEEE International Symposium on New Frontiers in Dynamic Spectrum Access Networks*, Baltimore, MD, USA, Nov 2005, pp. 131–136
24. H.-B. Yılmaz, Cooperative spectrum sensing and radio environment map construction in cognitive radio networks. Ph.D. Thesis, Institute for Graduate Studies in Science and Engineering, Bogazici University, Istanbul, Turkey, 2012
25. S.-M. Mishra, A. Sahai, and R.W. Brodersen, Cooperative sensing among cognitive radios, in *Proceeding of IEEE International conference on Communications* (ICC), Istanbul, June 2006, pp. 1658–1663
26. J. Unnikrishnan, V.-V. Veeravalli, Cooperative sensing for primary detection in cognitive radio. IEEE J. Sel. Top. Sig. Process. **2**(1), 18–27 (2008)
27. I.-F. Akyildiz, B.-F. Lo, R. Balakrishnan, Cooperative spectrum sensing in cognitive radio networks: a survey. Phys. Commun. **4**(1), 40–62 (2011)
28. G. Zhang, K. Yang, J. Song, Y. Li, Fair and efficient spectrum splitting for unlicensed secondary users in cooperative cognitive radio networks. Wireless Pers. Commun. **71**(1), 299–316 (2013)
29. K. Letaief, W. Zhang, Cooperative communications for cognitive radio networks. Proc. IEEE **97**(5), 878–893 (2009)
30. C.-H. Lee, W. Wolf, Energy efficient techniques for cooperative spectrum sensing in cognitive radios, in *Proceedings of IEEE Conference of Consumer Communications and Networking* (CCNC), Las Vegas, NV, 10–12 Jan 2008, pp. 968–972
31. C. Sun, W. Zhang, K.-B. Letaief, Cooperative spectrum sensing for cognitive radios under bandwidth constraints, in *Proceeding of IEEE Wireless Communications and Networking Conference* (WCNC), Kowloon, 11–15 March 2007, pp. 1–5
32. C. Sun, W. Zhang, K.-B. Letaief, Cluster-based cooperative spectrum sensing in cognitive radio systems, in *Proceedings of IEEE International Conference on Communications* (ICC), Glasgow, 24–28 June 2007, pp. 2511–2515
33. P. Flajolet, G.N. Martin, Probabilistic counting algorithms for data base applications. J. Comput. Syst. Sci. **31**(2), 182–209 (1985)
34. X. Zhu, L. Shen, T.-P. Yum, Analysis of cognitive radio spectrum access with optimal channel reservation. IEEE Commun. Lett. **11**(4), 304–306 (2007)

35. W.-Y. Lee, I.-F. Akyildiz, Spectrum-aware mobility management in cognitive radio cellular networks. IEEE Trans. Mobile Comput. **11**(4), 529–542 (2012)
36. Y.-S. Chen, C.-H. Cho, I. Yu, H.-C. Chao, A cross-layer protocol of spectrum mobility and handover in cognitive LTE networks. Simul. Model. Practices Theor **19**(8), 1723–1744 (2011)
37. E. Hossain, D. Niyato, D.-In Kim, Evolution of future trends of research in cognitive radio: a contemporary survey. Wireless Commun. Mobile Comput. **15**(11), 1530–1564 (2015)
38. Z. Quan, S. Cui, A.-H. Sayed, Optimal linear cooperation for spectrum sensing in cognitive radio networks. IEEE J. Sel. Top. Sig. Process. **2**(1), 28–40 (2008)
39. J. Ma, Y.-G. Li, Soft combination and detection for cooperative spectrum sensing in cognitive radio networks, in *Proceedings of IEEE Global Telecommunications Conference* (GLOBECOM), Washington, DC, 26–30 Nov 2007, pp. 3139–3143
40. C.-W. Chang, C.-C. Kuo, An interweave cognitive radio system based on the hierarchical 2D-spread MC-DS-CDMA, in *Proceedings of IEEE Vehicular Technology Conference Fall* (VTC 2010-Fall), Ottawa, Canada, 6–9 Sept 2010, pp. 1–5
41. A. Soysal, S. Ulukus, C. Clancy, Channel estimation and adaptive M-QAM in cognitive radio links, in *Proceedings of IEEE International Conference on Communications*, Beijing, 2008, pp. 4043–4047
42. M.-B. Pursley, T.-C. Royster, Low-complexity adaptive transmission for cognitive radios in dynamic spectrum access networks. IEEE J. Sel. Areas Commun. **26**(1), 83–94 (2008)
43. D.-I. Kim, L.-B. Le, E. Hossain, Joint rate and power allocation for cognitive radios in dynamic spectrum access environment. IEEE Trans. Wireless Commun. **7**(12), 5517–5527 2008
44. S. Srinivasa, S.-A. Jafar, The throughput potential of cognitive radio: a theoretical perspective, in *Proceedings of 40th Asilomar Conference on Signals, Systems and Computers*, Pacific Grove, CA, 29 Oct–1 Nov 2006, pp. 221–225
45. M. Taki, F. Lahouti, Spectral efficiency optimized adaptive transmission for interfering cognitive radios, in *Proceedings of IEEE International Conference on Communications Workshops* (ICC), Dresden, Germany, June 2009, pp. 1–6
46. M. Costa, Writing on dirty paper. IEEE Trans. Inform. Theor. **29**(3), 439–441 (1983)
47. S. Pandit, G. Singh, An overview of spectrum sharing techniques in cognitive radio communication system. *Wireless Networks* (Online), Dec. 2015
48. C. Ghosh, Innovative approaches to spectrum selection, sensing, and sharing in cognitive radio networks. Ph.D. Thesis, University of Cincinnati, Ohio, April 2009
49. M. Dashti, P. Azmi, K. Navaie, S.-M. Razavizadeh, Ergodic sum rate maximization for underlay spectrum sharing with heterogeneous traffic. Wireless Pers. Commun. **71**(1), 586–610 (2013)
50. X. Kang, Y.-C. Liang, H.-K. Garg, L. Zhang, Sensing based spectrum sharing in cognitive radio networks. IEEE Trans. Veh. Technol. **58**(8), 4649–4654 (2009)
51. D. Xu, Z. Feng, P. Zhang, On the impacts of channel estimation errors and feedback delay on the ergodic capacity for spectrum sharing cognitive radio. Wireless Pers. Commun. **72**(4), 1875–1887 (2013)
52. Y. Chen, G. Yu, Z. Zhang, H. Chen, P. Qiu, On cognitive radio networks with opportunistic power control strategies in fading channels. IEEE Trans. Wireless Commun. **7**(7), 2752–2761 (2008)
53. S. Pandit, G. Singh, Channel capacity in fading environment with CSI and interference power constraints for cognitive radio communication system. Wireless Netw. **21**(4), 1275–1288 (2015)
54. X. Kang, Optimal power allocation for fading cognitive multiple access channels: a two-user case. IEEE Wireless Commun. Lett. **2**(6), 683–686 (2013)
55. G. Bansal, Md-J Hossain, V.-K. Bhargava, Adaptive power loading for OFDM based cognitive radio systems with statistical interference constraint. IEEE Trans. Wireless Commun. **10**(9), 2786–2791 (2011)

56. S. Wang, F. Huang, C. Wang, Adaptive proportional fairness resource allocation for OFDM-based cognitive radio networks. Wireless Netw. **19**(3), 273–284 (2013)
57. L.-B. Le, E. Hossain, Resource allocation for spectrum underlay in cognitive radio networks. IEEE Trans. Wireless Commun. **7**(12), 5306–5315 (2008)
58. A. Benjebbour, Y. Saito, Y. Kishiyama, A. Li, A. Harada, T. Nakamura, Concept and practical considerations of non-orthogonal multiple access (NOMA) for future radio access, in *Proceedings of International Symposium on Intelligent Signal Processing and Communications Systems* (ISPACS), Naha, Japan, 12–15 Nov 2013, pp. 770–774
59. Y. Saito, A. Benjebbour, Y. Kishiyama, T. Nakamura, System-level performance evaluation of downlink non-orthogonal multiple access (NOMA), in *Proceedings of IEEE International Symposium on Personal Indoor and Mobile Radio Communications* (PIMRC), London, UK, 8–11 Sept. 2013, pp. 611–615
60. Y. Saito, Y. Kishiyama, A. Benjebbour, T. Nakamura, A. Li, and K. Higuchi, “Non-orthogonal multiple access (NOMA) for cellular future radio access,” *Proc. of IEEE Vehicular Technology Conference*, Dresden, June 2013, pp. 1–5
61. W.-C. Pao, Y.-F. Chen, Adaptive gradient based method for adaptive power allocation in OFDM based cognitive radio networks. IEEE Trans. Veh. Technol. **63**(2), 836–848 (2014)
62. Y. Zhang, C. Leung, An efficient power loading scheme for OFDM based cognitive radio systems. IEEE Trans. Veh. Technol. **59**(4), 1858–1864 (2010)
63. D. Li, Efficient power allocation for multi user cognitive radio networks. Wireless Pers. Commun. **59**(4), 589–597 (2011)
64. Z. Chen, X.-D. Zhang, Power and time allocation between multiple channels in cognitive radio networks. Wireless Pers. Commun. **64**(4), 783–794 (2012)
65. Y. Teng, F.-R. Yu, Y. Wei, L. Wang, Y. Zhang, Behavior modeling for spectrum sharing in wireless cognitive networks. Wireless Netw. **18**(8), 99–947 (2012)
66. G. Singh, Optimization of spectrum management issues for cognitive radio. J. Emerg. Technol. Web Intell. **3**(4), 263–267 (2011) (Invited paper)
67. S. Guo, C. Dang, X. Liao, Distributed algorithm for resource allocation of physical and transport layer in wireless cognitive ad-hoc networks. Wireless Netw. **17**(2), 337–356 (2011)
68. Y. Xi, E.-M. Yeh, Distributed algorithm for spectrum allocation, power control, routing and congestion control in wireless networks, in *Proceeding of the ACM International Symposium on Mobile Ad hoc Networking and Computing*, Canada, 9–14 Sept 2007, pp. 180–189
69. S. Guo, C. Dang, X. Liao, Distributed resource allocation with fairness for cognitive radios in wireless mobile ad-hoc networks. Wireless Netw. **17**(6), 1493–1512 (2011)
70. D. Niyatoand, E. Hossain, Spectrum trading in cognitive radio networks: a market equilibrium based approach. IEEE Wirel. Commun. **15**(6), 71–80 (2008)
71. Q. Yu, A survey of cooperative games for cognitive radio networks. Wireless Pers. Commun. **73**(3), 949–966 (2013)
72. N. Nie, C. Comaniciu, Adaptive channel allocation spectrum etiquette for cognitive radio networks. J. Mobile Netw. Appl. **11**(6), 779–797 (2006)
73. D. Niyato, E. Hossain, A noncooperative game-theoretic framework for radio resource management in 4G heterogeneous wireless access networks. IEEE Trans. Mobile Comput. **7** (3), 332–345 (2008)
74. S. Pandit, G. Singh, Spectrum sharing in cognitive radio using game theory, in *Proceedings of IEEE International Advance Computing Conference* (IACC-2013), India, 22–23 Feb 2013, pp. 1503–1506
75. H.-B. Chang, K.-C. Chen, Auction based spectrum management of cognitive radio networks. IEEE Trans. Veh. Technol. **59**(4), 1923–1935 (2010)
76. S. Maharjan, Y. Zhang, S. Gjessing, Economic approaches for cognitive radio networks: a survey. Wireless Pers. Commun. **57**(1), 33–51 (2011)
77. Z. Ji, K.-J.-R. Liu, Dynamic spectrum sharing: a game theoretical overview. IEEE Commun. Mag. **45**(5), 88–94 (2007)
78. D. Niyato, E. Hossain, Competitive spectrum sharing in cognitive radio networks: a dynamic game approach. IEEE Trans. Wireless Commun. **7**(7), 2651–2660 (2008)

79. O. Aftab, Economic mechanisms for efficient wireless coexistence. MIT Technical Report MIT-LCSTR-876 (Massachusetts Institute of Technology Press, Cambridge, MA) Aug 2002
80. G. Iosifidis, I. Koutsopoulos, Challenges in auction theory driven spectrum management. IEEE Commun. Mag. **49**(8), 128–135 (2011)
81. M.-O. Jackson, J.-M. Swinkels, Existence of equilibrium in single and double private value auctions. Econometrica **73**(1), 93–139 (2005)
82. M. Peters, S. Severinov, Internet auctions with many traders. J. Econ. Theor. **130**(1), 220–245 (2006)
83. Y. Teng, Y. Zhang, C. Dai, F. Yang, M. Song, Dynamic spectrum sharing through double auction mechanism in cognitive radio networks, in *Proceedings of IEEE Wireless Communications and Networking Conference* (WCNC), Cancun, Quintana Roo, 28–31 March 2011, pp. 90–95
84. K. Deshmukh, A.-V. Goldberg, J.-D. Hartline, A.-R. Karlin, Truthful and competitive double auctions, in *Proceedings of the 10th European Symposium on Algorithms* (ESA '02), Rome, Italy, Sept 2002, pp. 361–373
85. G.-S. Kasbekar, S. Sarkar, Spectrum auction framework for access allocation in cognitive radio networks. IEEE/ACM Trans. Network. **18**(6), 1841–1854 (2010)
86. Y. Wu, Q. Zhu, J. Huang, D.-H.-K. Tsang, Revenue sharing based resource allocation for dynamic spectrum access networks. IEEE J. Sel. Areas Commun. **32**(11), 2280–2296 (2014)
87. R. Zhang, Y.-C. Liang, Exploiting multi-antennas for opportunistic spectrum sharing in cognitive radio networks. IEEE J. Sel. Top. Sig. Process. **2**(1), 88–102 (2008)
88. Y. Xu, X. Zhao, Y.-C. Liang, Robust power control and beamforming in cognitive radio networks: a survey. IEEE Commun. Surv. Tutorials 99, (2015) doi:10.1109/COMST.2015.2425040
89. O. Bakr, M. Johnson, R. Mudumbai, K. Ramchandran, Multi-antenna interference cancellation techniques for cognitive radio applications, in *Proceedings of IEEE Wireless Communications and Networking Conference* (WCNC 2009), Budapest, 5–8 April 2009, pp. 1–6
90. M. Chiani, M.-Z. Win, A. Zanella, On the capacity of spatially correlated MIMO Rayleigh-fading channels. IEEE Trans. Inf. Theor. **49**(49), 2363–2371 (2003)
91. W.-C. Jeong, J.-M. Chung, D. Liu, Characteristic-function-based analysis of MIMO systems applying macroscopic selection diversity in mobile communications. J. Electron. Telecommun. Res. Inst. **30**(3), 335–364 (2008)
92. H.-I. Ihan, I. Altunbas, M. Uysal, Moment generating function-based performance evaluation of amplify-and-forward relaying in N^* Nakagami fading channels. IET Commun. **5**(3), 253–263 (2011)
93. V.K. Dwivedi, G. Singh, Moment generating function based performance analysis of maximal ratio combining diversity receivers in the generalized-K fading channels. Wireless Pers. Commun. **77**(3), 1959–1975 (2014)
94. V.K. Dwivedi, G. Singh, A novel marginal MGF based analysis of the channel capacity over correlated Nakagami-m fading with maximal-ratio combining diversity. Prog. Electromagn. Res. B **41**, 333–356 (2012)
95. V.K. Dwivedi, G. Singh, A novel MGF based analysis of channel capacity of generalized-K fading with maximal ratio combining diversity. Prog. Electromagn. Res. C **26**, 153–165 (2012)
96. V.K. Dwivedi, G. Singh, Analysis of channel capacity of generalized-K fading with maximal ratio combining diversity receivers, in *Proceedings International Conference on Communication System and Networking technology* (CSNT-2011), India, 3–5 June 2011, pp. 550–553
97. Y. Wang, D.-W. Yue, Capacity of MIMO Rayleigh fading channels in the presence of interference and receive correlation. IEEE Trans. Veh. Technol. **58**(8), 4398–4405 (2009)
98. L. Bixio, G. Oliveri, M. Ottonello, M. Raffetto, Cognitive radios with multiple antennas exploiting spatial opportunities. IEEE Trans. Sig. Process. **58**(8), 4453–4459 (2010)

99. M.-S. Kang, B.-C. Jung, D.-K. Sung, W. Choi, A pre-whitening scheme in a MIMO-based spectrum-sharing environment. IEEE Commun. Lett. **12**(11), 831–833 (2008)
100. K.-T. Phan, S.-A. Vorobyov, N.-D. Sidiropoulos, C. Tellambura, Spectrum sharing in wireless networks via QoS-aware secondary multicast beamforming. IEEE Trans. Sig. Process. **57**(6), 2323–2335 (2009)
101. S. Sridharan, S. Vishwanath, On the capacity of a class of MIMO cognitive radios. IEEE J. Sel. Top. Sig. Process. **2**(1), 103–117 (2008)
102. M.-G. Adian, H. Aghaeinia, Optimal resource allocation in heterogeneous MIMO cognitive radio networks. Wireless Pers. Commun. **76**(1), 23–39 (2014)
103. S. Rini, A. Goldsmith, On the capacity of the multiantenna Gaussian cognitive interference channel. IEEE J. Sel. Areas Commun. **32**(11), 2252–2267 (2014)
104. E. Telatar, Capacity of multi-antenna Gaussian channels. Eur. Trans. Telecommun. **10**, 585–598 (1999)
105. G.J. Foschini, M.J. Gans, On limits of wireless communications in a fading environment when using multiple antennas. Wireless Person. Commun. **6**, 311–335 (1998)
106. S. Sfar, L. Dai, K.B. Letaief, Optimal diversity-multiplexing tradeoff with group detection for MIMO systems. IEEE Trans. Commun. **53**, 1178–1190 (2005)
107. F. Rashid-Farrokhi, L. Tassiulas, K. Liu, Joint optimal power control and beamforming in wireless networks using antenna arrays. IEEE Trans. Commun. **46**, 1313–1323 (1998)
108. R. Zhang, Y.C. Liang, Exploiting multi-antennas for opportunistic spectrum sharing in cognitive radio networks, in *Proceedings of IEEE 18th International. Symposium on Personal, Indoor, and Mobile Radio Communications (PIMRC*' 07), Athens, Greece, Sept. 2007, pp. 1–5
109. L. Zhang, Y.C. Liang, Y. Xin, Robust cognitive beamforming with partial channel state information, in *Proceedings of CISS 2008*, pp. 890–895, March 2008
110. M.H. Islam, Y.C. Liang, A.T. Hoang, Joint beamforming and power control in the downlink of cognitive radio networks, in *Proceedings of IEEE WCNC*, Hong Kong, March 2007, pp. 21–26
111. K. Hamdi, W. Zhang, K.B. Letaief, Opportunistic spectrum sharing in cognitive MIMO wireless networks. IEEE Trans. Wireless Commun. **8**(8), 4098–4109 (2009)
112. R. Manna, R.H. Louie, Y. Li, B. Vucetic, Cooperative spectrum sharing in cognitive radio networks with multiple antennas. IEEE Trans. Sig. Process. **59**(11), 5509–5522 (2011)
113. B. Maham, P. Popovski, Cognitive multiple-antenna network with outage and rate margins at the primary system. IEEE Trans. Veh. Technol. **64**(6), 2409–2423 (2015)
114. L. Sboui, H. Ghazzai, Z. Rezki, M.S. Alouini, Achievable rate of spectrum sharing cognitive radio multiple-antenna channels. IEEE Trans. Wireless Commun. **14**(9), 4847–4856 (2015)
115. A. Maharshi, L. Tong, A. Swami, Cross-layer designs of multichannel reservation MAC under Rayleigh fading. IEEE Trans. Sig. Process. **51**(8), 2054–2067 (2003)
116. J. Wang, L. Li, S.-H. Low, J.-C. Doyle, Cross-layer optimization in TCP/IP networks. IEEE/ACM Trans. Netw. **13**(3), 582–595 (2005)
117. H. Kim, K.-G. Shin, Efficient discovery of spectrum opportunities with MAC-layer sensing in cognitive radio networks. IEEE Trans. Mobile Comput. **7**(5), 533–545 (2008)
118. C. Stevenson, G. Chouinard, Z.-D. Lei, W.-D. Hu, S. Shellhammer, W. Caldwell, IEEE 802.22: The first cognitive radio wireless regional area network standard. IEEE Commun. Mag. **47**(1), 130–138 (2009)
119. S. Bhattarai, J.-M. Park, B. Gao, K. Bian, W. Lehr, An overview of dynamic spectrum sharing: Ongoing initiatives, challenges, and a roadmap for future research. IEEE Trans. Cognitive Commun. Netw. **2**(2), pp. 110–128 (2016)
120. A. Fragkiadakis, V. Siris, N. Petroulakis, Anomaly-based intrusion detection algorithms for wireless networks, in *WWIC 10* Wired/Wireless Internet Communications, vol. 6074 of the Series Lecture Notes in Computer Science, pp. 192–203
121. A. Fragkiadakis, E. Tragos, T. Tryfonas, I. Askoxylakis, Design and performance evaluation of a lightweight wireless early warning intrusion detection prototype. EURASIP J. Wireless Commun. Netw. **2012**(73), 1–18 (2012)

122. M. Thamilarasu, S. Mishra, R. Sridhar, A cross-layer approach to detect jamming attacks in wireless ad hoc networks, in *Proceedings of IEEE Military Communications Conference* (MILCOM 2006), Washington, DC, pp. 1–7
123. M. Raya, I. Aad, J. Hubaux, Alaeddine El Fawal, DOMINO: detecting MAC layer greedy behavior in IEEE 802.11 Hotspots. IEEE Trans. Mobile Comput. **5**(12), 1–15 (2006)
124. A. Cardenas, S. Radosavac, J. Baras, Evaluation of detection algorithms for mac layer misbehavior: theory and experiments. IEEE/ACM Trans. Netw. **17**, 605–617 (2009)
125. W. Wang, Y. Sun, H. Li, Z. Han, Cross-layer attack and defense in cognitive radio networks, in *Proceedings of IEEE Global Telecommunications Conference* (GLOBECOM 2010), 6–10 Dec 2010, pp. 1–6
126. A. Toledo, X. Wang, Robust detection of selfish misbehavior in wireless networks. IEEE J. Sel. Areas Commun. **25**, 1124–1134 (2007)
127. V. Giri, N. Jaggi, MAC layer misbehavior effectiveness and collective aggressive reaction approach, in *Proceedings of IEEE Sarnoff Symposium*, 12–14 April 2010, pp. 1–5
128. C. Cormio, K. Chowdhury, A survey on mac protocols for cognitive radio networks. Ad Hoc Netw. **7**, 1325–1329 (2009)
129. J. H′andez-Serrano, O. Le′on, M. Soriano, Modeling the lion attack in cognitive radio networks. EURASIP J. Wireless Commun. Network. **2011**, 1–10 (2011)
130. F. Digham, M. Alouini, M. Sinon, On the energy detection of unknown signals over fading channels. IEEE Trans. Commun. **55**, 21–24 (2007)
131. H. Kim, K. Shin, In-band spectrum sensing in IEEE 802.22 WRANs for incumbent protection. IEEE Trans. Mobile Comput. **9**(10), 1766–1779 (2010)
132. S. Gong, W. Liu, W. Yuan, W. Cheng, S. Wang, Threshold-learning in local spectrum sensing of cognitive radio, in *Proceedings of 69th IEEE Vehicular Technology Conference* (VTC Spring 2009), Barcelona, 26–29 April 2009, pp. 1–6
133. C. Clancy, J. Hecker, E. Stuntebeck, Applications of machine learning to cognitive radio networks. IEEE Wireless Commun. **14**(4), 47–52 (2007)
134. K. Bian, J. Park, Enabling fair spectrum sharing: Mitigating selfish misbehaviors in spectrum contention. IEEE Netw. **27**(3), 16–21 (2013)
135. P. Steenkiste, D. Sicker, G. Minden, D. Raychaudhuri (eds.), Future Directions in Cognitive Radio Network Research, in *NSF Workshop Report*, 9–10 March 2009.[Online] (Available at: http://www.cs.cmu.edu/ ~ prs/NSF_CRN_Report_Final.pdf)
136. E. Biglieri, A.J. Goldsmith, L.J. Greenstein, N. Mandayam, H.V. Poor, *Principles of Cognitive Radio* (Cambridge University Press, Cambridge, 2012)

Chapter 2
Spectrum Sensing in Cognitive Radio Networks: Potential Challenges and Future Perspective

2.1 Introduction

With the vast number and diversity of wireless devices and technologies, exponential increase in the number of wireless subscribers, the emergence of new applications, and the continuous demand for higher data rates, RF spectrum is becoming increasingly crowded. These developments in the communications market demand systems and devices which are aware of their RF environment and can facilitate flexible, efficient, and reliable operation and utilization of available spectral resources. Therefore, spectrum sensing and its ability to identify underutilized spectrum is becoming progressively more important to current and future wireless communication systems to identify underutilized spectrum with characterizing interference and consequently, achieving reliable and efficient operation. The cognitive radio is an intelligent radio that is aware of its surrounding environment, capable of learning and adapting its behaviour and operation to provide a better match to its surrounding environment as well as to the user's needs as extensively presented in Chap. 1. In order to exploit spectrum in a dynamic fashion, cognitive radios must have a sensing mechanism for identifying spectrum opportunities and avoiding interference with licensed primary users. In addition to dynamic spectrum access, spectrum sensing techniques are important for both civilian and military spectrum management operations [1].

The operation of a cognitive radio for dynamic spectrum access involves two main components: spectrum sensing and spectrum opportunity exploitation. Due to hardware limitations and energy constraints, a cognitive radio may be unable to sense the entire spectrum simultaneously. Hence, a sensing policy that defines when and which frequency band to sense must be implemented either individually or collaboratively. In addition, we must assume that the sensing periods have already been synchronized among different cognitive radios, because simultaneous transmission and sensing on the same frequency band is generally inefficient. Such a

S. Pandit and G. Singh, *Spectrum Sharing in Cognitive Radio Networks*,
DOI 10.1007/978-3-319-53147-2_2

policy defines whether a cognitive radio performs sensing in a given period and, if so, which channel or channels it senses. Collaborative sensing policies are generally expected to offer benefits over individually selected policies. However, individual sensing policies have been proposed [2–4] that use a decision theory approach by formulating the design of an optimal sensing policy as a partially observable Markov decision process (POMDP). Myopic sensing policies that seek to maximize an immediate reward are analyzed in [5, 6]. Cognitive radios must also determine their access policy in order to exploit available spectral opportunities after they have been detected. An access policy addresses when and on which channels to transmit, or whether to transmit at all if conserving the energy of battery-operated terminals is necessary or channel quality is low. Access policy, like sensing policy, must be determined individually or collectively. Interference management is an integral part of spectrum exploitation. The cognitive radio system must ensure that its combined interference to the primary systems stays within the bounds set by regulatory authorities. Sensing and access policies are closely interwoven, and both are areas where cognition most naturally comes into play. In dynamic signal environment techniques such as reinforcement learning [7, 8], the potential for achieving the most efficient utilization of the available resources is significant [9, 10]. Feedback from past decisions and actions may be used to ascertain the state of the environment and thus enable better decisions in the future. Several other issues must also be resolved, including the modulation formats, transmit powers, and routing issues, as discussed in Chap. 1. Moreover, in addition to technological challenges, regulatory challenges must be met. Regulatory policies defining the rules for opportunistic spectrum access must be established to ensure that cognitive radios conform to the rules. An overview of spectrum sensing methods and algorithms for cognitive radios is presented in the following sections.

2.2 Spectrum Sensing Techniques

Spectrum sensing enables a cognitive radio to measure, learn, and be aware of its operating environment—for instance, spectrum availability and interference status. When a certain frequency band is detected as underutilized by the primary/licensed user at a particular time in a specific position, the secondary users can utilize the spectrum, i.e., a spectrum opportunity exists. Therefore, spectrum sensing can be performed across the domains of time, frequency, and space. With the recent development of beamforming technology, multiple users can utilize the same channel/frequency at the same time in the same geographical location. Thus, if a primary user (PU) is not transmitting in all directions, spectrum opportunities can be created for secondary users in the directions not in service, and spectrum sensing must also take into account the angle of arrivals [9]. The primary users can also use their assigned bands by means of spread-spectrum or frequency hopping, and secondary users can then transmit in the same band simultaneously without severe disruption to the primary users, provided they adopt an orthogonal code in relation to the codes adopted by the primary users [10]. This creates spectrum opportunities in code domain, but requires detection of the codes used by the primary users as well as multipath parameters. Because detecting primary users that are receiving data is generally very difficult, many studies on spectrum sensing have focused on primary transmitter detection based on the local measurements of secondary users Spectrum sensing and channel probing to acquire real-time spectrum/channel information required by the cognitive MAC layer are also critical components of cognitive radio networks. In general, spectrum sensing performs the following tasks [11]: (1) detection of spectrum holes, (2) determination of spectral resolution for each spectrum hole, (3) estimation of the spatial directions of an incoming interfering signal, and (4) signal classification. Among these, the detection of spectrum holes is probably the most important task, and is explored through a binary hypothesis-testing problem. Therefore, detection of spectrum holes on a narrow frequency band is usually referred to as spectrum sensing, which detects the presence or absence of primary users in the underlying band.

Spectrum sensing techniques can be divided into two main categories: non-cooperative/transmitter detection and cooperative detection (Fig. 2.1). Transmitter detection approaches are based on the detection of signals transmitted from a primary system through the local observations of cognitive radio users. Transmitter, or non-cooperative, detection techniques are generally based on the assumption that the location of the primary transmitter is unknown to the cognitive device. Therefore, cognitive users should rely only on the detection of weak primary transmitter signals and use only local observations to perform spectrum sensing. A cognitive device does not have complete knowledge of spectrum occupancy in its coverage area. As a consequence, it is not possible to completely avoid harmful interference with primary users. Moreover, transmitter detection cannot prevent a hidden terminal problem. Three schemes are usually employed for primary transmitter detection: matched filter detection, energy detection, and features detection. These schemes are discussed in detail in Sect. 2.3.

A cognitive user (CU) may have a good line of sight with a primary receiver, but may not be able to detect the presence of a primary transmitter (hidden terminal) as a result of the shadowing phenomenon, which is very common in urban/indoor

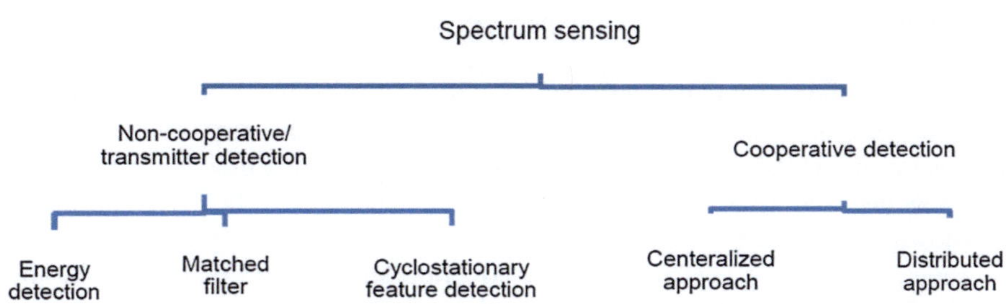

Fig. 2.1 Spectrum sensing techniques

environments. Cooperative detection strategies are implemented to mitigate this problem. Cooperative detection refers to spectrum sensing methods that enable multiple cognitive radios to share their local sensing information for more accurate primary transmitter detection [9, 12]. Cooperative detection can be implemented in either a centralized or a distributed manner. In the centralized method, a central unit collects sensing information from cognitive devices, identifies the available spectrum bands, and broadcasts this information to other cognitive radios [9]. In a distributed approach, there is no central node, and the sensing information is shared among the cognitive devices [9]. Distributed detection is easier to implement and does not require a backbone infrastructure, while centralized detection is more accurate and can effectively mitigate both multi-path fading and shadowing effects. The central node can also assign a specific weight to each spectrum sensing result to mitigate fading phenomena [13]. Cooperative detection techniques can be also classified as a soft or hard combination, according to the nature of the information shared among cognitive users. The soft combination refers to a cooperative strategy in which each node senses a certain frequency band and then sends the results of its measures—i.e., the energy of the received signal—to the central node [14–17]. Conversely, in hard combination strategies, each node decides whether a primary user is present, and then reports to the central node only the results of its decision [14–17]. Soft detection is usually more accurate and can implement macro-diversity techniques, as signals received from distant nodes tend to be uncorrelated. Hard detection is not as accurate but requires less information exchange between nodes. If a cognitive device is equipped with multiple antennas, sophisticated sensing strategies can be implemented, exploiting spatial, time, and/or frequency coding. Such cooperative spectrum sensing is discussed in detail in [18], and the authors demonstrate that the probability of false alarms can be reduced through the use of space, time, and frequency transmit diversity. Relay diversity can be further employed to compensate for the reduced sensing diversity order when some nodes in a cooperative spectrum sensing system cannot report directly to the central node (i.e., due to shadowing phenomenon).

Generally, spectrum sensing is performed using simple signal detection methods to detect unoccupied frequencies as quickly as possible. However, these simple techniques cannot achieve reliable and accurate sensing results in low-SNR and deep fading environments [9, 19]. Various methods have been proposed to enhance the reliability and accuracy of spectrum detection including fusion of multiple local detection decisions and cooperative spectrum sensing [20, 21]. The selection of the most suitable detection method for local spectrum sensing is a major challenge, because detection techniques differ in their performance. For example, the energy detector (ED) is unable to detect signals with low SNR. This can be achieved with the cyclostationary feature detector (CSFD), but with added time and complexity. The matched filter (MF) is the optimal detection technique if the PU's information is known. In contrast to the matched filter and cyclostationary feature detector, however, the energy detector requires no prior knowledge of the primary user signal. These observations raise the question of whether it would be possible to enhance sensing performance through collaboration among different detection techniques for local spectrum sensing, and if so, at what cost. Recent studies have proposed a two-stage spectrum sensing model, with a simple detection method is used in the first stage, and a more powerful one is used in the second stage [22, 23]. To achieve optimal performance, spectrum sensing techniques must be able to identify spectrum holes and any change in frequency-in-use status in a quick, secure, accurate, and reliable manner. Figure 2.2 shows potential requirements for spectrum sensing. However, developing a cognitive radio with spectrum sensing capability that meets all these requirements is impeded by several challenges. Detection results have a dramatic effect on the accuracy of the other cognitive radio components. Spectrum sensing is thus a critical issue in cognitive radio, and has recently received the attention of many researchers.

Cognitive radio can interact with its radio environment to acquire important information about its surroundings, including the presence of primary users and appearance of spectrum holes during spectrum sensing [1]. It is only with this information that it can adapt its transmitting and receiving parameters, such as transmission power, frequency, and modulation schemes, in order to achieve efficient spectrum utilization. Therefore, spectrum sensing and analysis is the first critical step toward dynamic spectrum management. In this chapter, we discuss three aspects of spectrum sensing: (1) spectrum hole detection, for determining additional available spectrum resources, including a comparison of several detection techniques; (2) cooperative sensing, which involves cooperation among

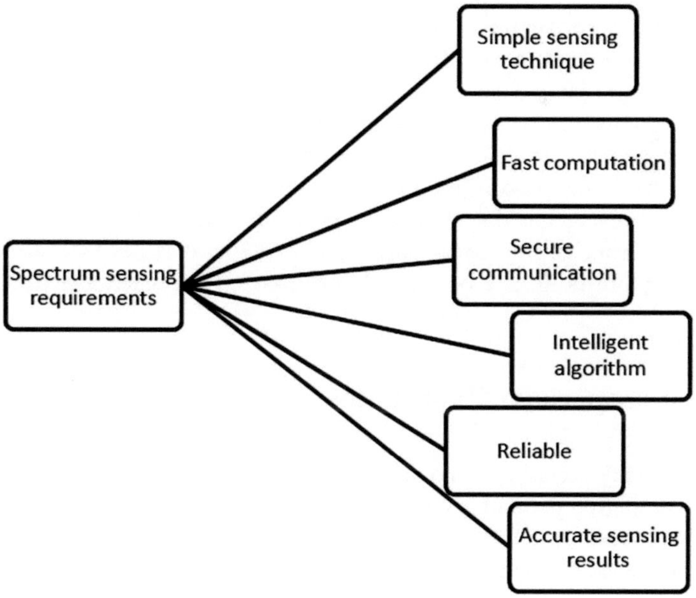

Fig. 2.2 Potential requirements of spectrum sensing

multiple cognitive users; and (3) interference temperature detection, which measures the interference level observed at a receiver and is used to protect licensed primary users from harmful interference due to unlicensed secondary users.

The model for transmitter detection can be described as a classical hypothesis testing approach, where H_0 is the null hypothesis, which states that there is no primary signal in a certain band, and H_1 is the alternative hypothesis (i.e., presence of the primary user). A testing variable is compared with a specific threshold to discriminate between the two hypotheses. System performance is evaluated in terms of probability of detection P_d (the probability of detecting the presence of a primary user) and probability of false alarm P_f (the probability of declaring the presence of a primary user in bands that are actually empty). Let us assume that the hypothesis model of the signal received at a cognitive radio user is:

$$y(t) = \begin{cases} h.s(t) + w(t) & H_1 : if\ PU\ is\,present \\ w(t) & H_0 : if\ PU\ is\,absent \end{cases} \tag{2.1}$$

where $y(t)$ is the received signal, h and $s(t)$ are the channel gain and primary user's signal to be detected at the secondary user (SU), which is assumed to be a Gaussian random process with variance σ_s^2, and $w(t)$ is the additive white Gaussian noise (AWGN) with zero-mean and variance σ_n^2. H_0 is a null hypothesis, meaning there is no primary user present in the band, while H_1 indicates the primary user's presence.

In the above-mentioned binary hypothesis test, there are two types of errors: type I and type II. A type I error, often called the probability of false alarm, is made if H_1 is accepted when H_0 is true. In spectrum sensing, the probability of a false alarm is an important design parameter for a detector, because it causes spectral opportunities to be overlooked. A type II error, on the other hand, occurs if H_0 is accepted when H_1 is true, known as a missed detection, which leads to collisions with primary user transmission and reduced data rates for both the primary and secondary user systems. In general, a cognitive radio system should satisfy the constraints of both the probability of false alarm and the probability of missed detection. However, the detection rule presents a trade-off between these two probabilities. From an implementation point of view, it is desirable to have algorithms whose threshold may be set and performance evaluated analytically.

2.3 Non-cooperative/Transmitter Detection

Spectrum sensing techniques requiring prior knowledge about the primary user's signal for comparing particular signal features to the cognitive user's received signal are called coherent signal detection techniques. Non-coherent detection techniques compare the received signal to a threshold defined on the basis of features that are independent of primary signal knowledge. Alternatively, spectrum sensing techniques can also be classified from a bandwidth perspective into wideband and narrowband detection techniques. Non-cooperative/transmitter detection is so named because cognitive radio sensing only detects a transmitted signal from a primary user transmitter [19]. Transmitter detection is classified as follows:

2.3.1 Energy Detection

Energy detection is the most commonly used spectrum sensing technique for determining the presence or absence of a primary user signal without requiring any information regarding the nature of the primary user signal. Energy detection is robust to the variation in the primary signal because it does not need any a priori knowledge of the primary signal. In the energy detection technique, shown in Fig. 2.3, the energy of a received signal is used to detect a primary user signal, and the presence of a signal in the channel is detected if the energy present is significantly greater than only noise [23]. Initially, the energy detector filters out the undesired signal from the unwanted frequency band [24]. The resulting output samples from the filter are then squared and summed, basically computing the signal energy. Finally, the output is compared with a threshold λ [25] to determine whether a licensed user is present or not as shown in Fig. 2.3. Setting the proper threshold is a challenging task, as it must differentiate between the signal and noise. Energy detection is the simplest method of detection. However, a priori knowledge of noise energy level is necessary, as its uncertainty degrades detector performance [26].

In addition, energy detection does not involve complicated signal processing and has low complexity that is especially suitable for wideband spectrum sensing. In this case, the simultaneous sensing of a number of sub-bands can be realized by simply scanning the power spectral density (PSD) of the received wideband signal. However, it is preferable to complete wideband spectrum sensing via the following two stages:

(1) Low-complexity energy detection is applied to search for possible idle sub-bands.
(2) More advanced spectrum sensing techniques with a higher detection sensitivity, and therefore higher complexity, are applied for accurate idle band detection.

Further, in a cognitive radio network, sensing time and periodic sensing intervals are optimized to maximize sensing accuracy or cognitive user throughput. In the energy detector, sensing time influences detector performance in terms of the probability of false alarm and the probability of missed detection. Moreover, when

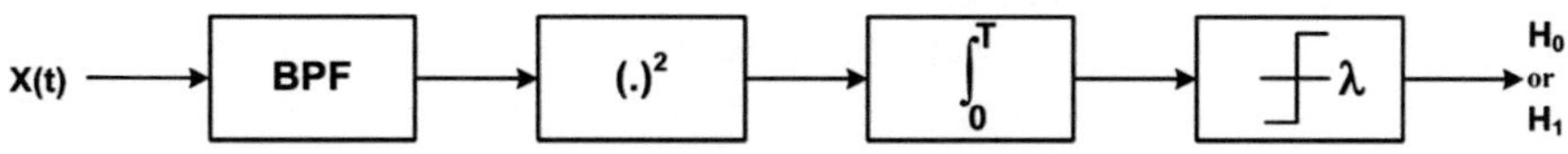

Fig. 2.3 The energy detection technique [9]

periodic sensing [9, 11, 12] is adopted, the periodic sensing interval affects the ability of the detector to grasp the spectrum opportunities and utilize them. If we consider optimizing the sensing time and the periodic sensing interval for each channel in the PU spectrum, then the objective would be to achieve the highest possible detector performance and opportunity utilization in that channel. For a multichannel system, this objective will still hold true, with a different interpretation of opportunity utilization, reflecting the utilization for all available opportunities in all channels rather than each channel individually.

However, energy detection is limited, as follows: (1) the energy detector cannot distinguish among the primary user signals, secondary user signals, and interference; (2) energy detection is susceptible to uncertainty in noise power; (3) prior knowledge of noise power or a reliable estimate of it is needed to obtain best performance; and (4) noise level uncertainty renders robust detection below a certain SNR impossible [20, 23]. To constrain the resulting false alarm rate, the detection threshold has to be set based on the worst case noise level uncertainty. As a result, energy detector performance depends heavily on the accuracy and reliability of the noise level estimate. The noise level may be estimated from guard bands or the detection may be performed in the frequency domain using a channelized radiometer [27, 28], which divides the total frequency band into smaller channels and then integrates energy from each channel separately using a radiometer. If the noise bandwidth is significantly larger than the signal bandwidth, a reasonably accurate noise level estimate may be obtained. In addition, collaboration among secondary users that employ energy detection mitigates the effects of noise uncertainty when users are experiencing independent and identically distributed (i.i.d.) fading or shadowing [29–31]. A review of the literature on energy-based detection is provided in [28]. Constant false alarm rate (CFAR) strategies for the channelized radiometer, such as cell averaging as discussed in [32], are considered in [28], and recent performance analyses of energy detection in fading channels are carried out in [21, 29, 33–37]. Experimental measurements of energy detection performance with noise uncertainty have been provided [36, 37]. If the signal power is below a certain threshold, called the SNR wall, the energy detector cannot distinguish the signal from a slightly larger noise power, regardless of the detection time [26]. Further, energy detection is suitable for random signal detection, and it does not require any assumptions about the primary signal. Unfortunately, this also means that energy detection cannot distinguish among different signals or interference. Ultimately, therefore, energy detection is not a suitable sensing approach if efficient spectral opportunity utilization is desired.

In the given flow diagram (Fig. 2.4), the probability of detection (P_d) and the probability of false alarm (P_f) are computed in order to analyze the effect of the fading channels on the performance metrics for detection. Further, in order to maximize the probability of detection, the threshold value is lowered. The detection statistics of an energy detector can be defined as the average (or total) energy of N observed samples. The energy of the received signal, which is the decision statistic, is given by [38]:

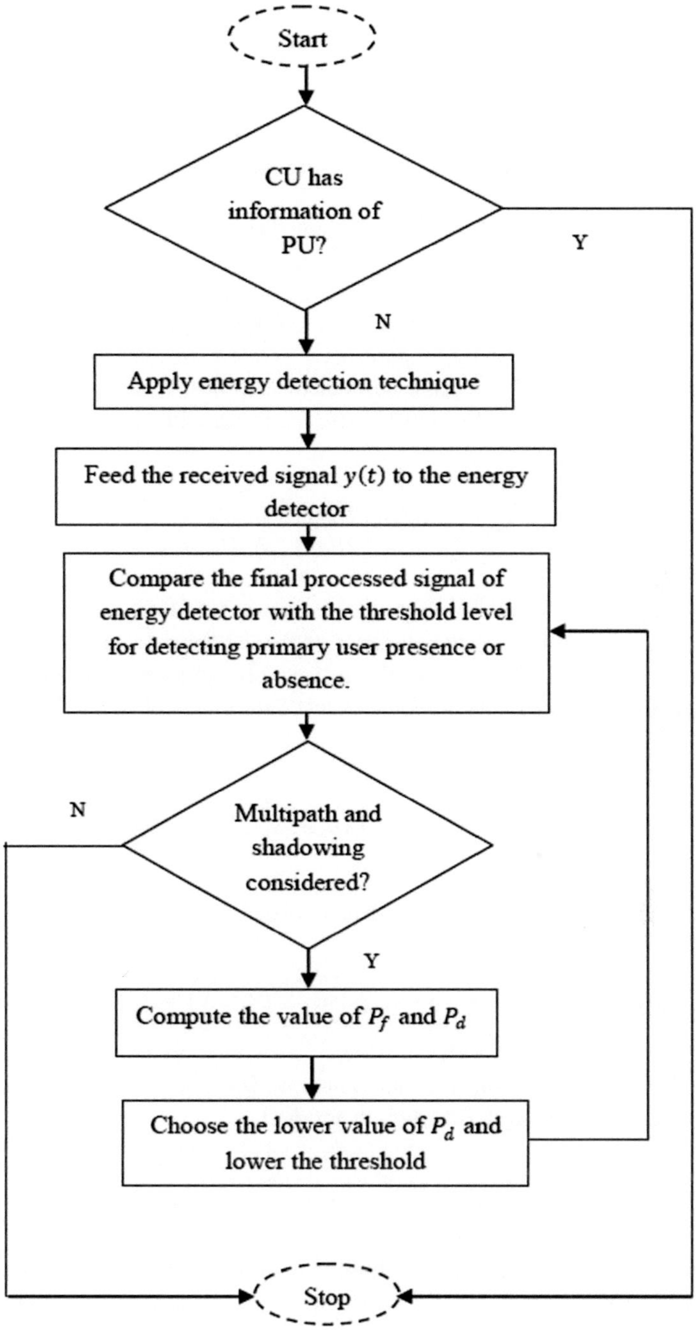

Fig. 2.4 Flow sequence of the energy detection technique

$$T = \frac{1}{N}\sum_{t=1}^{N} y^2(t) \tag{2.2}$$

where N is the number of samples considered. The energy detector model for cognitive radio can be formulated as the following binary hypothesis [38]:

$$d_{ED} = \begin{cases} +1 & \textit{if } H_1 \textit{ is declaired } (T \geq \lambda) \\ -1 & \textit{if } H_0 \textit{ is declaired } (T < \lambda) \end{cases} \tag{2.3}$$

The decision on whether the spectrum is being occupied by the primary user is made by comparing the detection statistics T (chi-square distribution) with a predetermined threshold λ. For a large number of samples, T can be approximated to Gaussian distribution using the central limit theorem, with test statistics as follows [38]:

$$T \sim \begin{cases} N(L\sigma_n^2, 2L\sigma_n^4) & \textit{if } T \geq \lambda \\ N(L\sigma_t^2, 2L\sigma_t^4) & \textit{if } T < \lambda \end{cases} \tag{2.4}$$

where $\sigma_t^2 = \sigma_n^2 + \sigma_s^2$. The probability of false alarm, detection, and missed detection (P_m) are given in [38]. The performance of the detector is characterized by two probabilities: the probability of false alarm P_f and the probability of detection P_d. P_f denotes the probability that the hypothesis test determines H_1 while it is actually H_0 that is [38, 39]:

$$P_f(ED) = P(T > \lambda/H_0) = \Gamma(u, \lambda/2)/\Gamma(u) = Q\left(\frac{\lambda - \sigma_n^2}{\sqrt{2L\sigma_n^4}}\right) \tag{2.5}$$

and P_d denotes the probability that the test correctly decides H_1.

$$P_d(ED) = P(T > \lambda/H_1) = Q_u\left(\sqrt{2SNR}, \sqrt{\lambda}\right) = Q\left(\frac{\lambda - \sigma_t^2}{\sqrt{2L\sigma_t^4}}\right) \tag{2.6}$$

where $\Gamma(.)$ is the incomplete gamma function, $Q_u(.)$ is the generalized Marcum Q-function, and u is the time-bandwidth product,

and

$$P_m(ED) = P(T > \lambda/H_1) = (1 - P_d) \tag{2.7}$$

A robust detector should ensure a high detection probability P_d and a low false alarm probability P_f, or it should optimize the spectrum usage efficiency while guaranteeing a certain level of primary user protection. To this end, various approaches have been proposed to improve energy detector efficiency for spectrum sensing. As detection performance is very sensitive to the noise power estimation error [40], an adaptive noise level estimation approach is proposed [41] in which the multiple signal classification algorithms are used to decouple the noise and signal subspaces and estimate the noise floor. A well-chosen detection threshold can minimize spectrum sensing error, provide the primary user with adequate protection, and fully enhance spectrum utilization. In [42], the detection threshold is optimized iteratively to satisfy the requirement on false alarm probability. Threshold optimization subject to spectrum sensing constraints is investigated in [43], where an optimal adaptive threshold level is developed by utilizing the spectrum sensing error function. Forward methods for energy detection have been proposed [44], where the noise power is unknown and is adaptively estimated. A localization algorithm based on double-thresholding (LAD) has been proposed [45] for finding and localizing narrowband signals, where the use of two thresholds can provide signal separation and localization. The LAD method with normalized thresholds can reduce computational complexity without performance loss by combining adjacent clusters, enabling more accurate estimation of the number of narrowband signals. The sensing throughput trade-off of energy detection is studied in [46], where the sensing period duration in a time slot is optimized to maximize the achievable throughput for the secondary users under the constraint that the primary users are sufficiently protected. A novel wideband spectrum sensing technique based on energy detection has been introduced in [47], in which joint detection of signal energy levels over multiple frequency bands improves the opportunistic throughput of CRs and reduces their interference with the primary systems. Analysis in [48] shows that detection of narrowband transmission using energy detection over multiband orthogonal frequency-division multiplexing (OFDM) is feasible, and can be further extended to cover more complex systems. Further, with the noise power level constantly changing over time, the determination of a detection threshold becomes challenging. Even if the threshold is set adaptively, the presence of any in-band interference would confuse the energy detector. In addition, in frequency-selective fading, it is not clear how the threshold is set with respect to channel notches. Because the energy detector cannot recognize the interference, it cannot benefit from adaptive signal processing for canceling the interferer. Furthermore, the spectrum policy for use of the band is limited to primary users, so a cognitive user should treat noise and other secondary users differently. Lastly, an energy detector is not effective for spread-spectrum signals—direct sequence and frequency hopping signals—for which more sophisticated signal processing algorithms must to be devised.

2.3.2 Matched Filter Detection

The matched filter is a coherent detection technique that employs a correlator matched to the signal of interest or to specific parts of it such as pilot and training sequences. Coherent detection processing provides very good performance under nominal conditions. With this technique, the received signal is matched with the PU signal, and the presence or absence of PU can thus be determined. Matched filter detection assumes that Gaussian noise exists, for which matched filtering is the optimal detection technique [49]. However, with the matched filter detection, the cognitive user needs to be fully synchronized with the PU, a capability that is not possible in most cases, particularly with low SNRs. The matched filter method detects a signal by computing the correlation between the received signal and a known copy of the signal. As the optimal detection technique, however, it requires perfect information regarding the primary user's signal, such as the operating frequency, bandwidth, modulation type and order, pulse shape, and packet format. In addition, if incorrect information is used for matched filtering, detection performance will be degraded. On the other hand, most wireless communication systems exhibit certain patterns, such as pilot tones, preambles, midambles, and spreading codes, which are used for purposes of control, equalization, synchronization, continuity, or reference. Even when perfect knowledge of a primary user's signal is not attainable, if a certain pattern is known from the received signals, coherent detection can be used to determine whether a primary user is transmitting [50] (Fig. 2.5).

Matched filter is the optimal detection method [51, 52] when the secondary user has a priori information on the primary user's signal. A matched filter can correlate a previously identified primary signal with the received signal to detect the presence of the primary user, maximizing the SNR in the presence of additive stochastic noise. An advantage of a matched filter, which needs fewer received signal samples, is the short time it requires to achieve acceptable detection performance such as a low probability of missed detection or false alarm [53]. However, the required number of signal samples also grows as the received SNR decreases, so there also exists a SNR wall [26] for a matched filter. Further, the matched filter needs receivers for all types of signals and corresponding receiver algorithms to be executed, resulting in excessive implementation complexity and power consumption [54]. There are two hypotheses in coherent detection:

$$y(t) = \begin{cases} \sqrt{\varepsilon}x_p(t) + \sqrt{1-\varepsilon}x(t) + w(t) & H_1 : \textit{if PU is present} \\ w(t) & H_0 : \textit{if PU is absent} \end{cases} \tag{2.8}$$

where $x_p(t)$ is a known pilot tone, ε is the fraction of energy allocated to the pilot tone, and $x(t)$ is the desired signals assumed to be orthogonal to the pilot tone. The

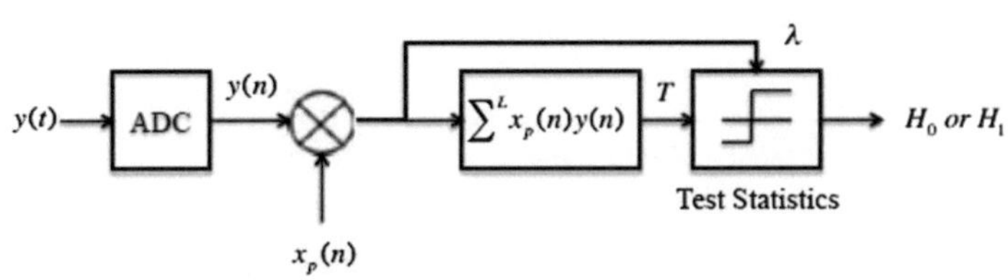

Fig. 2.5 Schematic of matched filter [50]

test statistics of the coherent detection are defined as the projected received signal in the pilot direction that is:

$$T = \frac{1}{N}\sum_{t=1}^{N} y(t)\hat{x}_p(t) \tag{2.9}$$

with $\hat{x}_p(t)$ representing a normalized unit vector in the direction of the pilot tone. As N increases, test statistics T under hypothesis H_1 are much greater than those under H_0. By comparing T with a predetermined detection threshold, one can determine the presence of a primary user. Coherent detection can also be performed in the frequency domain [55]. One can express the binary hypothesis test using the power spectrum density of the received signal $S_Y(\omega)$, and distinguish between H_0 and H_1 by exploiting the unique spectral signature exhibited in $S_X(\omega)$. Coherent detection is robust to noise uncertainty and not limited by the SNR wall [56], as N is large enough. Moreover, coherent detection outperforms energy detection in sensing convergence time [57, 58], because the sensing time of energy detection increases quadratically with SNR reduction, while that of coherent detection increases only linearly [58]. However, information about waveform patterns is a prerequisite for implementing coherent detection; the more precise the information that a coherent detector has, the better the sensing performance will be. The matched filter is Gaussian in nature and works on the principle of maximizing the received SNR. However, the main advantage of matched filter detection is that, because of coherency, it requires less time to achieve high processing gain. The flow sequence of matched filter detection, beginning with the hypothesis model, is demonstrated in Fig. 2.6.

The CU receives the signal $y(t)$. In order to apply the matched filter technique, the CU must have the PU signal information. The main advantage of the matched filter is that, because of coherency, it requires less time to achieve high processing gain, since only $O(1/\text{SNR})$ samples are needed to meet a given probability of detection constraint. However, a significant drawback of the technique is that the cognitive radio would need a dedicated receiver for every primary user class.

2.3.3 Cyclostationary Feature Detection

Feature detection relies on identification of primary signals based on their deterministic or statistical properties. Since feature detection is based on extracted signal features, it can distinguish signals with different features. In general, feature detection has higher computational complexity than energy detection or matched filtering. One important subclass of feature detectors is the cyclostationarity-based detectors, which is more robust against noise uncertainty than that of the energy based detection because the noise is typically not cyclostationary. However, cyclostationarity-based detection can be very sensitive to synchronization errors,

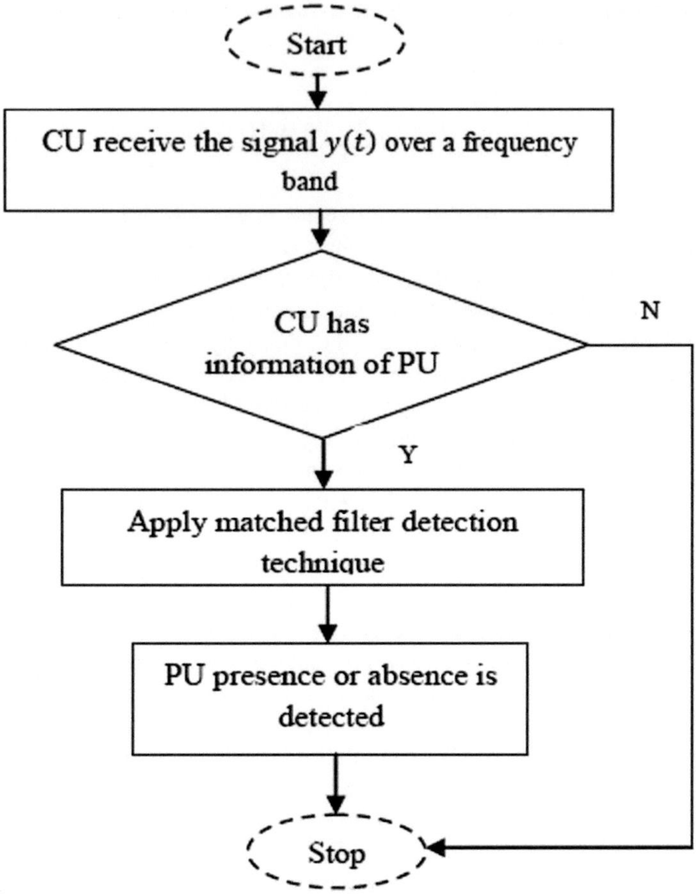

Fig. 2.6 Flow sequence of matched filter detection

resulting in carrier frequency and sampling clock frequency offsets. The cyclostationary feature detection technique used in cognitive radio is a very attractive spectrum sensing scheme because it is capable of differentiating the primary signal from interference and noise [59]. This spectrum sensing technique relies on periodic redundancy introduced into the signal by modulation and sampling because modulated signals are, in general, coupled with sine wave carriers, pulse trains, spreading sequences, or cyclic prefixes, causing periodicity in the transmitted signal [60, 61]. The cyclostationary feature detector uses these non-random periodic statistics of signals for detection by observing the mean and autocorrelation of the received signal. If the mean and autocorrelation vary periodically in time, then the received signal is associated with the primary user, otherwise it is noise, which lacks periodicity. As a result, cyclostationary feature detectors can operate successfully in extremely low-SNR environments and can differentiate between the primary user signal and noise [61]. This detector has demonstrated enhanced detection capability, especially in the presence of noise power uncertainty, and is suitable when the pilot signal of the primary user is known. However, a matched filter detector is more suitable when the period of the primary signal is known. Probability-based spectrum sensing techniques have recently been proposed, utilizing statistical information on primary user activity. The more a cognitive user knows about the primary signal, the better the detector works. These types of detectors exploit certain PU signal properties, such as pilots or cyclostationary features to perform the detection. However, this type of detection requires a very accurate synchronization which is difficult to maintain under low-SNR conditions [62]. A schematic of cyclostationary feature detection is shown in Fig. 2.7.

There are specific features associated with the information transmission of a primary user. For instance, the statistics of transmitted signals in many communication paradigms are periodic because of inherent periodicities such as the modulation rate and carrier frequency. Such features are typically viewed as cyclostationary, based upon which a detector can distinguish cyclostationary signals from stationary noise. In a more general sense, the features can refer to any intrinsic characteristics associated with a primary user's transmission, as well as cyclostationary features. For example, center frequencies and bandwidths [63] extracted from energy detection can also serve as reference features for classification and determining a primary user's presence. As in most communication systems, transmitted signals are modulated signals coupled with sine wave carriers, pulse trains, hopping sequences, or cyclic prefixes, while additive noise is generally wide-sense stationary (WSS) with no correlation. Therefore, cyclostationary feature detectors can differentiate noise from primary users' signals [57, 64, 65] and can distinguish among different types of transmissions and primary systems [66]. A cyclostationary feature detector differs from an energy detector, which uses time-domain signal energy as test statistics however cyclostationary feature detector perform a transformation from the time-domain into the frequency feature domain, followed by conducting a hypothesis test in the new domain. Specifically, the cyclic autocorrelation function (CAF) of the received signal is defined as:

$$R_y^\alpha = E[Y(t+\tau)Y^*(t-\tau)e^{j2\pi\alpha t}] \quad (2.10)$$

where $E[.]$ is the expectation operation, * denotes the complex conjugation, and α is the cyclic frequency. Given that periodicity is a common property of wireless modulated signals, while noise is WSS, the received signal's CAF also demon-

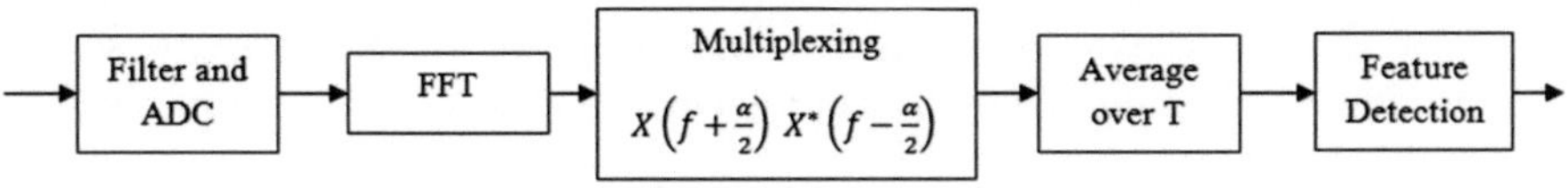

Fig. 2.7 Schematic of cyclostationary feature detection

strates periodicity when the primary signal is present. Thus, we can represent CAF using its Fourier series expansion, called the cyclic spectrum density (CSD) function, expressed as [54]:

$$S(f, \alpha) = \sum_{\tau=-\infty}^{\infty} R_Y^{\alpha}(\tau) e^{-j2\pi f \tau} \tag{2.11}$$

The CSD function has peaks when the cyclic frequency α equals the fundamental frequencies of the transmitted signal *x*(*t*), i.e., $\alpha = (k/T_x)$ with T_x being the period of *x*(*t*). Under the hypothesis H_0, the CSD function does not have any peaks, as the noise comprises non-cyclostationary signals. A peak detector [67] or a generalized likelihood ratio test [57] can be further used to distinguish between the two hypotheses. Different primary communication systems using different air interfaces (modulation, multiplexing, coding, etc.) can also be differentiated by their different cyclostationarity properties. Indeed, in comparison to energy detectors, which are prone to high false alarm risk due to noise uncertainty and are unable to detect weak signals in noise, cyclostationary detectors represent an attractive alternative, as they can differentiate noise from the primary user's signal and have more robust detection in a low-SNR regime. A spectrum sensing method based on maximum cyclic autocorrelation selection was proposed in [55] in which the peak and non-peak values of the cyclic autocorrelation function were compared to determine whether the primary signal was present. This method does not require noise variance estimation, and is robust against noise uncertainty and interference signals. Frequency-selective fading and uncertain noise impair the robustness of cyclostationary signal detection in low SNR environments. Run time noise calibration has been considered in [9, 56] in order to improve detector robustness. The method exploits in-band measurements at frequencies where a pilot is absent in order to calibrate the noise statistics at the pilot frequencies. Generalized feature detection refers to a detection and classification process that extracts feature information other than the cyclostationarity due to the modulated primary signals, such as the transmission technologies used by a primary user, the amount of energy and its distribution across different frequencies [68, 69], channel bandwidth and its shape [45, 48], power spectrum density [45], center frequency [48], and fast Fourier transform (FFT)-type features [60]. Primary users can be identified by matching the features extracted from the received signal to a priori information on primary user transmission characteristics. Information on the location of the primary signal is also an important feature that can be used to distinguish a primary user from other signal sources. Under a primary user emulation attack, a malicious secondary user transmits signals whose characteristics emulate those of the primary signals. A transmitter verification scheme is proposed in [70] to secure trustworthy spectrum sensing based on verification of primary user location.

In conclusion, the cyclostationary technique is implemented in order to differentiate between the primary user signal and noise signal by exploiting the unique nature of the received signal *y*(*t*). This is performed by the modulation of the received signal using periodic sequences and then computing the spectral correlation function to detect the correlation. If a correlation exists, then the primary user signal is assumed to be present; otherwise, it is noise, and further action is required according the detection results, as shown in Fig. 2.8. These modulated signals are characterized as cyclostationary [22], since their mean and autocorrelation exhibit periodicity. Such features are detected by analyzing a spectral correlation function. The main advantage of this function is its ability to differentiate noise energy from modulated signal energy, which is a primary need. As a result, since the noise is a wide-sense stationary signal with no correlation [23], modulated signals are cyclostationary, with spectral correlation due to the embedded redundancy of signal periodicity. Therefore, a cyclostationary feature detector can perform better than an energy detector in discriminating against noise because of its robustness to the uncertainty in noise power. The flow sequence of a cyclostationary detection technique with the hypothesis model is demonstrated in Fig. 2.8.

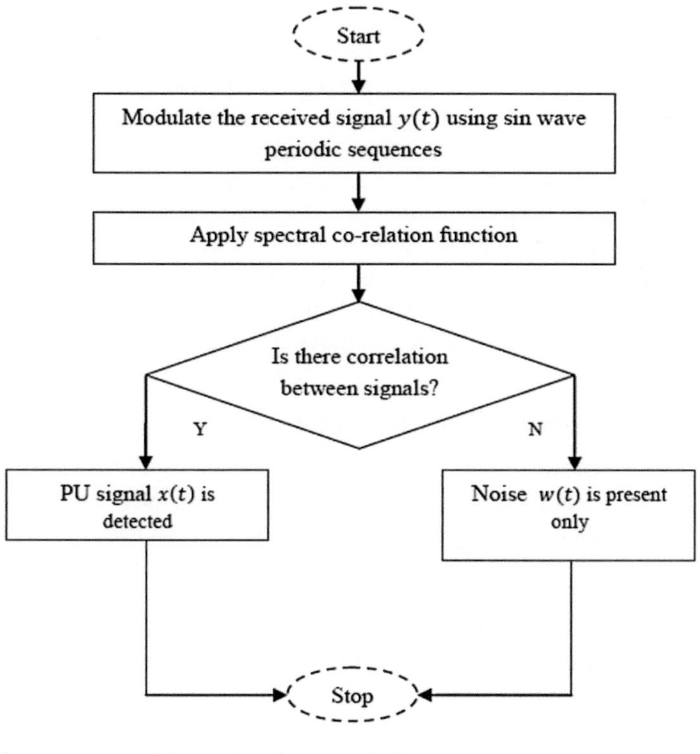

Fig. 2.8 Flow sequence of the cyclostationary technique

2.4 Cooperative Detection

The hidden terminal problem is a critical issue in spectrum sensing. It occurs when cognitive radio is shadowed and, because of the very low SNR of the received signal, cannot reliably sense the presence of the primary user. This cognitive radio assumes that the observed channel is vacant, and begins to access the channel while the primary user is still in operation, resulting in interference. As discussed in the preceding sections, several challenges are inherent in spectrum sensing which can negatively impact sensing reliability. In addition, each of the local spectrum sensing techniques has its own strengths and limitations, and no any optimal scheme exists for all applications and scenarios. Various spectrum sensing studies have proposed that cooperation among several spatially distributed cognitive users is needed to mitigate the issues with local spectrum sensing techniques. Therefore, multiple cognitive radios can perform spectrum sensing in a coordinated and cooperative manner. Several recent works have shown that cooperative spectrum sensing can greatly increase the probability of detection in fading channels [21]. Cooperative sensing in cognitive radio networks is analogous to distributed decision making in wireless sensor networks, where each sensor makes a local decision, and those decision results are reported to a fusion center (FC) to produce a final decision according to a certain fusion rule [71]. The main difference between these two applications lies in the wireless environment. Compared to wireless sensor networks, cognitive radios and the FC (or common receiver) are distributed over a larger geographic area. This creates a much greater challenge for cooperative spectrum sensing, because sensing channels (from the primary user to cognitive radios) and reporting channels (from cognitive radios to the FC or common receiver) are normally subject to fading or heavy shadowing. Therefore, cooperative spectrum sensing aims to utilize the variation in cognitive user locations to ultimately produce one global decision for all cognitive users [72]. Based on the methods used by cognitive users to share their sensing information, cooperative spectrum sensing techniques can be categorized into two main classes: centralized and distributed [73].

- Centralized Cooperative Spectrum Sensing

In this class, all cognitive users sense a band of interest using the same or different sensing techniques, and ultimately send their local decisions, either hard or soft, through a control channel to a central unit. Subsequently, all received data are fused to arrive at one final or global decision regarding the PU's current status [73, 74]. Interestingly, centralized cooperative spectrum sensing can be organized into both centralized and distributed types. If the fusion process is performed at a central base station, the cooperative system is recognized as a centralized model. In cognitive radio ad hoc networks (CRAHNs), on the other hand there is no base station and one of the cooperating nodes coordinates the synchronization and fusion processes [21, 75]. Several fusion models that rely on various factors to make their final decision have been suggested in the literature.

- Distributed Cooperative Spectrum Sensing

Instead of relying on a central FC, cognitive nodes exchange sensing information and eventually converge to make one global decision after trading information several times. Distributed cooperative spectrum sensing systems might cost less than other models because their establishment does not require any infrastructure. Several algorithms have been employed in cooperative spectrum sensing to coordinate the sensed data at different cognitive nodes. A discrete time gossip protocol has been employed in which a secondary user senses a band of interest during a certain time slot, and later sends its observations to a set of neighboring cognitive users selected at random [76]. Similarly, a dissemination strategy for sensing information among cognitive users has also been proposed [77], where a small group of cooperating cognitive users exchange their local decisions during a particular time slot, after which a cognitive user within this group sends all received data to a randomly selected neighbor that serves as the designated user in the next time slot, and so on, until all cognitive users receive the sensing information.

However, for traditional cooperative spectrum sensing algorithms such as AND, OR, and majority fusion rules, if most or all of the cognitive nodes are located in low-SNR environments, the cooperation between these nodes provides no advantage, and can even degrade the overall sensing accuracy. This is largely because those cooperative spectrum sensing techniques involve sensing information acquired blindly by different unlicensed users, without specific consideration for the surrounding contexts (e.g., SNR values) of these secondary users. In this study, the SNR value of every secondary user is considered within the contextual data in the fusion process. In fact, the value of the SNR for each cognitive node implicitly works as a weighting factor for the SU's local detection information.

The entire centralized cooperative spectrum sensing process consists of three steps: local spectrum sensing, transmission of the results of local spectrum sensing, and information fusion. We will now briefly describe these three steps of cooperative spectrum sensing and highlight the problems we have considered in each of these.

Step 1: Local spectrum sensing
Every cognitive radio performs local spectrum measurements independently, utilizing detection algorithms such as energy detection, matched filter detection or cyclostationary detection, and then makes a binary decision. Because energy detection is a simple and facile method, as discussed in the previous section, many studies have used this technique to assess local spectrum sensing performance [78–80]. When this method is used in local spectrum sensing, each cognitive user transmits the detected energy signal or decision results to the destination node.

Step 2: Transmission of the results of local spectrum sensing
In centralized cooperative spectrum sensing, each cognitive user sends detected signals to the FC through the reporting channel. Many researchers have studied cooperative spectrum sensing performance when the reporting channels (the channels from the SUs to the FC) are additive white Gaussian noise (AWGN) [29,

73, 74, 81, 82]. The hidden terminal problem also exists in the reporting channels—for example, shadowing between the cognitive user and FC. Thus the data transmitted from the cognitive user to the FC will be impacted by channel fading, which may result in transmission error. The literature [83, 84] has shown that fading of the reporting channel will also affect the performance of cooperative spectrum sensing. At present, research on the performance of cooperative spectrum sensing under both imperfect sensing and reporting channels is still in the initial stages. Further, cognitive radios forward their binary decisions to a common receiver, which is an access point in a wireless LAN or a base station in a cellular network.

Step 3: Information fusion at the FC
In centralized cooperative spectrum sensing, the FC combines all of the information from each cognitive user and makes a final decision to infer the presence or absence of the cognitive user in the observed channels. There are different procedures for information fusion, and a variety of methods have been studied in the literature [15–17, 85]. We can conclude that the major fusion methods include a soft and hard combination. In the soft combination method, the cognitive user is weighted before sending information to the FC, so that the channel state information can be used to improve the accuracy of the combined information, whereas with the hard combination, the cognitive user sends the information directly to the FC with no preprocessing. Fusion methods can also be divided into data fusion and decision fusion according to the data format transmitted by the cognitive user. From step 1 to step 2, when each cognitive user performs local spectrum sensing, it can either send the detected primary user information directly to the FC or make a judgment first and then send the result to the FC—the former constituting data fusion and the latter decision fusion. Afterwards, the common receiver combines those binary decisions and makes a final decision to infer the absence or presence of the primary user in the observed band.

Cooperative spectrum sensing uses two successive channels: the sensing channel (from the primary user to cognitive radios) and the reporting channel (from cognitive radios to the common receiver). A simple decision fusion method is typically used to conserve the control channel bandwidth. Each cognitive user makes a binary decision based on its local observation, indicating the presence of the primary user if the local decision result is 1, and the absence of the primary user if the decision is 0. The benefit of cooperative spectrum sensing lies primarily in the achievable space diversity afforded by the independent sensing channels, or sensing diversity gain, provided by multiple cognitive radios. Even if one cognitive radio fails to detect the signal of the primary user, many detection opportunities remain for other cognitive radios. With the increased number of cooperative cognitive radios, the probability of missed detection for all users is extremely small. As the number of cooperating cognitive users participating in cooperative spectrum sensing increases, so does sensing diversity order and sensing performance. Another merit of cooperative spectrum sensing is the mutual benefit of communicating with each other to improve sensing performance [86]. When a cognitive radio is far removed from the primary user, the received signal may be too weak to detect. However, by employing a cognitive radio located near the primary user, as a relay, the signal of the primary user can be detected reliably by a distant user.

Improved cognitive user performance through user collaboration was investigated [73, 74] in the case of AWGN sensing channels, which presented methods of cooperation between two users as well as multiple users based on periodic spectrum sensing. Others studies have investigated the effect of sensing diversity order on cooperative spectrum sensing performance when the sensing channel experienced AWGN and fading channel, respectively [29, 81]. The results illustrate significantly improved performance by cooperative spectrum sensing with an increase in sensing diversity. Furthermore, it has been theoretically proven that cooperative spectrum sensing can reduce the demand of the average SNR of sensing channels compared with single user spectrum sensing. However, these investigations are based on periodic spectrum sensing, in which the sensing time and sensing performance are contradictory: a longer sensing duration can improve sensing performance, but results in a longer waiting time for the SUs to access the channel, causing serious interference for the PU [84]. Therefore, studies have been undertaken to determine optimal sensing duration to improve the system performance [87–90].

In decision based cooperative spectrum sensing, the control bandwidth can be greatly reduced by one-bit quantization compared with data fusion and multiple bits quantization method. However, when the number of sensing users is very large, the total number of sensing bits transmitted to the FC remains significant, and use of the larger control bandwidth also creates a potential problem. Further, the influence of reporting channel fading on sensing performance is related to sensing diversity order [83]. Therefore, establishing an appropriate trade-off between reporting channel fading and sensing diversity order must be further considered. System performance can be effectively improved through the soft combination method versus the hard combination in cooperative spectrum sensing. At present, research on soft combination-based cooperative spectrum sensing is largely focused on the data fusion method, in which the SU can provide relatively detailed and effective local detection information for the FC. In [14], the authors proposed an optimal soft combination scheme, demonstrating that cooperative spectrum sensing performance increased as the number of sensing users grew. However, infinite bits are required, and this will result in a large communication bandwidth for many cognitive users, leading to substantial waste of communication bandwidth.

The elements of cooperative spectrum sensing are shown in Fig. 2.9 and are briefly described as follows:

- Cooperation models consider how cognitive users cooperate to perform spectrum sensing. The popular parallel fusion network models [71] and the recently developed game theory models [91, 92] have been considered for achieving optimal detection performance. Most existing models for cooperative spectrum sensing are centered on detection performance in terms of cooperative gain, and the modeling of cooperation overhead is still an open issue.
- Sensing techniques are used for the RF environment, taking observation samples and employing signal processing techniques for detecting a primary user signal

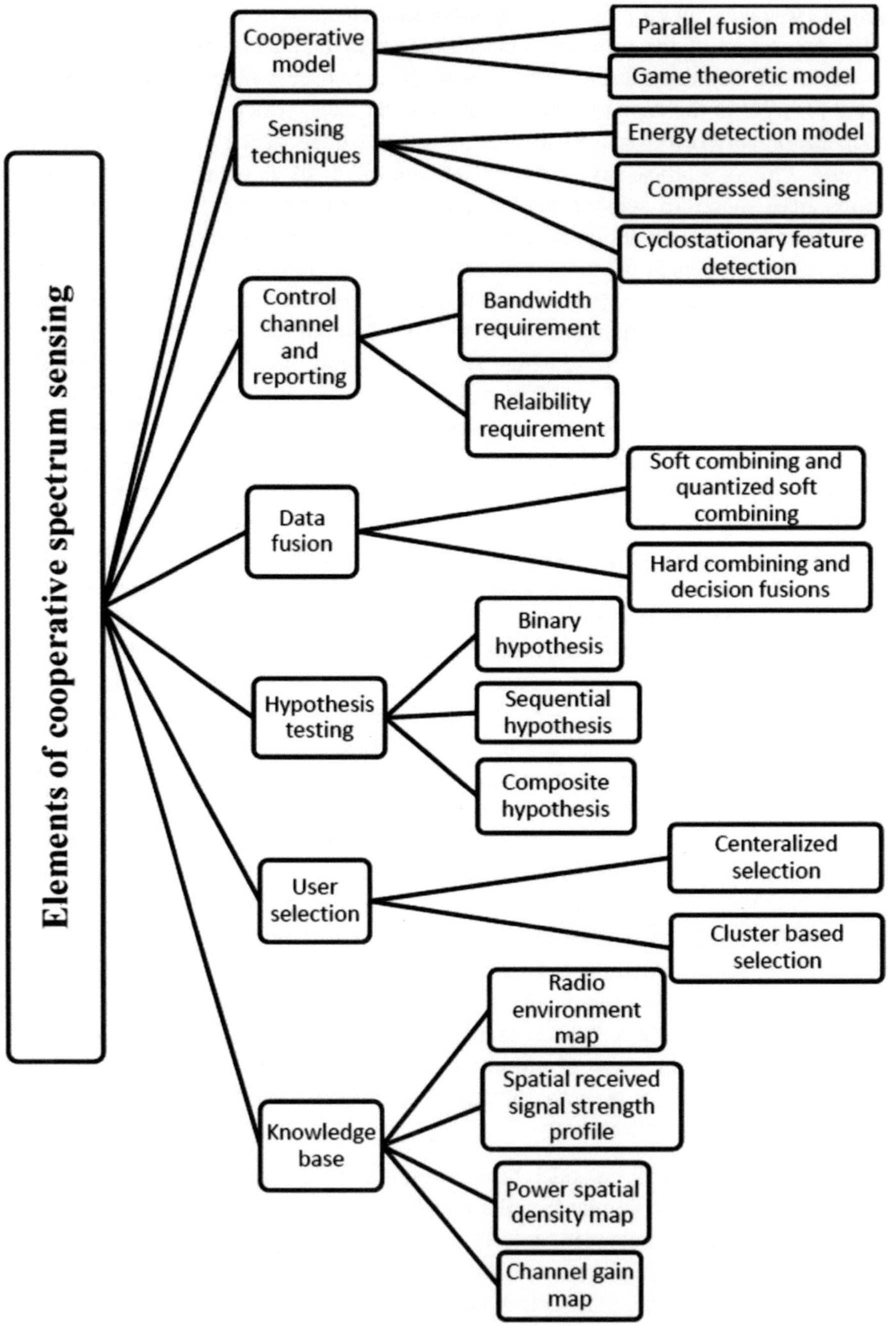

Fig. 2.9 Potential elements of the cooperative spectrum sensing technique

or available spectrum. The choice of sensing technique influences how cognitive radio users cooperate with each other. The process of cooperative spectrum sensing begins with local spectrum sensing at each cooperating cognitive user. Sensing techniques are crucial in cooperative spectrum systems, because the sensing, sampling, and processing of primary signals is strongly dependent on cognitive user cooperation. However, due to sub-Nyquist rate sampling and insufficient number of samples, a weak primary user signal with a nearby strong signal may not be properly reconstructed for detection in a wideband spectrum. In such a scenario, it may be challenging to achieve detection sensitivity by compressed sensing in a wideband spectrum.

- Control and reporting channels are concerned with how the sensing results obtained by cooperating cognitive users can be efficiently and reliably reported to the FC or shared with other cognitive users via the bandwidth-limited and fading-susceptible control channel. In cooperative spectrum sensing, a common control channel (CCC) [93, 94] is generally used by cognitive users to report local sensing data to the FC or for sharing sensing results with neighboring nodes. A MAC scheme is generally used by all cooperating cognitive users to access the control channel. From the perspective of the physical layer, a physical point-to-point link from a cooperating cognitive user to the FC is called a reporting channel. The availability of a perfect control channel in cooperative sensing is unrealistic, but recent studies suggest that imperfect control channels for influencing cooperative sensing performance should be considered as reasonable alternatives. However, the design of a control channel that is resilient to channel impairments, robust to primary user activity, and bandwidth-efficient for delivering sensing data is not a trivial task. Most existing cooperative sensing schemes assume a dedicated control channel for data reporting. In certain applications where the control channel must be dynamic allocated according to primary user activity, channel availability, and network topology, this allocation scheme significantly increases the difficulty for cognitive user cooperation and data reporting in cooperative sensing.
- Data fusion is the process of combining the reported or shared sensing results for making a cooperative decision. Depending on their data type, sensing results can be combined by signal combining techniques or decision fusion rules. In cooperative sensing, data fusion is a process for combining local sensing data for hypothesis testing. Depending on the control channel bandwidth requirement, reported sensing results may be of different forms, types, and sizes. In general, sensing results reported to the FC or shared with neighboring users can be combined in three different ways in descending order of control channel bandwidth performance: (1) soft combining, where cognitive users can transmit all local sensing samples or the complete local test statistics for a soft decision; (2) quantized soft combining, in which cognitive users can quantize local sensing results and send only the quantized data for soft combining in order to alleviate control channel communication overhead; and (3) hard combining, where cognitive users make a local decision and transmit the one-bit decision for hard combining. The use of soft combining at the FC clearly achieves the best

detection performance among the three, at a cost of greater control channel overhead, while quantized soft and hard combinations require much less control channel bandwidth, although degradation of performance is possible due to the loss of information from quantization.

- Hypothesis testing is a statistical test to determine the presence or absence of a primary user. This test can be performed individually by each cooperating user for local decisions or performed by the FC for a cooperative decision. However, large numbers of samples are needed to reach a decision during extended sensing time, which is a challenging task [95].
- User selection facilitates optimal selection of the cooperating cognitive users and determines the proper cooperation footprint/range to maximize cooperative gain and minimize cooperation overhead. The selection of cognitive users for cooperative sensing plays a key role in determining the performance of cooperative sensing because it can improve cooperative gain and address the overhead issues. For example, when cooperating cognitive users experience correlated shadowing, selecting independent cognitive users for cooperation can improve the robustness of sensing results, indicating that user selection is a critical issue for cooperation performance [20]. Potential challenges are summarized as follows:

  (1) Cooperation footprint [20] is the area where cognitive users cooperate with one another. As cooperative gain is obtained from spatial diversity, the cooperation footprint is an important parameter for evaluating performance and overhead in cooperative sensing. Thus, in addition to the distance between CR users, the selection of user schemes should consider the distribution of cognitive users and the area covered by their cooperation. However, deriving the exact footprint of cooperation from user selection is a challenge.
  (2) User selection and overhead: User selection is strongly related to every type of cooperative sensing overhead ranging from control channel bandwidth to energy efficiency, to security issues, among others. A trade-off exists between detection performance and the various types of overhead. Because attempting to address all overhead issues within the user selection scheme is challenging, most user selection schemes target one or two of these issues to address.

- The knowledge base stores information and facilitates the cooperative sensing process to improve detection performance. The stored information is either a priori knowledge or knowledge accumulated through user experience. The knowledge may include PU and CR user locations, PU activity models, and received signal strength profiles. The performance of cooperative sensing schemes largely depends on the knowledge of PU characteristics such as traffic patterns, location, and transmission power. PU information, if available in a database, can facilitate PU detection. The knowledge base is an indispensable element of cooperative sensing, because it can be utilized to assist, complement, or even replace cooperative sensing for detecting PU signals and identifying available spectrum. In addition, it serves two roles in cooperative sensing: (1) enhancing detection performance by utilizing the accumulated knowledge and learned experience, such as statistical models, in the database; and (2) alleviating the burden of cooperative sensing by retrieving spectrum information (e.g., a list of PU-occupied channels) from the database. To address security issues in cooperative sensing, the database should include other types of knowledge such as the behavior model of CR users and the model for jammer identification. Although cooperatively establish accurate statistical models for security purposes is challenging, the knowledge derived from these models can significantly improve security in cooperative sensing. In addition, because a recent U.S. Federal Communications Commission (FCC) ruling [96] has removed spectrum sensing requirement in TV white space, CR devices are able to access PU activity and spectrum information from a remote spectrum database. This ruling gives rise to new challenges for on-demand and web-based processing applications such as cloud computing [97, 98] in providing CR users with fast, secure, scalable, and energy-efficient access to a remote knowledge base.

2.5 Interference Temperature

The interference temperature limit is defined as the amount of additional interference that a receiver could tolerate [99], but a potential issue with this approach is the calibration of the limit itself. However, the conventional approach is based on the worst-case assumption of various primary users transmitting simultaneously. Severe constraints are imposed on the transmission power of cognitive users that should operate below the noise floor of primary systems. Various spectrum sensing methods based on interference level are reported in the literature [99, 100]. In dynamic spectrum access, cognitive users need to detect the primary user's appearance and decide, according to different metrics, which portion of the spectrum is available. The traditional approach is to limit the transmitter power of interfering devices such that the transmitted power should be no higher than a prescribed noise floor at a certain distance from the transmitter. However, constraints on transmitter power become more problematic as the mobility and variability of RF emitters increases, potentially revealing new, unpredictable sources of interference. The FCC Spectrum Policy Task Force [101] proposed a new metric on interference assessment, the interference temperature, to enforce an interference limit perceived by receivers. The interference temperature is a measure of the RF power available at a receiving antenna that is then to be delivered to a receiver, reflecting the power generated by other emitters and noise sources [102]. The purpose of the metric was to expose and remove the subjectivity that regulatory agencies might use to analyze interference. The development of an interference metric is critical if more intensive, dynamic use of the spectrum is desired. Interference-based detection is an underlay approach based on an estimation of the interference level at the primary receiver. Although interference is regulated by the transmitter, it actually occurs at the receivers. In interference-based approaches, a cognitive user transmits only if the new interference introduced by its own transmission is below a specific threshold, or the interference temperature limit. Using the interference temperature parameter, two crucial controls can be defined: (1) the upper threshold, above which the channel is declared to be occupied, and (2) the lower threshold, below which the channel can be declared empty or available for another user.

More specifically, it is defined as the temperature equivalent to the RF power available at a receiving antenna per unit bandwidth [103]: $T_I(f_c, B) = (P_I(f_c, B)/kB)$, where $P_I(f_c, B)$ is the average interference power in watts centered at f_c, covering bandwidth B measured in Hertz, and Boltzmann's constant is $k = 1.38 \times 10^{-23}$ J/K. Any unlicensed secondary transmitter using the licensed band must ensure that their transmission, plus the existing noise and interference, does not exceed the interference temperature limit at a licensed receiver. Any transmission in the licensed band is viewed as harmful if it increases the noise floor above the interference temperature limit. Thus, the receiver needs a reliable spectral estimate of the interference temperature. This requirement can be met by using the multi-taper method to estimate the power spectrum of the interference temperature with a large number of sensors [15]. If a regulatory body sets an interference temperature limit for a particular frequency band, then the secondary transmitters must keep the average interference below this level. Thus the interference temperature serves as a cap placed on potential RF energy that could appear on that band. Previous efforts have shown how to implement efficient spectrum allocation within the interference temperature limit. Spectrum shaping has been proposed as a method to improve spectrum efficiency [104] in cognitive radio networks. More specifically, using interference fitting, a cognitive radio senses the shape of the interference power spectrum and creates spectra inversely shaped to the current interference environment to take advantage of gaps between the noise floor and the cap on the interference temperature limit. A comprehensive analysis is presented in [25], which quantifies how interference temperature limits should be selected and how those choices affect the range of licensed signals. The FCC received input from external parties commenting that the interference temperature approach is not workable, as it would increase interference in the frequency bands where it would be used. Therefore, in May 2007, the FCC terminated the rule making process for implementing the interference temperature model.

2.6 The Spectrum Sensing Hybrid Model

The hybrid model for non-cooperative spectrum detection is the combination of all three techniques: matched filter, energy detection, and cyclostationary feature detection. Under this approach, the proper channelization of these techniques and add-on functionalities are used for opportunistic detection of idle spectrum bands. Let us now consider an area where spectrum sensing through a non-cooperative technique must be implemented, as shown in Fig. 2.10.

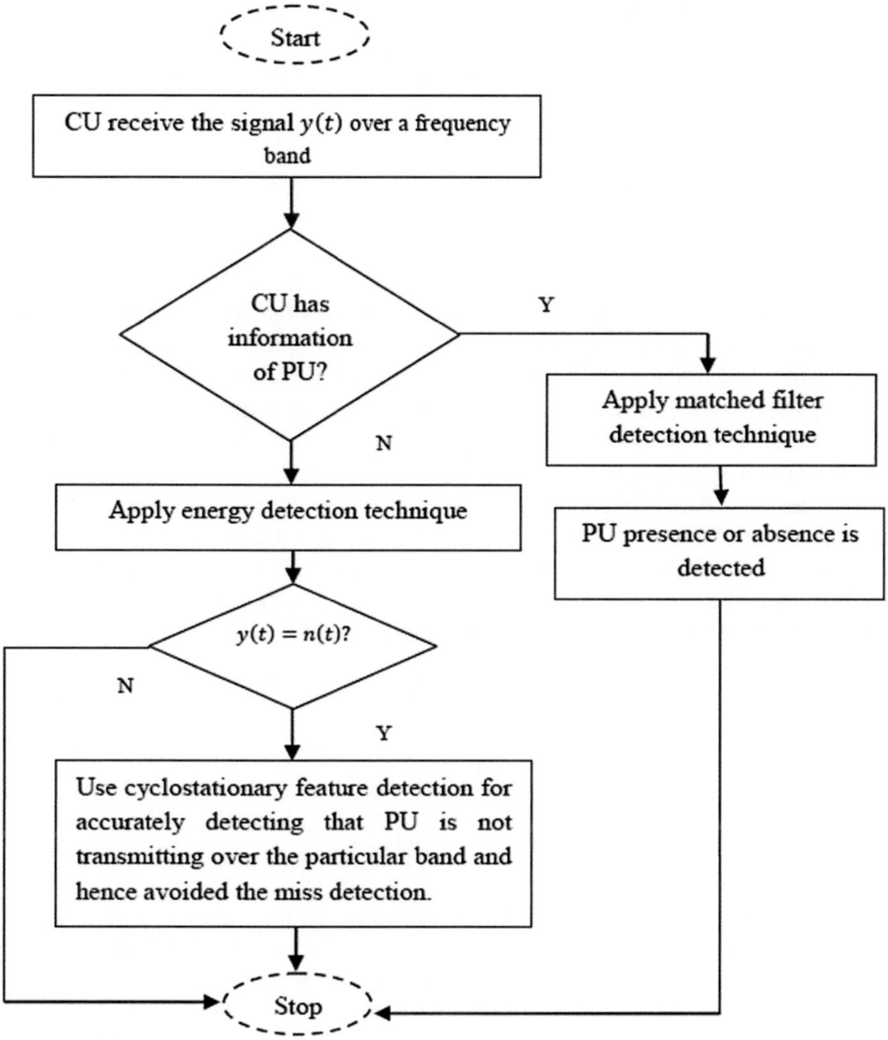

Fig. 2.10 Flowchart of the hybrid model for transmitter sensing

Step 1: The cognitive user receives the signal from the designated frequency band. The receiver will then determine whether the PU signal characteristics are present at the CU. If the signal is present, then directly matched filter detection is used for locating the licensed user on that band at that particular moment.

Step 2: If the cognitive user does not have any knowledge of the PU signal characteristics, an energy detection technique is employed. However, energy detection is not a highly accurate detection method and may result in sensing error, causing interference to licensed users. Therefore, hybrid model energy detection can be used as a fast sensing method. The idle channel sensed by the energy detector is again sensed by the cyclostationary feature detector to avoid missed detection. Cyclostationary feature detection provides higher sensing accuracy than the energy detector, but at the cost of greater processing time. The hybrid sensing method thus provides a shorter sensing time and accurate sensing results.

2.7 Threshold Setting

Setting an optimal threshold—the value needed to meet detection performance requirements—is one of the most important challenges to implementing detection techniques. Under optimal conditions, the probability of false alarm must be as low as possible, and the probability of detection as high as possible. A low probability of false alarm increases spectrum utilization, while a high probability of detection ensures the presence or absence of a primary user and reduces the probability of interference. The threshold can be set as either fixed or dynamic; two principles can be used to set a fixed threshold: constant false alarm rate (CFAR) and (2) constant detection rate (CDR) [105]. In CFAR, the threshold is set to meet a target P_f, and the obtained threshold is then used to compute the corresponding P_d, whereas in CDR, a certain P_d is used to set the threshold. For energy detection, the threshold can be computed based on these two principles [105]:

$$\lambda_f = \sigma_n^2\left(L + Q(P_f)\sqrt{2L}\right) \tag{2.12}$$

where λ_f is the threshold based on CFAR.

$$\lambda_d = \sigma_t^2\left(L + Q(P_d)\sqrt{2L}\right) \tag{2.13}$$

where λ_d is the threshold based on CDR. From Eqs. (2.12) and (2.13), in contrast to CDR, CFAR does not need the signal power of a PU to set the threshold; therefore, CFAR is more commonly used. However, constantly setting P_f to a small value such as 0.1 means that the corresponding threshold will be high. Consequently, it is difficult to detect low-power signals, and interference may occur. Therefore, a fixed threshold based on CFAR is not optimal. An optimal threshold setting can be archived if each cognitive user dynamically sets its threshold according to its channel states. In this context, the concept of constant false alarm is utilized to compute the energy detection threshold, and that value is then compared to the decision statistic to identify the primary user's current status (active/idle). The CFAR is used to compute the threshold value. The false alarm probability (P_f) is swept through a set of values in the range [0, 1], and the corresponding threshold is simultaneously computed using Monte Carlo simulations for each threshold value. Noise variance, as a significant parameter used to compute the threshold, is also varied. For each value of noise variance, the false alarm probability is updated through different values to observe the impact of this variation on the energy detection threshold. A low probability of missed detection and false alarm must always be jointly maintained to optimize detection performance in an SNR-varying environment. Minimizing the probability of missed detection affords greater protection to the PU against potential cognitive user transmissions, whereas minimizing the false alarm probability allows cognitive users to efficiently utilize the unused bands of spectrum. Therefore, the decision threshold must be adaptively adjusted to satisfy these two conflicting requirements for various channel conditions. The overall performance objective of the entire CRN can also be put into a single optimization problem of minimizing the total sensing error, which is discussed in detail in [106].

In cooperative spectrum sensing, local decisions are obtained by an energy detector based on CFAR. The authors in [107] proposed an optimal threshold method based on minimizing the total error rate, which is the summation of the probability of false alarm and missed detection, as follows:

$$P_e = P_m + P_f \tag{2.14}$$

By substituting Eqs. (2.5) and (2.7) in Eq. (2.14), the total error of energy detection is:

$$P_{e(ED)} = 1 - Q\left(\frac{\lambda - \sigma_t^2}{\sqrt{2L\sigma_t^4}}\right) + Q\left(\frac{\lambda - \sigma_n^2}{\sqrt{2L\sigma_n^4}}\right) \tag{2.15}$$

The optimal threshold (λ_{opt}) is the value that gives the minimum total error rate, which is obtained by solving the next optimization problem:

$$\lambda_{opt} = \arg \lambda \min P_e \tag{2.16}$$

The solution to this problem is as follows [107]:

$$\lambda_{opt} = \frac{-B - \sqrt{B^2 - 4AC}}{2A} \tag{2.17}$$

where $A = \frac{-1}{2L}\left(\frac{1}{\sigma_t^2} + \frac{1}{\sigma_n^2}\right)$, $B = \left(\frac{\sigma_s^2}{\sigma_t^2 \sigma_n^2}\right)$ and $C = -2\ln\left(\frac{\sigma_n^2}{\sigma_t^2}\right)$

This threshold setting approach is employed for the energy detector in [107]. However, this method can be applied to both the matched filter and cyclostationary feature detector. Further, for the dynamic threshold setting for the matched filter, the total error is [108]:

$$P_{e(MF)} = 1 - Q\left(\frac{\lambda - \varepsilon}{\sqrt{\varepsilon\sigma_n^2}}\right) + Q\left(\frac{\lambda}{\sqrt{\varepsilon\sigma_n^2}}\right) \tag{2.18}$$

where $\varepsilon = \sum_1^L x_p^2$. Using the dynamic threshold setting scheme [107], the optimal threshold is:

$$\lambda_{opt(MF)} = \arg\lambda \min\left[1 - Q\left(\frac{\lambda - \sigma_t^2}{\sqrt{2L\sigma_t^4}}\right) + Q\left(\frac{\lambda - \sigma_n^2}{\sqrt{2L\sigma_n^4}}\right)\right] \tag{2.19}$$

The solution to this minimization problem is the threshold value that makes the derivative of the total error equal to zero.

$$\frac{\partial P_{e(MF)}}{\partial\lambda} = 0 = -\frac{\partial}{\partial\lambda}\int\limits_{\frac{\lambda-\varepsilon}{\sqrt{\varepsilon\sigma_n^2}}}^{\infty} e^{-t^2/2}dt + \frac{\partial}{\partial\lambda}\int\limits_{\frac{\lambda}{\sqrt{\varepsilon\sigma_n^2}}}^{\infty} e^{-t^2/2}dt = 0 \tag{2.20}$$

Using Leibniz's integral rule, the above Equation becomes:

$$\frac{e^{-\frac{(\lambda-\varepsilon)^2}{\varepsilon\sigma_n^2}}}{\sqrt{\varepsilon\sigma_n^2}} - \frac{e^{-\frac{\lambda^2}{\varepsilon\sigma_n^2}}}{\sqrt{\varepsilon\sigma_n^2}} = 0 \tag{2.21}$$

$$(\lambda - \varepsilon)^2 = \lambda^2 \tag{2.22}$$

The optimal threshold of the matched filter is: $\lambda_{opt(MF)} = \varepsilon/2$. Similarly, in dynamic threshold setting for cyclostationary feature detection, the total error is given by [108]:

$$P_{e(CSFD)} = 1 - Q_1\left(\frac{S}{\sigma_1}, \frac{\lambda}{\sigma_1}\right) + e^{-\frac{\lambda^2}{\sigma_0^2}} = 1 - \int\limits_{\frac{\lambda}{\sigma_1}}^{\infty} xe^{\left(-\frac{\left(x^2 + \frac{s^2}{\sigma_1^2}\right)}{2}\right)} I_0\left(\frac{S.x}{\sigma_1}\right).dx + e^{\frac{-\lambda^2}{\sigma_0^2}} \tag{2.23}$$

The objective function is to find the optimal threshold that minimizes $P_{e(CSFD)}$. This problem is defined as: $\lambda_{opt(CSFD)} = \arg\lambda \min P_{e(CSFD)}$. The solution to this minimization problem is the threshold value that makes the derivative of the total error equal to zero; thus the solution lies in finding the value of λ that solves Eq. (2.24).

$$\frac{\partial P_{e(CSFD)}}{\partial \lambda} = \lambda^2\left(\frac{1}{\sigma_0^2} - \frac{1}{2\sigma_0^2}\right) + \ln\left(I_0\left(\frac{S_{xN'}^{\alpha_0}(n,k_0)N\lambda}{\sigma_1^2}\right) - \ln(2\sigma 1)\right) - \frac{\left(S_{xN'}^{\alpha_0}(n,k_0)N\right)^2}{\sigma_1^2} = 0 \tag{2.24}$$

Using a numerical methods such as the Newton-Raphson method, Eq. (2.24) can be solved with respect to λ. The detection and false alarm probabilities depend on the threshold λ, and hence it is necessary to choose an appropriate value that meets specific requirements. Detection probability also depends on signal power and the time-bandwidth product, whereas the false alarm probability depends only on the time-bandwidth product apart from the threshold. Therefore, one approach to choosing the threshold for a given time-bandwidth product is to select λ to meet the desired false alarm probability.

2.8 Potential Spectrum Sensing Challenges

Designing an efficient spectrum-sensing technique is the most fundamental yet problematic functionality in the cognitive radio paradigm because the levels of complexity, accuracy, reliability, computational cost, and sensing time of spectrum sensing fluctuate. Indeed, it is difficult for any given spectrum sensing technique to achieve high performance for all these spectrum sensing requirements; thus a trade-off among these requirements is necessary to achieve overall satisfactory spectrum-sensing results. Several potential challenges that make spectrum sensing an exigent task are shown in Fig. 2.11. Wideband spectrum sensing for cognitive radio applications requires a high sampling rate, high-resolution analog-to-digital converter (ADC) with a large dynamic range, multiple analog front-end circuitry, and high speed signal processors [109], all of which demand potential hardware, software, or algorithms/approaches. In traditional receiver design, noise variance or interference temperature estimation over the transmission of desired narrowband signals has been commonly used for optimal receiver designs such as channel estimation and soft information generation, as well as for improved handoff, power control, and channel allocation techniques. The noise/interference estimation problem is easier for these purposes, as receivers are tuned to receive signals that are transmitted over a desired bandwidth. Moreover, receivers are generally capable of processing the narrowband baseband signals with reasonably low complexity and low-power processors. However, in cognitive radio, the terminals are required to process transmission over a much wider band to sense an opportunity. Cognitive radio should thus be able to capture and analyze a relatively large band for

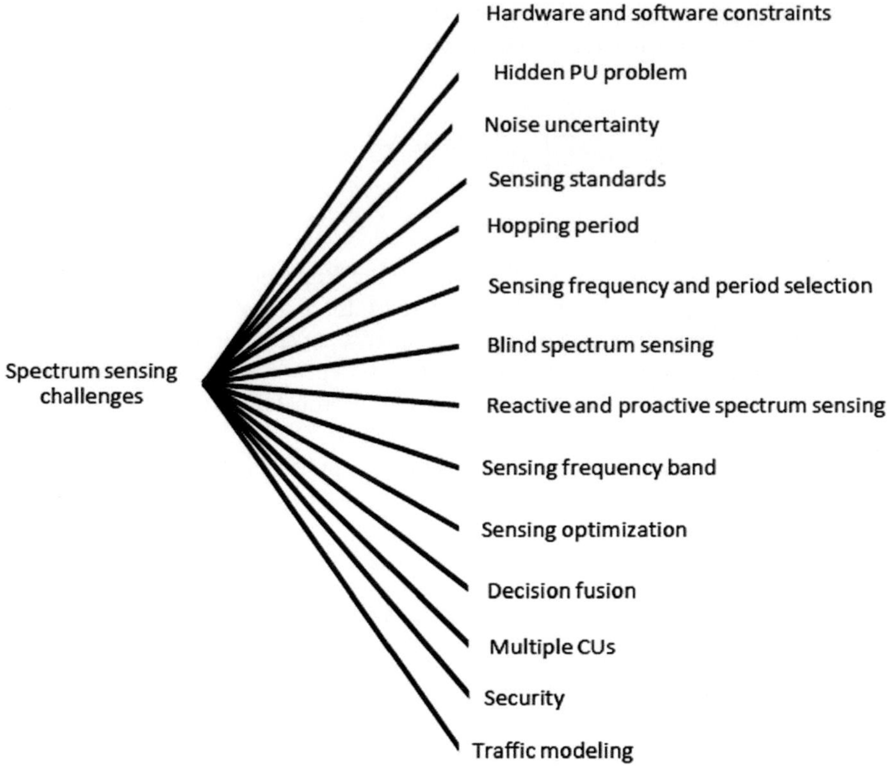

Fig. 2.11 Potential spectrum sensing challenges

identifying spectrum opportunities. Further, the high-speed processing units or field-programmable gate arrays (FPGAs) are needed for performing computationally demanding signal processing tasks with relatively short delay. The sensing can be performed via two architectures: single-radio and dual-radio [109, 110]. A specific time slot is allocated for spectrum sensing in the single radio architecture, and minimum accuracy can be guaranteed for spectrum sensing results. Moreover, spectrum efficiency decreases as some portion of the available time slot is used for sensing instead of data transmission. The obvious advantage of single-radio architecture is its simplicity and lower cost. However, in the dual-radio sensing architecture, one radio chain is dedicated to data transmission and reception, while the other chain is dedicated to spectrum monitoring. The potential limitation of such an approach is increased power consumption and hardware cost.

The level of noise power is required to estimate SNR, but it is difficult to measure the exact level of the noise power that is the noise uncertainty. In several studies, noise power is assumed to be known and fixed, but in fact it varies in time, requiring real-time measurements to determine its exact value. By considering noise uncertainty in performing spectrum sensing, it was shown that primary users' signals could not be detected under a certain a SNR value even over an extended sensing period [26]. This value is called the SNR wall, and its exact value depends on the detection technique used. The SNR wall is expressed as [31] $SNR_{wall} = 10\log_{10}\left[10^{x/10} - 1\right]$, where x is noise uncertainty in dB. In addition, in order to provide a promising security level and low probability of detection and interference, the wireless communication systems uses a spread spectrum technique or frequency hopping that utilizes spread frequencies with a wide bandwidth. Due to these characteristics, hopping is one of the main concerns in PU detection, requiring prior knowledge of PU hopping patterns [9]. Another crucial design element in cognitive radio spectrum sensing is the identification of the sensing period and how often it should be performed (sensing frequency). During the sensing period, data transmission is suspended, thus reducing network throughput and increasing end-to-end delay. Thus the sensing time chosen should be as short as possible. However, short sensing times may negatively affect detection performance, and sensing must be repeated frequently to ensure that the channel usage status for primary users is accurate. In other words, sensing must be active most of the time, which affects network performance. Hence, the selection of a suitable detection time must weigh these considerations.

Another fundamental design parameter of spectrum sensing is related to the frequency bands. Sensing a wide frequency band guarantees identification of more frequency opportunities, at the expense of time and hardware cost. A parallel sensing mechanism has been proposed [111] whereby cognitive users sense different frequencies simultaneously, and subsequently send their estimations to a FC, an approach that could enable rapid sensing of wider frequency bands. Another potential issue is determining the most effective frequency bands for a given cognitive radio environments to provide high QoS for both primary and cognitive users. Cognitive radio not only inherits the security concerns of wireless communication, but also raises new security concerns, such as primary user emulation and belief manipulation attacks [112, 113]. These malicious actions may degrade the performance of spectrum sensing and other cognitive radio functionalities. However, most proposed spectrum sensing techniques have not adequately addressed such security concerns [112, 114], and thus this important issue in cognitive radio will require significant attention. Furthermore, there is a high possibility that multiple secondary user networks competing for the same licensed bands will increase the likelihood of interference; thus coordination among SUs will be necessary [59].

An additional important consideration for cognitive radio networks aimed at maximizing performance is a sensing policy that addresses decisions about when, how long, and which frequency bands to sense. Sensing policies should be coordinated among cognitive users, and sensing periods must be synchronized among cognitive radios. Ideally, a cognitive radio user wants to minimize the amount of time required for identifying spectral opportunities in order to maximize the time available for transmission. Opportunistic spectrum access and/dynamic spectrum access are still in their infancy, and several complex technical, economical, and regulatory issues must be addressed before its potential can be fully assessed and realized. Potential research efforts within the signal processing community are particularly important in providing technical data for crafting of spectrum regulatory policies.

Moreover, the potential importance and challenges of spectrum sensing have been presented, however the present study has been emphasized to explore the possibilities to enhance the local sensing results in low-SNR environments. Further, experimental investigations are also needed to assess the effects of fading/shadowing (composite fading) on sensing results. However, an important initial step is determining whether collaboration between different detection techniques can significantly enhance sensing performance. A collaborative spectrum sensing model must be able to utilize various detection techniques to support reliable detection decisions. Much of the recent research in this area has focused on multistage spectrum sensing [83, 84, 115]. All detection techniques require an estimate of the noise power to compute SNR, but measuring noise uncertainty is problematic because this parameter changes with time. Therefore, it is important to evaluate spectrum sensing under certain noise uncertainty scenarios. MAC layer sensing schemes in cognitive radio networks generally consider both reactive and proactive sensing. In proactive sensing, adapted and non-adapted sensing period schemes are also assessed, via two performance metrics: available spectrum utilization and idle channel search delay. Simulation results show that the best performance is achieved with proactive sensing and adapted periods, but with observable overhead computational tasks to be performed by the network nodes.

2.9 Summary

In this chapter, we have provided a comprehensive survey on the fundamentals of cognitive radio spectrum sensing and the major research challenges, focusing on non-cooperative and cooperative spectrum sensing perspectives. The fundamental tasks of cognitive radio include spectral estimation of a radio frequency (RF) spectrum, hole identification, extraction of channel state estimation, and transmitter power control. Efficient utilization of radio spectrum by the cognitive radio transmitter can be achieved only with spectral information of the radio environment and spectrum hole identification in the neighborhood of a receiver, as well as information on the evolution of spectrum holes This information can be used by cognitive radio transmitter, for example, to select the appropriate modulation and coding format and transmission power level. The basic objective of the transmit power control function problem is to determine the transmit power levels for cognitive radio transmitters so that their data transmission rates can be maximized under the constrained interference limits in the frequency bands.

We have presented system models for selected detection techniques—the energy detector, matched filter, and cyclostationary feature detector, and compared them with fixed and dynamic threshold setting methods. Hybrid spectrum sensing techniques were also used to improve sensing performance through the proper channelization of detection techniques in a non-cooperative environment. However, as discussed, a first step in the spectrum sensing process could use energy detection or spectral estimation to provide a quick, coarse sensing in order to narrow the set of potentially available frequency bands, which would then be checked using more computationally complex feature detectors or matched filters. The main purpose of this initial step is to determine whether the power level at a given frequency band is below a specified threshold to enable secondary user transmissions. Regardless of the spectrum sensing algorithm employed, each algorithm provides a trade-off between the probability of false alarm and the probability of missed detection. These probabilities further depend on the number of collaborating users, the fusion rule employed, and the number of samples. However, selection of a proper detection threshold is a cross-layer optimization problem. The MAC layer protocols define the bounds for the physical layer algorithms for obtaining a desired trade-off between false alarms and missed detections. Physical layer algorithms whose thresholds can be set analytically to obtain a desired trade-off are preferred for their simplicity and predictability. For a multiuser distributed cognitive radio network, self-organization can be achieved with the help of two basic mechanisms—cooperation and competition. With cooperation (via either a distributed or centralized approach), the cognitive nodes can share network information among one another to achieve coordinated and efficient spectrum management. However, synchronization among the nodes may be required, resulting in a more complex network design. Conversely, a competitive (or non-cooperative) approach may simplify the network design, but at the expense of network performance. However, cooperative spectrum sensing techniques, with their advantages and limitations, have also been presented. Various parameters required for the design of cognitive radio models were explored, including hardware, spectrum sensing techniques, reasoning agent, and spectrum model.

In order to ensure low-interference operation for primary users, the detection sensitivity of cognitive users must be very high. However, spectrum sensing that relies on opportunistic access is not possible without tolerance of significant interference. The sensing problem is typically formulated to detect the primary transmitters instead of primary receivers. In practice, this is the only feasible option if the primary receivers are passive. There are multiple means of improving the detection sensitivity of a cognitive radio network. These include RF front-end sensitivity, designing and employing powerful signal processing algorithms well-suited to the task, and exploiting spatial diversity through collaborative sensing among multiple cognitive radios.

In summary, each of the two major classes of spectrum sensing (non-cooperative and cooperative) has its advantages and disadvantages. The selection and design of a proper detection algorithm is highly dependent on the application and primary user system. An algorithm best suited for every application may not exist. Hence, the use of a library of different sensing algorithm—for example, both energy and feature detectors—may be the most viable strategy. The spectrum sensing approach should be primarily system-oriented in order to maximize the probability of spectral opportunity detection. Therefore, feature detection or matched filter methods should be used whenever a desired performance must be achieved, with the aid of a computationally feasible algorithm; alternatively, energy detection may be used.

References

1. Q. Zhao, B.M. Sadler, A survey of dynamic spectrum access. IEEE Signal Process. Mag. **24** (3), 79–89 (2007)
2. Q. Zhao, L. Tong, A. Swami, Y. Chen, Decentralized cognitive MAC for opportunistic spectrum access in ad hoc networks: a POMDP framework. IEEE J. Sel. Areas Commun. **25** (3), 589–600 (2007)
3. H.-S. Chen, W. Gao, D.G. Daut, Spectrum sensing for wireless microphone signals, in *Proceedings of the 5th IEEE Annual Communications Society Conference on Sensor, Mesh and Ad Hoc Communications and Networks. SECON Workshops '08*, 16–20 June 2008, pp. 1–5
4. Y. Chen, Q. Zhao, A. Swami, Distributed spectrum sensing and access in cognitive radio networks with energy constraint. IEEE Trans. Signal Process. **57**(2), 783–797 (2009)
5. Q. Zhao, B. Krishnamachari, K. Liu, On myopic sensing for multi-channel opportunistic access: structure, optimality, and performance. IEEE Trans. Wireless Commun. **7**(12), 5431–5440 (2008)
6. S.H.A. Ahmad, M. Liu, T. Javidi, Q. Zhao, B. Krishnamachari, Optimality of myopic sensing in multichannel opportunistic access. IEEE Trans. Inf. Theory **55**(9), 4040–4050 (2009)
7. R.S. Sutton, A.G. Barto, *Reinforcement Learning: An Introduction* (MIT Press, Cambridge, MA, 1998)
8. S. Haykin, *Neural Networks: A Comprehensive Foundation*, 2nd edn. (Prentice-Hall, Inc., 1999)
9. T. Yucek, H. Arslan, A survey of spectrum sensing algorithms for cognitive radio applications. IEEE Commun. Surv. Tutorials **11**(1), 116–130, First Quarter (2009)
10. W.D. Horne, Adaptive spectrum access: using the full spectrum space, in *Proceedings of the Annual Telecommunications Policy Research*, Arlington, VA, Oct 2003
11. S. Haykin, D.J. Thomson, J.H. Reed, Spectrum sensing for cognitive radio. Proc. IEEE **97** (5), 849–877 (2009)
12. C. Cormiob, K.R. Chowdhury, A survey on MAC protocols for cognitive radio networks. Ad Hoc Netw. **7**(7), 1315–1329 (2009)
13. A. Sabharwal, P. Schniter, D. Guo, D.W. Bliss, S. Rangarajan, R. Wichman, In-band full-duplex wireless: challenges and opportunities. IEEE J. Sel. Areas Commun. **32**(9), 1637–1652 (2014)
14. B. Shen, K.S. Kwak, Soft combination schemes for cooperative spectrum sensing in cognitive radio networks. ETRI **31**(3), 263–273 (2009)
15. M. Kam, Q. Zhu, W.S. Gray, Optimal data fusion of correlated local decisions in multiple sensor detection systems. IEEE Trans. Aerosp. Electron. Syst. **28**(3), 916–920 (1992)
16. B. Chen, R.X. Jiang, T. Kasetkasem, P.K. Varshney, Channel aware decision fusion in wireless sensor networks. IEEE Trans. Signal Process. **52**(12), 3454–3458 (2004)
17. Z. Chair, P.K. Varshney, Optimal data fusion in multiple sensor detection systems. IEEE Trans. Aerosp. Electron. Syst. **22**(1), 98–101 (1986)
18. X. Chen, H.H. Chen, W. Meng, Cooperative communications for cognitive radio networks—from theory to applications. IEEE Commun. Surv. Tutorials **16**(3), 1180–1192, Third Quarter (2014)

19. B. Wang, K.J. Ray Liu, Advances in cognitive radio networks: a survey. IEEE J. Sel. Top. Signal Process. **5**(1), 5–23 (2011)
20. S.M. Mishra, A. Sahai, R.W. Brodersen, Cooperative sensing among cognitive radios, in *Proceedings of the IEEE International Conference on Communications (ICC'06)*, Istanbul, June 2006, vol 4, pp. 1658–1663 (2006)
21. A. Ghasemi, E.S. Sousa, Collaborative spectrum sensing for opportunistic access in fading environments, in *Proceedings of the 1st IEEE International Symposium on New Frontiers in Dynamic Spectrum Access Networks (DySPAN 2005)*, Baltimore, MD, USA, 8–11 Nov 2005, pp. 131–136
22. W. Yue, B. Zheng, A two-stage spectrum sensing technique in cognitive radio systems based on combining energy detection and one-order cyclostationary feature detection, in *Proceedings of the International Symposium on Web Information Systems and Applications (WISA09)*, China, 22–24 May 2009, pp. 327–330
23. P.R. Nair, A.P. Vinod, A.K. Krishna, A fast two stage detector for spectrum sensing in cognitive radios, in *Proceedings of the IEEE Vehicular Technology Conference (VTC Fall)*, San Francisco, CA, 5–8 Sept 2011, pp. 1–5
24. P.R. Nair, AP Vinod, A.K. Krishna, An adaptive threshold based energy detector for spectrum sensing in cognitive radios at low SNR, in *Proceedings of the IEEE International Conference on Communication Systems (ICCS, 2010)*, pp. 574–578
25. T.C. Clancy, On the use of interference temperature for dynamic spectrum access. Ann. Telecommun. **64**(7), 573–585 (2009)
26. R. Tandra, A. Sahai, SNR walls for signal detection. IEEE J. Sel. Top. Signal Process. **2**(1), 4–17 (2008)
27. D.J. Torrieri, *Principles of Military Communication Systems* (Artech, Dedham, MA, 1981)
28. J. Lehtomäki, Analysis of energy based signal detection, Ph.D. thesis, Faculty of Technology, University of Oulu, 2005
29. A. Ghasemi, E.S. Sousa, Opportunistic spectrum access in fading channels through collaborative sensing. J. Commun. **2**(2), 71–82 (2007)
30. J. Ma, Y. Li, Soft combination and detection for cooperative spectrum sensing in cognitive radio networks, in *Proceedings of the IEEE Global Communications Conference (GLOBECOM)*, 26–30 Nov 2007, pp. 3139–3143
31. H. Kim, K.G. Shin, In-band spectrum sensing in cognitive radio networks: energy detection or feature detection? in *Proceedings of the 14th ACM International Conference on Mobile Computing and Networking (MobiCom)*, 14–19 Sept 2008, pp. 14–25
32. J.J. Lehtomäki, M. Juntti, H. Saarnisaari, CFAR strategiesor channelized radiometer. IEEE Signal Process. Lett. **12**(1), 13–16 (2005)
33. V.I. Kostylev, Energy detection of a signal with random amplitude, in *Proceedings of the IEEE International Conference on Communications (ICC)*, vol 3, 28 Apr–2 May 2002, pp. 1606–1610
34. F.F. Digham, M.-S. Alouini, M.K. Simon, On the energy detection of unknown signals over fading channels, in *IEEE International Conference on Communications (ICC)*, vol 5, 11–15 May 2003, pp. 3575–3579
35. A. Ghasemi, E.S. Sousa, Impact of user collaboration on the performance of sensing based opportunistic spectrum access, in *Proceedings of the 64th IEEE Vehicular Technology Conference (VTC-2006 Fall)*, 25–28 Sept 2006, pp. 1–6
36. D. Cabric, A. Tkachenko, R. Brodersen, Spectrum sensing measurements of pilot, energy, and collaborative detection, in *Proceedings of the Military Communications Conference (MILCOM)*, 23–25 Oct 2006, pp. 1–7
37. D. Cabric, Addressing the feasibility of cognitive radios. IEEE Signal Process. Mag. **25**(6), 85–93 (2008)
38. Z. Xuping, P. Jianguo, Energy-detection based spectrum sensing for cognitive radio", in *Proceedings of the IET Conference on Wireless, Mobile and Sensor Networks (CCWMSN07)*, Shanghai, 12–14 Dec 2007, pp. 944–947
39. H.V. Trees, *Detection, Estimation, and Modulation Theory* (Wiley, New York, 1968)

40. R. Tandra, A. Sahai, Fundamental limits on detection in low SNR under noise uncertainty, in *Proceedings of the Wireless Communication*, June 2005, pp. 464–469
41. M.P. Olivieri, G. Barnett, A. Lackpour, A. Davis, A scalable dynamic spectrum allocation system with interference mitigation for teams of spectrally agile software defined radios, in *Proceedings of the IEEE International Symposium on New Frontiers in Dynamic Spectrum Access Networks (DySPAN)*, Baltimore, MD, Nov 2005, pp. 170–179
42. F. Weidling, D. Datla, V. Petty, P. Krishnan, G. Minden, A framework for RF spectrum measurements and analysis, in *Proceedings of the IEEE International Symposium on New Frontiers in Dynamic Spectrum Access Networks (DySPAN)*, Baltimore, MD, Nov 2005, pp. 573–576
43. D.-C. Oh, Y.-H. Lee, Energy detection based spectrum sensing for sensing error minimization in cognitive radio networks. Int. J. Commun. Netw. Inf. Secur. **1**(1) (2009)
44. J. Lehtomaki, J. Vartiainen, M. Juntti, H. Saarnisaari, Spectrum sensing with forward methods, in *Proceedings of the IEEE Military Communication Conference*, Washington, D. C., Oct 2006, pp. 1–7
45. J. Vartiainen, H. Sarvanko, J. Lehtomki, M. Juntti, M. Latva-aho, Spectrum sensing with LAD-based methods, in *Proceedings of the 18th Annual IEEE International Symposium on Personal, Indoor, Mobile Radio Communication (PIMRC)* 2007, pp. 1–5
46. Y.-C. Liang, E.P.Y. Zeng, A.T. Hoang, Sensing-throughput tradeoff for cognitive radio networks, in *Proceedings of the IEEE International Conference on Communication (ICC)*, Glasgow, UK, June 2007, pp. 5330–5335
47. Z. Quan, S. Cui, A.H. Sayed, H.V. Poor, Wideband spectrum sensing in cognitive radio networks, in *Proceedings of the IEEE International Conference on Communication*, Beijing, China, May 2008, pp. 901–906
48. M. Wylie-Green, Dynamic spectrum sensing by multiband OFDM radio for interference mitigation, in *Proceedings of the 1st IEEE International Symposium on New Frontiers in Dynamic Spectrum Access Networks (DySPAN)*, 2005, pp. 619–625
49. S. Kapoor, S.V.R.K. Rao, G. Singh, Opportunistic spectrum sensing by employing matched filter in cognitive radio network, in *Proceedings of the IEEE International Conference on Communication Systems and Network Technologies (CSNT)*, 2011, India, pp. 580–583
50. J. Ma, G.Y. Li, B.H. Juang, Signal processing in cognitive radio. Proc. IEEE **97**(5), 805–823 (2009)
51. L. Lu, X. Zhou, U. Onunkwo, G.Y. Li, Ten years of research in spectrum sensing and sharing in cognitive radio. EURASIP J. Wireless Commun. Netw. **2012**(1), 28 (2012)
52. D. Cabric, A. Tkachenko, R.W. Brodersen, Spectrum sensing measurements of pilot, energy, and collaborative detection, in *Proceedings of the Military Communications Conference (MILCOM)*, 2006, pp. 1–7
53. S. Kapoor, G Singh, Non-cooperative spectrum sensing: a hybrid model approach, in *Proceedings of International Conference on Devices and Communications (ICDeCom)*, 2011, India, pp. 1–5
54. S.M. Kay, *Fundamentals of Statistical Signal Processing*, 5th edn. (Prentice Hall, 2004), pp. 487–488
55. K. Muraoka, M. Ariyoshi, T. Fujii, A novel spectrum-sensing method based on maximum cyclic autocorrelation selection for cognitive radio system, in *Proceedings of the 3rd IEEE Symposium on New Frontiers in Dynamic Spectrum Access Networks (DySpan)*, 2008, pp. 1–7
56. R. Tandra, A. Sahai, Noise calibration, delay coherence and SNR walls for signal detection, in *Proceedings of the 3rd IEEE International Symposium on New Frontiers in Dynamic Spectrum Access Networks* (2008), pp. 1–11
57. M. Öner, F. Jondral, Air interface recognition for a software radio system exploiting cyclostationarity, in *Proceedings of the 15th IEEE International Symposium on Personal, Indoor, Mobile Radio Communication (PIMRC)*, vol 3 (2004), pp. 1947–1951

58. A. Fehske, J.D. Gaeddert, J.H. Reed, A new approach to signal classification using spectral correlation and neural networks, in *Proceedings of the IEEE Symposium on New Frontiers in Dynamic Spectrum Access Networks (DySPAN)*, Baltimore, MD, Nov 2005, pp. 144–150
59. A. Ghasemi, E.S. Sousa, Spectrum sensing in cognitive radio networks: requirements, challenges and design trade-offs. IEEE Commun. Mag. **46**(4), 32–39 (2008)
60. W.A. Gardner, Signal interception: performance advantages of cyclic-feature detectors. IEEE Trans. Commun. **40**, 149–159 (1992)
61. A. Sahai, N. Hoven, R. Tandra, Some fundamental limits on cognitive radio, in *Allerton Conference on Communication, Control, and Computing* (2004), pp. 1662–1671
62. D. Cabric, Addressing feasibility of cognitive radios. IEEE Signal Process. Mag. **25**(6), 85–93 (2008)
63. T. Yücek, H. Arslan, Spectrum characterization for opportunistic cognitive radio systems, in *Proceedings of the IEEE Military Communication Conference* (2006), pp. 1–6
64. D. Cabric, R.W. Brodersen, Physical layer design issues unique to cognitive radio systems, in *Proceedings of the IEEE 16th International Symposium on Personal, Indoor, Mobile Radio Communication (PIMRC)*, vol 2 (2005), pp. 759–763
65. N. Shankar, C. Cordeiro, K. Challapali, Spectrum agile radios: utilization and sensing architectures, in *Proceedings of the 1st IEEE International Symposium on New Frontiers in Dynamic Spectrum Access Networks (DySPAN)* (2005), pp. 160–169
66. J. Lunden, V. Koivunen, A. Huttunen, H. Poor, Spectrum sensing in cognitive radios based on multiple cyclic frequencies, in *Proceedings of the 2nd International Conference on Cognitive Radio Oriented Wireless Networks and Communications (CrownCom)* (2007), pp. 37–43
67. L.P. Goh, Z. Lei, F. Chin, Feature detector for DVB-T signal in multipath fading channel, in *Proceedings of the 2nd International Conference on Cognitive Radio Oriented Wireless Networks and Communications (CrownCom)*, 2007
68. G. Vardoulias, J. Faroughi-Esfahani, G. Clemo, R. Haines, Blind radio access technology discovery and monitoring for software defined radio communication systems: Problems and techniques, in *Proceedings of the 2nd International Conference on 3G Mobile Communication Technology* (2001), pp. 306–310
69. M. Mehta, N. Drew, G. Vardoulias, N. Greco, C. Niedermeier, Reconfigurable terminals: an overview of architectural solutions. IEEE Commun. Mag. **39**(8), 82–89 (2001)
70. R. Chen, J. Park, Ensuring trustworthy spectrum sensing in cognitive radio networks, in *Proceedings of the 1st IEEE Workshops on Networking Technologies* for *Software Defined Radio (SDR) Networks* (2006), pp. 110–119
71. P.K. Varshney, C.S. Burrus, *Distributed Detection and Data Fusion* (Springer, New York, 1997)
72. D. Cabric, S.M. Mishra, Robert W. Brodersen, Implementation issues in spectrum sensing for cognitive radios, in *Proceedings of Asilomar Conference on Signals, Systems, and Computers*, vol 1 (2004), pp. 772–776
73. G. Ganesan, Ye Li, Cooperative spectrum sensing in cognitive radio, part I: two user networks. IEEE Trans. Wireless Commun. **6**(6), 2204–2213 (2007)
74. G. Ganesan, Ye Li, Cooperative spectrum sensing in cognitive radio, part II: multiuser networks. IEEE Trans. Wireless Commun. **6**(6), 2214–2222 (2007)
75. R. Peterson, E. Visotsky, S. Kuffner, On collaborative detection of tv transmissions in support of dynamic spectrum sharing, in *Proceedings of the 1st International Symposium on New Frontiers in Dynamic Spectrum Access Networks (DySPAN 2005)*, 2005, pp. 338–345
76. A. Dobra, D. Kempe, J. Gehrke, Gossip-based computation of aggregate information, in *Proceeding of IEEE Symposium on Foundations of Computer Science (FOC)*, Oct 2003, pp. 482–491
77. D.N.C. Tse, J.N. Laneman, G.W. Wornell, Cooperative diversity in wireless networks: efficient protocols and outage behavior. IEEE Trans. Inf. Theory **50**(12), 3062–3080 (2004)
78. J. Hillenbrand, T. Weiss, F.K. Jondral, Calculation of detection and false alarm probabilities in spectrum pooling systems. IEEE Commun. Lett. **9**(4), 349–351 (2005)

79. F.F. Digham, M.S. Alouini, M.K. Simon, On the energy detection of unknown signals over fading channels. IEEE Trans. Wireless Commun. **55**(1), 21–24 (2007)
80. H. Urkowitz, Energy detection of unknown deterministic signals. Proc. IEEE **55**(4), 523–531 (1967)
81. G. Amir, S.S. Elvino, Spectrum sensing in cognitive radio networks: the cooperation-processing tradeoff. Wireless Commun. Mobile Comput. **7**(9), 1049–1060 (2007)
82. G. Scutari, D. Palomar, S. Barbarossa, Cognitive MIMO radio. IEEE Signal Process. Mag. **25**(6), 46–59 (2008)
83. W. Zhang, K.B. Letaief, Cooperative spectrum sensing with transmit and relay diversity in cognitive radio networks. IEEE Trans. Wireless Commun. **7**(12), 4761–4766 (2008)
84. K.B. Letaief, W. Zhang, Cooperative communications for cognitive radio networks. Proc. IEEE **97**(5), 878–893 (2009)
85. R. Viswanathan, V. Aalo, On counting rules in distributed detection. IEEE Trans. Acoust. Speech Signal Process. **37**(5), 772–775 (1989)
86. G. Ganesan, Y.G. Li, Agility improvement through cooperation diversity in cognitive radio, in *Proceedings of the IEEE Global Communications Conference*, St Louis, Missouri, USA, 28 Nov–2 Dec 2005, vol 5, pp. 2505–2509
87. Y.C. Liang, Y.H. Zeng, E.C.Y. Peh, Anh Tuan Hoang, Sensing-throughput tradeoff for cognitive radio networks. IEEE Trans. Wireless Commun. **7**(4), 1326–1337 (2008)
88. J.Y. Shen, T. Jiang, S.Y. Liu, Zhongshan Zhang, Maximum channel throughput via cooperative spectrum sensing in cognitive radio networks. IEEE Trans. Wireless Commun. **8** (10), 5166–5175 (2009)
89. E.C.Y. Peh, Y.C. Liang, Y.L. Guan, Yonghong Zeng, Optimization of cooperative spectrum sensing in cognitive radio networks: a sensing-throughput tradeoff view. IEEE Trans. Veh. Technol. **58**(9), 5294–5299 (2009)
90. Y.H. Zeng, Y.C. Liang, Spectrum-sensing algorithms for cognitive radio based on statistical covariances. IEEE Trans. Veh. Technol. **58**(4), 1804–1815 (2009)
91. W. Saad, Z. Han, M. Debbah, A. Hjorungnes, T. Basar, Coalitional games for distributed collaborative spectrum sensing in cognitive radio networks. Proc. IEEE INFOCOM **2009**, 2114–2122 (2009)
92. B. Wang, K. Ray Liu, T. Clancy, Evolutionary cooperative spectrum sensing game: how to collaborate? IEEE Trans. Commun. **58**(3), 890–900 (2010)
93. I.F. Akyildiz, W.-Y. Lee, M.C. Vuran, S. Mohanty, NeXt generation/dynamic spectrum access/cognitive radio wireless networks: a survey. Comput. Netw. **50**(13), 2127–2159 (2006)
94. B.F. Lo, I.F. Akyildiz, A.M. Al-Dhelaan, Efficient recovery control channel design in cognitive radio ad hoc networks. IEEE Trans. Veh. Technol. **59**(9), 4513–4526 (2010)
95. R. Blum, S. Kassam, H. Poor, Distributed detection with multiple sensors I. Advanced topics. Proc. IEEE **85**(1), 64–79 (1997)
96. FCC, Second memorandum opinion and order, ET Docket No. 10-174
97. M. Armbrust, A. Fox, R. Griffith, A.D. Joseph, R. Katz, A. Konwinski, G. Lee, D. Patterson, A. Rabkin, I. Stoica, M. Zaharia, A view of cloud computing. Commun. ACM **53**, 50–58 (2010)
98. V. Namboodiri, Towards sustainability in portable computing through cloud computing and cognitive radios" in *Proceedings of 39th International Conference on Parallel Processing Workshops (ICPPW 2010)*, 2010, pp. 468–475
99. Y. Xing, C.N. Mathur, M. Haleem, R. Chandramouli, K. Subbalakshmi, Dynamic spectrum access with QoS and interference temperature constraints. IEEE Trans. Mobile Comput. **6** (4), 423–433 (2007)
100. J. Bater, H.P. Tan, K.N Brown, L. Doyle, Modelling interference temperature constraints for spectrum access in cognitive radio networks, in *Proceeding of IEEE International Conference on Communications, ICC'07*, June 2007, pp. 6493–6498

101. Establishement of interference temperature metric to quantify and manage interference and to expand available unlicensed operation in certain fixed mobile and satellite frequency bands, FCC, 2003, FCC Doc. ET Docket 03-289
102. P.J. Kolodzy, Interference temperature: a metric for dynamic spectrum utilization. Int. J. Netw. Manag. **16**(2), 103–113 (2006)
103. T.C. Clancy, Formalizing the interference temperature model. J. Wireless Commun. Mobile Comput. **7**(9), 1077–1086 (2007)
104. A. Wagstaff, N. Merricks, A subspace-based method for spectrum sensing, in *Proceedings of SDR Forum Technical Conference* (2007)
105. Z. Ye, G. Memik, J. Grosspietsch, Energy detection using estimated noise variance for spectrum sensing in cognitive radio networks, in *Proceedings of the IEEE Wireless Communications and Networking Conference (WCNC 2008)*, 2008, pp. 711–716
106. D.R. Joshi et al., Gradient-based threshold adaptation for energy detector in cognitive radio systems. IEEE Commun. Lett. **15**(1), 19–21 (2011)
107. X. Zhai, H. He, G. Zheng, Optimal threshold and weighted cooperative data combining rule in cognitive radio network, in *Proceedings of the 12th IEEE International Conference on Communication Technology (ICCT)*, 2010, pp. 1464–1467
108. Y.M. Maatug, Spectrum sensing in cognitive radio: multi-detection techniques, MS dissertation, University of Waterloo, Ontario, Canada, 2012
109. S. Haykin, Cognitive radio: brain-empowered wireless communications. IEEE J. Sel. Areas Commun. **23**(2), 201–220 (2005)
110. H. Arslan, *Cognitive Radio, Software Defined Radio, and Adaptive Wireless Systems* (Springer, Berlin, 2007)
111. S. Xie, Y. Liu, Y. Zhang, R. Yu, A parallel cooperative spectrum sensing in cognitive radio networks. IEEE Trans. Veh. Technol. **59**(8), 4079–4092 (2010)
112. T.C. Clancy, N. Goergen, Security in cognitive radio networks: threats and mitigation, in *Proceedings of the 3rd International Conference on Cognitive Radio Oriented Wireless Networks and Communications (CrownCom 2008)*, Singapore, 15–17 May 2008, pp. 1–8
113. A. Fragkiadakis, E. Tragos, I. Askoxylakis, A survey on security threats and detection techniques in cognitive radio networks. IEEE Commun. Surv. Tutorials **15**(1), 428–445 (2013)
114. J.L. Burbank, Security in cognitive radio networks: the required evolution in approaches to wireless network security, in *Proceedings of the 3rd International Conference on Cognitive Radio Oriented Wireless Networks and Communications (CrownCom 2008)*, Singapore, 15–17 May 2008, pp. 1–7
115. H. Jiang, L. Lai, R. Fan, H. Poor, Optimal selection of channel sensing order in cognitive radio. IEEE Trans. Wireless Commun. **8**(1), 297–307 (2009)

Chapter 3
Medium Access Control Protocol for the Distributed Cognitive Radio Network

3.1 Introduction

As the sphere of technology continues to expand, spectrum scarcity has become a bottleneck for the development of wireless communication. In addition, the growing number of unlicensed wireless devices has led to overcrowding of the industrial-scientific-medical (ISM) band of the radio frequency spectrum. As a solution for alleviating spectrum utilization pressure on affected spectrum bands, cognitive radio is designed to constantly sense and access spectrum opportunities across the entire radio spectrum. A key challenge in cognitive radio networks is obtaining an efficient sensing and non-interfering spectrum access decision, enabling cognitive users to reserve chunks of the spectrum for certain periods of time. The modeling of variable bandwidth communication in cognitive radio is very complicated, and channel access policies must be defined for cognitive radio users. In this chapter, we propose a novel medium access control (MAC) protocol for the distributed cognitive radio network that defines the cognitive radio access policies for the unutilized spectrum.

Various MAC protocols for distributed cognitive radio networks have been proposed by researchers and scientists [1–13], and are discussed in detail in the next section of this chapter. Technical issues for some of the protocols proposed in [1–3, 8] for a distributed cognitive radio network include (1) a hidden terminal problem in the hardware constrained- MAC (HC-MAC) protocol [1], (2) the synchronization requirement among cognitive users in the cognitive MAC (C-MAC) protocol [2], (3) large communication overhead before data transmission in the cognitive radio-enabled multichannel (CREAM)-MAC protocol [3], and (4) the contention interval access scheme and its detrimental effect on the throughput of the SMC-MAC protocol [8]. These technical problems are rectified in the proposed MAC protocol discussed in this chapter, which provides significant throughput enhancement compared to the SMC-MAC [8]. We have implemented a backoff algorithm for contention solving among cognitive users, hence reserving the idle

S. Pandit and G. Singh, *Spectrum Sharing in Cognitive Radio Networks*,
DOI 10.1007/978-3-319-53147-2_3

licensed channels for data transmission. In the proposed multichannel cooperative MAC protocol for the distributed cognitive radio network, cognitive users share sensing results with one another over a control channel. Available licensed channels, along with the control channel, are divided into cycle time consisting of four intervals: (1) idle, (2) sensing-sharing, (3) contention, and (4) data transmission. However, in the reported SMC-MAC [8] protocol, a lower number of contention slots during the contention interval results in significantly more collisions, and the large contention slots increase successful cognitive users while decreasing the data transmission interval, since the total cycle time is fixed. Hence, less data transmission time results in lower throughput, which is a major limitation of the SMC-MAC protocol [8]. In addition, under this protocol, cognitive users who collide in the contention interval are unable to select another contention slot in the current cycle, and therefore must wait for the next cycle to succeed in transmitting data. The proposed method in this chapter applies a backoff mechanism to resolve this issue by allowing collided cognitive users to select a different contention slot in the same cycle time. We have also optimized the number of contention slots to enable all users to succeed in transmitting data, resulting in a significant increase in the volume of data transmitted.

The remainder of chapter is organized as follows: Section 3.2 presents the related work in the field of cognitive radio MAC protocols. In Sect. 3.3, the system model of the proposed MAC protocol is described. Section 3.4 introduces the proposed algorithm for contention solving among cognitive users in the cognitive radio network, and presents a performance analysis of the proposed MAC protocol. In Sect. 3.5, the numerical simulation results of the proposed MAC protocol are discussed. Finally, Sect. 3.6 provides a summary of the chapter and a discussion of future directions in the field.

3.2 Related Work

For dynamic spectrum access (DSA)-based cognitive radio networks, MAC protocols that have been designed for traditional wireless networks need to be modified to include spectrum sensing and spectrum access. The coexistence of cognitive users and licensed users greatly complicates the design of the MAC protocol for cognitive radio [14], which must achieve optimal spectrum utilization by accurately detecting all opportunities for accessing the spectrum in order to minimize collisions with other cognitive users. However, depending on channel quality, transmission parameters such as modulation and coding level can be adapted at the MAC layer. Various ideas have been discussed regarding the use of an optimization model [1, 14, 15] for spectrum sensing and spectrum access decisions. In [14], Kim and Shin describe a mechanism for optimizing the sensing period and reducing idle channel discovery delay among cognitive users. In [15], a partially observable Markov decision process (POMDP) is employed for accessing of licensed channels by cognitive users. The MAC protocol must select the best available channels for sensing, and the cognitive radio user then uses these results to determine the channel to access for data transmission. This decision is based on the objective of maximizing the transmission rate and constraints such as maintaining interference with a licensed user below a certain threshold. Taking into consideration hardware constraints such as the single radio, partial spectrum sensing, and spectrum aggregation limits, a hardware-constrained MAC (HC-MAC) [1] was proposed for efficient spectrum sensing and access decisions. While the model is applicable for both single and multiple channels/users, it suffers from a multichannel hidden terminal problem [1]. MAC protocols for a multichannel and multiuser cognitive radio system are discussed in [2–7].

The main objectives of these protocols are to perform negotiation among cognitive users for spectrum access in a multichannel environment and to avoid collisions due to simultaneous transmissions. In [2], the cognitive MAC (C-MAC) protocol is proposed for the distributed cognitive radio network in which there is no central entity such as a base station available for coordination among the cognitive radio terminals. In C-MAC [2], each available licensed channel is scheduled, which is divided into super-frames that consist of consecutive beacon and data transmission periods. A rendezvous channel (RC) is assumed to be available throughout the network operation, and provides the synchronization and coordination among the cognitive users through non-overlapping beacon periods. There is also a backup channel which is detected during sensing and is used to immediately provide an alternate choice of spectrum band to CR if a primary user appears. Each cognitive radio user visits the RC periodically in order to share load information of each band for (1) synchronization, (2) gathering information about primary and secondary user discovery, (3) avoiding the hidden node problem, and (4) exchanging schedules for beacon periods so that beacons are not transmitted simultaneously over all spectrum bands. In addition, each cognitive terminal seeking to send data to its intended receiver will first send a beacon signal during its designated beacon slot, then coordinate with other users, and once synchronized, transmit over the assigned channel. However, any spectrum change by the cognitive terminal occurring in C-MAC must first be announced over the RC so that other cognitive users will be aware of the change. Therefore, establishing an RC that is available throughout the cognitive network is very important. This protocol has certain technical issues, however, such as setting up non-overlapping beacon periods, quiet periods without a central entity, and RC availability [2]. In addition, network synchronization is needed in C-MAC, and the requirement for beacon control infrastructure makes it more complex. However, it is free from the hidden terminal problem found in HC-MAC [1]. The cognitive radio-enabled multichannel (CREAM) MAC protocol has also been discussed in [3], which is free from the hidden terminal problem and network synchronization; however, there is large communication overhead in this MAC protocol.

Opportunistic spectrum access–MAC (OSA-MAC) for distributed cognitive radio networks is proposed in [4]. This is somewhat similar to the architecture of the IEEE 802.11 ad hoc MAC protocol, but it functions differently from WLAN IEEE 802.11 MAC [16]. In OSA-MAC, there is one dedicated control channel for cognitive users for exchanging control information, which is owned by the cognitive user service provider. Channel time is also divided into beacon intervals, and all cognitive users are synchronized with the periodic beacon transmission. Each beacon interval consists of three phases: channel selection, sensing, and data transmission [4]. The cognitive user transmitter first sends an ad hoc traffic indication message (ATIM) over the control channel to its receiver, which contains the list of licensed idle channels for data transmission use. With agreement on the selected channel, the cognitive receiver sends the ATIM-ACK (acknowledgment) back to the transmitter over the control channel, after which the cognitive user switches to the selected channel and begins sensing it continuously during the sensing phase. However, if no primary user is detected on the selected channel, then data are transmitted during the data transmission phase. Otherwise, with the detection of the primary user, the cognitive radio switches back to the control channel. A major limitation of OSA-MAC is the large overhead before the actual data transmission, where the data of the cognitive user is transmitted after request-to-send (RTS) and clear-to-send (CTS) message exchange with the respective receiver, which is preceded by the amount of time at which the backoff timer has expired. There is also bandwidth wastage during the ATIM window in OSA-MAC.

An error-adaptive MAC protocol [5] has been proposed with switching between error recovery and dual transmission modes according to the channel status of the cognitive radio network. Additional channels detected during sensing are utilized for error recovery in poor channel conditions and for increasing the throughput in good channel states. However, this protocol increases the complexity of receiver systems because it requires precise channel estimation and more than one transceiver for utilizing a large number of idle channels. A self-scheduling multichannel cognitive radio-MAC (SMC-MAC) [8] protocol for distributed cognitive radio networks was recently proposed, in which the cooperation among the cognitive users minimizes the sensing time and enhances throughput. However, technical issues associated with this protocol must be addressed, including cognitive user collisions in the contention interval and bandwidth wastage over the licensed channels during the sensing-sharing and contention periods [17]. A dynamic common control channel (DCCC)-based MAC protocol was proposed [9] for a cellular cognitive radio network. In addition, several other MAC based protocols have been proposed recently, including an opportunistic matched filter-based MAC [10], prioritized cognitive radio MAC (PCR-MAC) [11], a cooperative access spectrum sharing protocol [12], distributed sequential-access MAC (DSA-MAC) [13], and cognitive adaptive MAC (CAMAC) [18], and a comparison of these is provided in Table 3.1.

The impact of selfish users on MAC protocol fairness is considered in [19] using Jain's fairness index [20]. Timmers et al. [21] explored CR-enabled networks with distributed control and concluded that distributed multichannel medium access control (MAC) protocols are the key enablers for these networks. In addition to spectrum scarcity, energy is rapidly becoming a major bottleneck in wireless operations, and must be considered as a key design criterion. The authors presented

Table 3.1 Performance comparisons of various cognitive radio MAC protocols

Protocol	MAC technique	Spectrum access technique	No. of transceivers	Dedicated control channel	Synchronization needed	Hidden terminal problem
HC-MAC [1]	Contention-based	Interweave/OSA	1	Yes	No	Yes
C-MAC [2]	Polling-based	Interweave/OSA	1	Yes	Yes	No
CREAM-MAC [3]	Contention-based	Interweave/OSA	1 with multiple sensors	Yes	No	No
OSA-MAC [4]	Contention-based	Interweave/OSA	1	No	Yes	No
Error-adaptive MAC [5]	Contention-based	Interweave/OSA	More than 1	No	No	Yes
SMC-MAC [8]	Contention-based	Interweave/OSA	1	Yes	No	No
PCR-MAC [11]	Contention-based	Interweave/OSA	2	Yes	No	No
Cooperate and access spectrum sharing protocol [12]	TDMA-based	Overlay	1	No	Yes	Yes
DSA-MAC [13]	Polling-based	Interweave/OSA	1	No	Yes	No
CAMAC [18]	Contention-based	Interweave/OSA	1	Yes	No	No
MMAC-CR [21]	Contention-based	Interweave/OSA	2	Yes	Yes	No

an energy-efficient distributed multichannel MAC protocol for CR networks (MMAC-CR), with simulation results showing that the proposed protocol significantly improved the performance through borrowing of licensed spectrum and protected primary users (PUs) from interference, even in hidden terminal situations [21]. The sensing costs were found to contribute only 5% to the total energy cost.

Since cognitive radio technology can significantly boost spectrum utilization by exploiting radio spectrum unoccupied by licensed users, it is rapidly gaining popularity and inspiring numerous innovations. However, many technical issues still need to be addressed for successful deployment of CR networks, especially in the MAC layer. Jha et al. [22] have focused on CR networks that have a distributed architecture because they offer ease of deployment, self-organizing capability, and flexibility in design. These networks are also believed to be more practical for future deployments compared to their centralized counterparts. The MAC protocols for distributed CR networks should consider the key features of these networks such as lack of a central unit to coordinate communication, dynamic topology, requirements to minimize interference with primary users, and variation in spectrum availability with time and location. To clarify the relevant research challenges and issues, we have provided a detailed study of the critical design issues and an overview of current state-of-the-art MAC protocols proposed for distributed cognitive radio networks. A classification of existing proposals is provided, and their salient features, advantages, and limitations are discussed in this chapter and summarized in Table 3.1. We then introduce and examine the proposed MAC protocol, which is better able to address some of the research issues than are existing solutions. We also highlight important research challenges that could drive future research in this area.

Xiang et al. [23] first reported the challenges in the design and implementation of CR-MAC protocols, provided a comprehensive survey of state-of-the-art CR-MAC protocols, and categorized them on the basis of spectrum sharing modes, i.e., overlay and underlay. Other classification metrics such as architecture (centralized or distributed), sharing behaviors (cooperative or non-cooperative), and access modes (contention-based or contention-free) were also considered. Through this study, we learn that most CR-MAC protocols are designed for the overlay mode. This is because CR-MAC protocols in underlay mode, while yielding higher spectrum utilization efficiency, bear the cost of more complicated power and admission control schemes. The centralized CR-MAC protocols are more suitable for spectrum sensing using quiet periods in the whole network, while the distributed CR-MAC protocols can be deployed more flexibly. Cormio and Chowdhury [24] provided an extensive survey, including the characteristic features, advantages, and limiting factors of the existing CR MAC protocols, for both infrastructure-based and ad hoc networks. An overview of the spectrum sensing is provided, which demonstrates that the channel access does not result in interference to the licensed users of the spectrum. A detailed classification of MAC protocols is also presented, considering the infrastructure support, integration of spectrum sensing functionalities, the need for time synchronization, and the number of radio transceivers. Cormio and Chowdhury present the main challenges and future research directions [24], highlighting the close coupling of the MAC protocol design with the other layers of the protocol stack. The MAC protocols exploit sensing stimuli to create a spectrum opportunity map, and available resources are scheduled to improve coexistence between the users of heterogeneous systems (dynamic spectrum sharing). In addition, the MAC protocols may allow cognitive users to vacate selected channels when their quality becomes unacceptable. The study performed by De Domenico et al. [25] exploited the fundamental role of the MAC layer, identifying its functionalities in a cognitive radio (CR) network. In addition, a review of C-MAC protocols, advantages, drawbacks, and further design challenges are discussed.

Operations such as data sharing in cooperative spectrum sensing, broadcasting spectrum-aware routing information, and spectrum coordination access rely on a control message exchange via a common control channel to improve spectrum efficiency. Thus, a reliable and continuously activated common control channel is indispensable. As the common control channel may be subject to primary user activity, its design in cognitive radio networks poses new challenges, in that cognitive radio users are unable to negotiate a new control channel when the original one is occupied by primary users. Lo [26] presented the problem of common control channel design by its classification, design challenges, design schemes, and its applications in network protocol layers. The issues of control channel saturation, robustness to primary user activity, limited control channel coverage, and control channel security are identified as design challenges. Major control channel design schemes including sequence-based, group-based, dedicated, and ultra-wideband approaches are also presented. The relationship of the common control channel with radio interface, cooperative sensing, medium access control, and routing are discussed as well. Krishna and Das [27] provided a comprehensive review of research performed at the MAC layer of OSA networks, especially with reference to ad hoc network design, sparking more research in this area. A classification of the MAC layer protocols used in OSA networks is provided, along with a thorough analysis and comparison of the essential features of the different MAC layer protocols for OSA networks and discussion of unresolved MAC layer research issues. Pawelczak et al. [28] presented an extensive comparison and analysis of opportunistic spectrum access (OSA) as a function of spectrum-sensing performance and licensed user activity of different control channel (CC) implementations for multichannel medium access control (MAC) algorithms. Their analysis is based on a discrete Markov chain model of a subset of representative multichannel OSA-MAC classes that incorporates physical layer effects, such as spectrum sensing and fading, complemented by extensive simulations. The major observations made by Pawelczak et al. [28] are as follows: (1) When the CC is implemented through a dedicated channel, sharing this channel with the licensed user does not significantly decrease the throughput achieved by the OSA network if the data packets are of sufficient size or the number of considered data channels is small. (2) Hopping OSA-MACs, where the CC is spread over all channels, are less susceptible to licensed user activity than those with a dedicated CC (in terms of both average utilization and on/off times). (3) Scanning efficiency has a significant impact on the achievable performance of licensed and OSA users for all analyzed protocols. (4) The multiple-rendezvous

MAC class, which has yet to be proposed in the OSA literature, outperforms all the multichannel MAC designs analyzed.

Chen et al. [29] have presented a MAC protocol design for random access cognitive radio networks by considering a two-level opportunistic spectrum access strategy. This strategy aims to optimize system performance of the secondary network and to protect the operation of the primary network. At the first level, secondary users (SUs) maintain a detection probability sufficient to avoid interference with primary users (PUs), and spectrum sensing time is optimized to control the total traffic rate of the secondary network which allows for random access when the channel is detected as available. At the second level, two MAC protocols, the slotted cognitive radio ALOHA (CR-ALOHA) and cognitive radio-based carrier-sensing multiple access (CR-CSMA), are developed to manage packet scheduling of the secondary network. Normalized throughput and average packet delay are considered as network metrics, and closed-form expressions have been derived to evaluate the performance of the secondary network for the proposed protocols. In addition, the authors use the interference and agility factors as performance parameters to measure the protective effects on the primary network. For various frame lengths and numbers of SUs, optimal performance of throughput and delay can be achieved during the same spectrum sensing period, and a trade-off exists between the achievable performance of the secondary network and protective effects on the primary network.

Park et al. [30] presented an analytical framework to assess the link layer throughput of multichannel OSA ad hoc networks, emphasizing an analysis of various combinations of collaborative spectrum sensing and MAC protocol abstractions. The authors decomposed collaborative spectrum sensing into layers, parameterized each layer, classified existing solutions, and proposed a new protocol, called truncated time division multiple access (TTDMA), that supports the efficient distribution of sensing results. In the case of multichannel MAC protocols, two approaches for control channel design, dedicated and hopping channels, are evaluated. Enhancements to these protocols are proposed that provide options for handling SU connections pre-empted by the PU through connection buffering until PU departure and connection switching to a vacant PU channel. Comparing and optimizing different design combinations demonstrates that it is generally better to buffer pre-empted SU connections than to switch them to PU vacant channels, and that TTDMA is a promising design option for a collaborative spectrum sensing process.

In [31], a novel medium access control (MAC) scheme is proposed for multichannel cognitive radio ad hoc networks that achieves high throughput while effectively protecting PUs. The PU signal may cover only a part of the network, and the nodes can produce variable sensing results for the same PU even if the same channel has been considered in the design of the MAC scheme. The proposed MAC scheme fully utilizes the spectrum access opportunity by allowing the nodes to use the channel on which the PU exists, provided their transmissions do not disturb the PU. The proposed MAC scheme [31] mitigates the hidden PU problem inherent in multichannel cognitive radio networks by adjusting the sensing priorities of channels at each node with the PU detection information of other nodes and by limiting the transmission power of the cognitive radio node to the maximum allowable for guaranteeing the quality-of-service (QoS) requirement of the PU. The performance of the proposed MAC scheme is evaluated using simulation, and the results reveal that the cognitive radio system with the proposed MAC achieves good performance in throughput and packet delay, while adequately protecting PUs [31].

There is significant capacity for devising protocols that adapt cognitive radio transmissions to the type of interferer. Updated performance metrics are needed that capture cognitive radio-specific improvements in order to evaluate different MAC protocols. We believe that MAC protocol design for cognitive radio is an area ripe for research that will be of interest to both industry and academia as this technology matures over the next few years.

3.3 MAC Protocol and System Design

3.3.1 System Model

In the proposed system model, we have considered a primary user network having N_{ch} number of licensed channels, and a cognitive radio network comprising N_{CU} number of cognitive users. The primary user network is assumed to be a cellular network, and the traffic within the cellular network is based on the Poisson distribution, according to research [32]. Cognitive users utilize licensed channels of the primary network for communication applications when the channels (licensed channels of the primary network) are idle. It is also assumed that the sensing performed by a cognitive user is perfect so that there are no probabilities of false alarm or missed detection [33] in the sensing results. In addition, a control channel is assumed to be always available to the cognitive network, and the cognitive user terminal is equipped with a single transceiver (full-duplex mode) that can change frequency among multiple channels. However, if a cognitive user wants to transmit/receive its data on/from different idle channels simultaneously, multiple transceivers must be available to the user. In addition, in order to increase the performance of the cognitive radio system, a cognitive radio user should sense as many licensed channels as possible. We know that there are different sensing techniques in a cognitive radio system, and that each technique requires some mathematical computation [34] of the received signals to detect the presence or absence of a primary user. Therefore, as an increasing number of licensed channels are sensed by a cognitive radio terminal, the complexity and power consumption of the terminal increases, resulting in a trade-off between number of sensed channels and complexity or power consumption. However, based on this consideration, we have attempted to limit the number of channels sensed by each terminal and have allowed the sharing of sensing results with other cognitive users, so that information is available for a greater number of licensed channels at each cognitive terminal in comparison to the channels it has sensed.

3.3.2 Proposed MAC Protocol

The proposed MAC protocol consists of a control channel on which cognitive users cooperate with one another, and N_{ch} licensed channels, as shown in Fig. 3.1a. Control channel cooperation among the cognitive users is performed by presenting all the sensing results of cognitive users for control channel activity, followed by

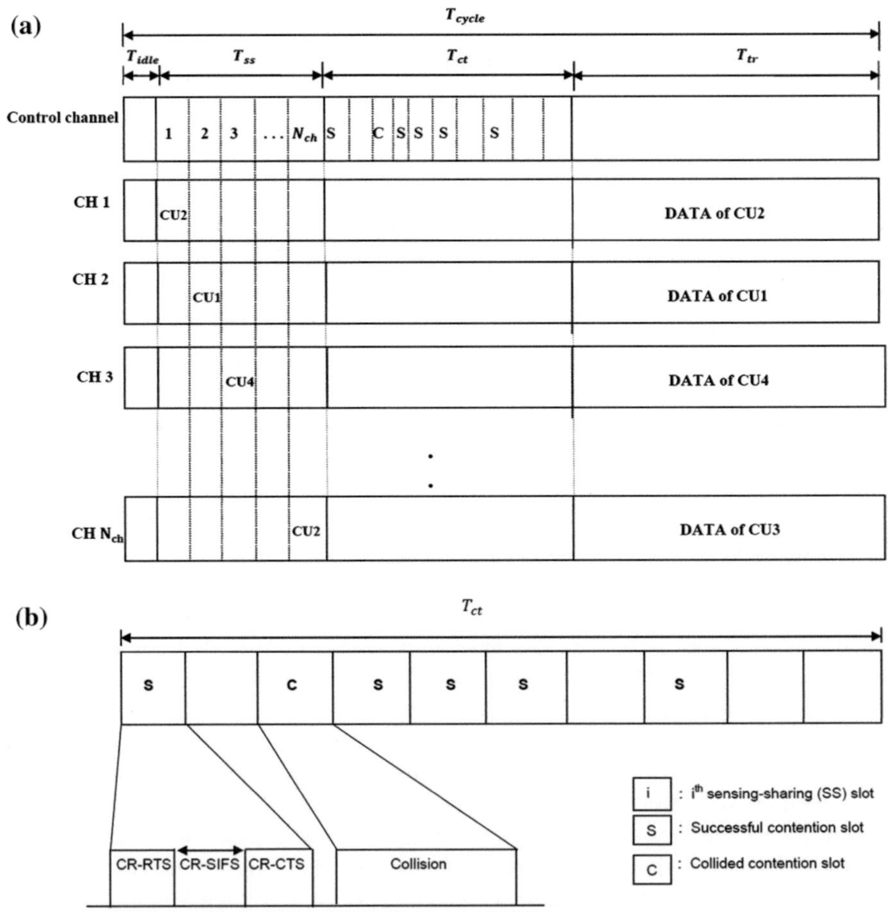

Fig. 3.1 The proposed distributed cognitive radio MAC protocol **a** system model consisting of multiple licensed channels and control channel for cooperation among the cognitive users, and **b** contention interval expansion of the control channel

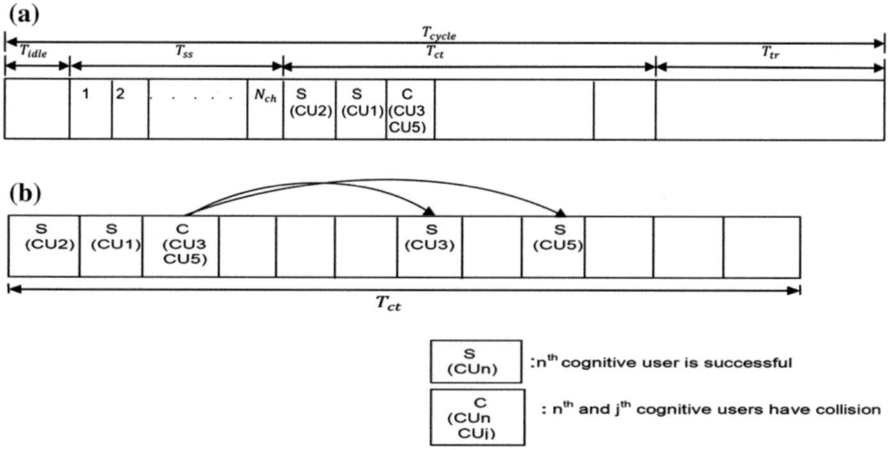

Fig. 3.2 The control channel structure of **a** the SMC-MAC protocol [8] without the backoff algorithm during the contention interval, and **b** the proposed scheme with the backoff algorithm during contention interval

cognitive user selection of idle channel/channels from the pool of available idle channels. Each channel is divided into cycle time, T_{cycle}, which is further divided into four intervals: idle interval (T_{idle}), sensing-sharing interval (T_{ss}), contention interval (T_{ct}), and data transmission interval (T_{tr}), as shown in Fig. 3.1a. It is assumed that all cognitive users are tuned to the control channel for the idle and sensing-sharing interval. In addition, cognitive users compete in the contention interval to reserve idle licensed channels and then tune to the selected idle channels. The sensing-sharing and contention intervals are further divided into a number of slots [8], as shown in Figs. 3.1 and 3.2. The sensing-sharing interval has a number of slots equal to the number of licensed channels, and each cognitive user randomly selects sensing-sharing slots in order to sense the selected slot number for the corresponding licensed channel during that slot period.

Let us assume that there are 20 licensed channels in the network, and each cognitive user can sense two licensed channels; therefore, there are 20 sensing-sharing slots, and for any of these 20 slots that are randomly selected by the cognitive users, the user will begin sensing to the selected slot number licensed channels. For example, as shown in Fig. 3.1, the second cognitive user has randomly selected the first and last slots; therefore, this user senses the first licensed channel during the first sub-slot, and the sensing information is then broadcast in the second and third sub-slots of the first slot, as depicted in Fig. 3.3. All other cognitive users are tuned to the control channel, which hears the broadcast sensing information in the first slot and thus stores the channel status information of the first licensed channel. Similarly, in the last sensing-sharing slot, the second cognitive user senses the last (20th) licensed channel, and shares the sensing information about the availability of this channel with the other cognitive users.

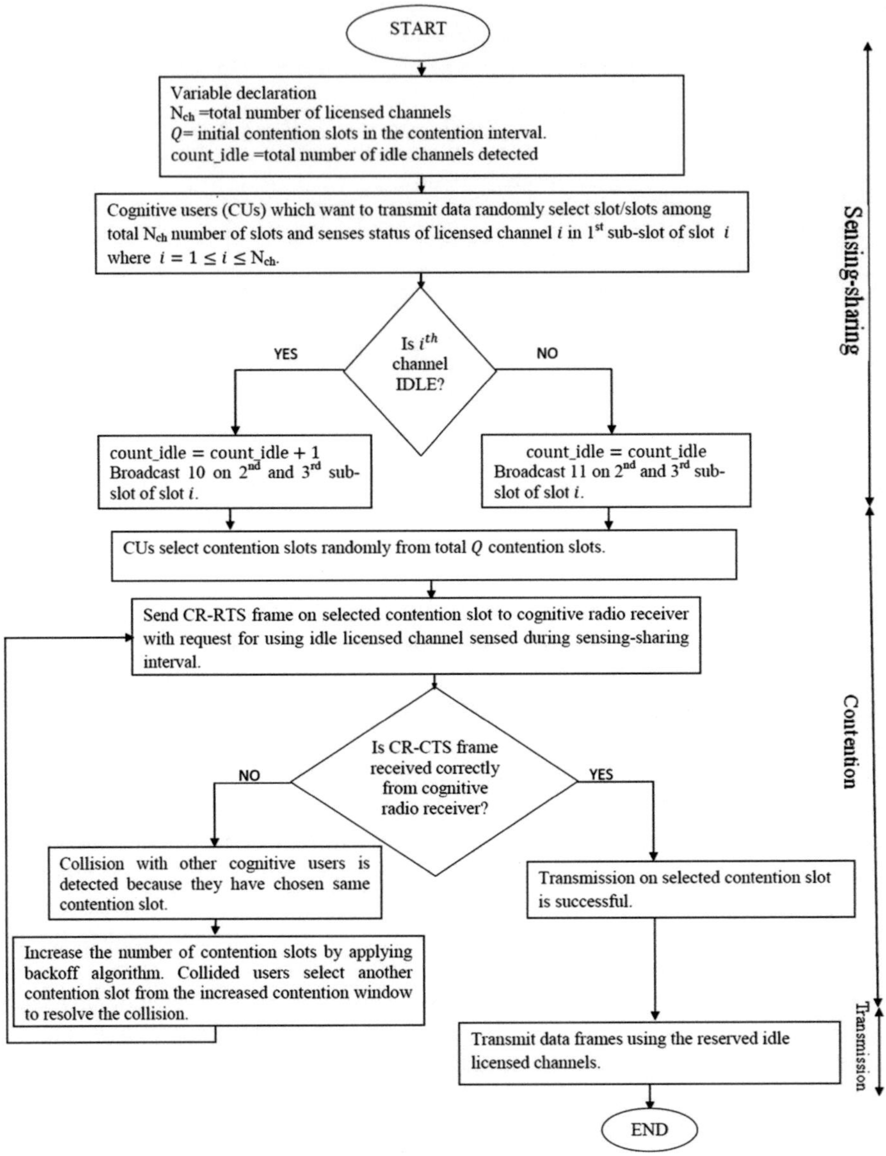

Fig. 3.3 The flow diagram of the proposed MAC protocol

Other cognitive users can also randomly pick two slots for sensing their respective licensed channels; hence, cognitive users cooperate by sharing the sensing results with one another in the sensing-sharing interval. It is also possible that more than one user senses the same licensed channel by selecting the same slot during the sensing-sharing interval. While sensing of the same licensed channel by two or more cognitive users is not a problem, simultaneously broadcasting the same information by users on the same channel corrupts sensing information. Therefore, we have considered that the cognitive user, after sensing a channel during the first sub-slot of the selected sensing-sharing slot, will wait for a random length of time during the second sub-slot before broadcasting sensing information. During this waiting period, if the cognitive user hears any transmission, it knows that another user has selected the same channel for sensing and is broadcasting sensing information over the control channel. As a result, the cognitive user would refrain from transmitting its own sensing information to avoid a collision, retrieving the channel sensing results from the information already broadcasted. However, the precise mechanism by which each user selects the random waiting period during the second sub-slot to avoid collision is beyond the scope of our proposed work. This sensing and sharing procedure is performed by all cognitive users during their selected slot, and hence each cognitive user has sensing information for both, the channels it has sensed and those senesd by other users, resulting in reduced sensing time.

Cognitive users compete to reserve idle licensed channels detected in the sensing-sharing interval by selecting a contention slot from the contention interval. A cognitive user can successfully send a frame in the transmission interval of the idle licensed channel only if that cognitive user is not colliding with other cognitive users in that slot, which is possible only if each transmitting cognitive user has selected a different contention slot in the contention interval. A collision by a cognitive user is detected by listening to the cognitive radio clear-to-send (CR-CTS) frame sent by the destination cognitive user in response to the cognitive radio ready-to-send (CR-RTS) frame transmitted by the source cognitive user. These frames are sent over the selected contention slot in the control channel, and it is obvious that if more than one source cognitive user has selected the same contention slot, they will not receive the CR-CTS frame correctly and detect collision. This probability of collision is significant if the number of contention slots is limited relative to the number of cognitive users. Although a large number of contention slots increases the success rate of cognitive users in the cognitive radio network, it simultaneously reduces the data transmission interval and hence the throughput of the cognitive network. Thus there is a contention slot-throughput trade-off problem in the SMC-MAC protocol [8]. Because a cognitive user occupying its contention slot knows the idle channels that other users have already reserved based on the exchange of CR-RTS and CR-CTS frames on the control channel, it will not ask to utilize those idle channels on its own CR-RTS frame. Hence, the possibility of reserving the same idle channel by more than one user is avoided by the cooperation over the control channel during the contention interval. On the CR-RTS frame, the source cognitive user sends a list of available idle channels to the destination cognitive user. However, it is possible that at the destination cognitive user location, those all channels are not idle, due hidden terminal problem. Therefore, the destination user sends a CR-CTS frame with a selected idle channel which is available at both the transmitter and receiver on which they will transmit data during the data transmission interval. The CR-RTS and CR-CTS frame structure with different fields are discussed in detail in [8]. In the proposed

MAC, the cooperation among cognitive users is shown in Fig. 3.1a, where the data of the third cognitive user (CU3) is transmitted on channel N_{ch} (CH N_{ch}), which is sensed as idle by the second cognitive user (CU2) during the sensing-sharing interval. This is because channels sensed by CU3 during the sensing-sharing interval are not detected as idle as that by CU2, which detected both channel 1 (CH 1) and channel N_{ch}(CH N_{ch}) as idle, and therefore CU3 utilized the extra idle channel of the second cognitive user for data transmission.

Figure 3.1b shows a detailed description of the contention interval. The inter-frame spacing between the CR-RTS and CR-CTS frames is given by CR-SIFS as that in IEEE 802.11 [16]. In SMC-MAC [8], it is proposed that each cognitive user randomly chooses a contention slot, increasing its vulnerability to collisions among cognitive users. Thus, in order to reduce the number of collisions, we have modified the control channel's contention interval, as shown in Fig. 3.2b, using the backoff algorithm in the contention interval. As one example, Fig. 3.2a shows that cognitive user 3 (CU3) and cognitive user 5 (CU5) collide during T_{ct} in SMC-MAC, preventing them from reserving the licensed channels during the current T_{cycle}. However, in the proposed method, performance can be improved by modifying the control channel, as shown in Fig. 3.2b, which allows cognitive users in collisions to select another contention slot in the same T_{cycle}. In Fig. 3.2b, following a collision, cognitive user 3 (CU3) and cognitive user 5 (CU5) select a contention slot from the contention window with the help of the backoff algorithm. If the selected contention slots are different, both cognitive users are successful and may have found an idle channel for transmitting data in the data transmission interval. However, if they both select the same slot from a wider contention window, another collision results, further increasing the contention window size and resulting in a repeat of the procedure, presented as a flow diagram in Fig. 3.3. We have considered the full-duplex capability of cognitive users that enables the cognitive nodes to simultaneously transmit and receive information/data. When the cognitive user selects a licensed idle channel during the contention interval, the cognitive node switches to the selected channel.

If the primary user signal has been sensed by the cognitive node on the selected licensed channel in the data transmission interval, the node stops transmission of its own signal to protect the primary user on that channel. Since the sensing is performed throughout most of the cycle by the cognitive node, however, during the sensing- sharing interval, the sensing results are also shared with other users to incorporate cooperation and enhance the performance of the cognitive network.

3.4 Performance Analysis

In this section, a numerical analysis of the proposed MAC protocol is performed and various parameters of the cognitive network are discussed. For a fixed number of channels sensed by each cognitive user, idle channels detected by cognitive users are computed in the sensing-sharing interval. The successful users after contention are computed, as well as the throughput of the cognitive users that have successfully reserved the idle channels for data transmission.

3.4.1 Sensing-Sharing Analysis

In [32], the authors discuss the behavior of cellular communication system subscribers, which follows a Poisson distribution and exponentially distributed arrival time between two calls. The Poisson distribution is a Markov process with state transitions limited to the next higher state or to the same state and having a constant transition rate. Therefore, for the given Poisson distribution of primary network cellular calls with an inter-arrival time T and average rate λ, the distribution of waiting times between successive calls is computed using the cumulative distribution function (CDF) [32]:

$$p_{\mathrm{i}} = P\left(T \leq T_{\mathrm{cycle}}\right) = 1 - P\left(T > T_{\mathrm{cycle}}\right) = 1 - exp\left(-\lambda T_{\mathrm{cycle}}\right)$$

where p_{i} is the given probability of cognitive user interfering with the primary user and T_{cycle} is the maximum interference time that a cognitive user is allowed to interfere with the primary user. Hence, the T_{cycle} is computed as:

$$T_{\mathrm{cycle}} = -\frac{ln(1 - p_{\mathrm{i}})}{\lambda}$$

Further, the ith licensed channel utilization is represented by the probability α_{i}, where $1 \leq i \leq N_{\mathrm{ch}}$, and we have assumed that on average the total utilization probability of each channel is: $\alpha = \frac{\sum_{i=1}^{N_{\mathrm{ch}}} \alpha_{\mathrm{i}}}{N_{\mathrm{ch}}}$. Therefore, the probability of l idle channels in the system follows the binomial distribution as given by [8]:

$$p(l) = \binom{N_{\mathrm{ch}}}{l} (1 - \alpha)^l \alpha^{N_{\mathrm{ch}} - l}, \quad 0 \leq l \leq N_{\mathrm{ch}} \tag{3.1}$$

where N_{ch} is the total number of licensed channels, and the average number of idle licensed channels present in the primary network is [8]:

$$E[L] = \sum_{l=0}^{N_{\mathrm{ch}}} lp(l) \tag{3.2}$$

where $p(l)$ is obtained from Eq. (3.1). Let us assume that the cognitive user senses limited Ch_{max} channels randomly among the total N_{ch} licensed channels. The probability distribution of the number of sensed idle channels m among the sensed licensed channel Ch_{max} by the single cognitive user is [8]:

$$p(m) = \binom{Ch_{\max}}{m}(1-\alpha)^m \alpha^{Ch_{\max}-m}, \quad 0 \le m \le Ch_{\max} \tag{3.3}$$

Thus, the average numbers of sensed idle channels by a cognitive user are:

$$E[M] = \sum_{m=0}^{Ch_{\max}} mp(m) \tag{3.4}$$

where $p(m)$ is achieved from Eq. (3.3). The probability of a cognitive user sensing a licensed channel is thus given by:

$$\mu = \frac{\text{Number of channels each cognitive user sense}}{\text{Total number of licensed channels}}$$

or

$$\mu = \frac{Ch_{\max}}{N_{\text{ch}}} \tag{3.5}$$

Since cognitive users choose and sense licensed channels independently, from Eq. (3.5) we can obtain the probability that a channel is not sensed by any N_{CU} number of cognitive users, which is given by:

$$p_{\text{nosensed}} = (1-\mu)^{N_{\text{CU}}} \tag{3.6}$$

However, from Eq. (3.6), the probability that a channel is sensed by at least one cognitive user is:

$$p_{\text{sensed}} = 1 - p_{\text{nosensed}} \tag{3.7}$$

The probability distribution of n detected idle channels among $E[L]$ idle licensed channels by N_{CU} cognitive users is determined using Eqs. (3.2) and (3.7) as:

$$p(n) = \binom{E[L]}{n} p_{\text{sensed}}^{n} (1 - p_{\text{sensed}})^{E[L]-n}, \quad 0 \le n \le E[L] \tag{3.8}$$

From Eq. (3.8), the average number of sensed idle channels by N_{CU} cognitive users is computed as:

$$E[N] = \sum_{n=0}^{E[L]} np(n) \tag{3.9}$$

where $p(n)$ is achieved from Eq. (3.8).

3.4.2 Contention Analysis

After sensing the licensed channels and sharing the results of sensing among N_{CU} cognitive users during the sensing-sharing interval, the cognitive users compete with each other for reserving the idle licensed channels in the contention interval. However, each cognitive user with data to send to its intended receiver randomly selects a contention slot among the total number of contention slots, Q, in the contention interval. Two cases are now considered, one in which the number of successful cognitive users is computed without any contention resolution, and the other in which a backoff algorithm is applied. The case without the backoff algorithm is for the existing SMC-MAC protocol discussed in [8].

- Case 1: Without backoff algorithm

Given that the selection of a contention slot by each cognitive user is random, it is possible that two or more cognitive users select the same contention slot, resulting in collisions among cognitive users who were unable to reserve idle licensed channels for data transmission during the data transmission interval. Successful contention slot selection occurs when a single cognitive user selects a contention slot and can transmit its data over the reserved idle licensed channel during the transmission interval. Since we have Q number of contention slots, the probability of selecting each contention slot is:

$$r = \frac{1}{Q}$$

The number of cognitive users selecting a given contention slot is denoted by random variable s, and follows binomial distribution as:

$$p(s) = \binom{N_{CU}}{s} r^s (1-r)^{N_{\mathrm{CU}}-s}, \quad 0 \le s \le N_{\mathrm{CU}} \tag{3.10}$$

The probability of a successful contention slot selection is determined from Eq. (3.10); When $s = 1$, the single cognitive user has selected a given contention slot. Therefore, the probability of success from Eq. (3.10) is [8]:

$$\begin{aligned} p_{\mathrm{success}} &= p(1) = \binom{N_{\mathrm{CU}}}{1} r^1 (1-r)^{N_{\mathrm{CU}}-1} \\ &= N_{\mathrm{CU}} r (1-r)^{N_{\mathrm{CU}}-1} \end{aligned} \tag{3.11}$$

If we consider t to be a random variable denoting the number of successful cognitive users, then the probability of t cognitive users being successful is [8]:

$$p(t) = \binom{Q}{t}(p_{\text{success}})^t(1 - p_{\text{success}})^{Q-t}, \quad 0 \le t \le Q \tag{3.12}$$

The average number of successful cognitive users is computed from Eq. (3.12) and is defined as:

$$E[T] = \sum\nolimits_{t=0}^{Q} tp(t) \tag{3.13}$$

From Eq. (3.13), the average number of cognitive users colliding is:

$$E[C] = N_{CU} - \sum\nolimits_{t=0}^{Q} tp(t) \tag{3.14}$$

where $p(t)$ is achieved from Eq. (3.12).

- Case 2: With backoff algorithm

In the proposed scheme, after the first collision is detected during the contention interval, the contention window size increases according to the backoff algorithm, and the cognitive user then selects another contention slot from the larger contention window. For each subsequent collision, the contention window size increases, and this process continues until there are no additional collisions and all cognitive users have successfully reserved a channel. It is evident that congestion is a function of the number of cognitive users, and alleviating congestion is possible by increasing the contention window size, and hence the number of contention slots in the contention interval. Therefore, the proposed backoff algorithm improves the flexibility of the contention interval based on the number of cognitive users in the network. The algorithm for the proposed scheme is described as follows:

Algorithm:
Step 1: Variable declaration
N_{CU} = *Number of cognitive users*
CW = *Number of initial contention slots*
= $2 \times N_{CU}$
CW_{new} = *$CW+2^4$, which is selected initially by cognitive users that experience their first collision during contention interval*
Count = *number of collided cognitive users*
Z = *Number of successful cognitive users*
Step 2: Count the number of collided cognitive users in the contention interval
N_{CU} = *Number of cognitive users that randomly select contention slots between 1 and* CW
IF N_{CU} *cognitive users have selected different contention slots*
N_{CU} *cognitive users are successful*

ELSE

Count = count the number of cognitive users that have selected the same contention slots

Z = N_{CU} - *Count*

END

Step 3: Solve contention among collided cognitive users with the help of the backoff algorithm
FOR *i = 1:10//taken by default*
Count *the number of cognitive users randomly select contention slot between* CW *and* CW_{new}
IF *Count the number of cognitive users have selected different contention slots*
All N_{CU} *cognitive users are successful*
break;
ELSE

X = *Number of cognitive users which have selected the same contention slot*

Z = Z + (Count-X)
Count = X

$CW = CW_{new}$
$CW_{new}=CW+2^i$

IF *Z =* N_{CU}
All cognitive users have become successful by selecting different contention slots.
break;

END

END

3.4.3 Data Transmission and Throughput Analysis

Successful cognitive users transmit their data in the data transmission interval on idle channels selected during the contention interval. The data transmission interval T_{tr} is defined by subtracting the idle time T_{idle}, the sensing-sharing time T_{ss}, and the contention time T_{ct} from the cycle time T_{cycle} [8]. This transmission interval is utilized for the computation of cognitive user throughput. However, the maximum achievable throughput is achieved when all detected licensed idle channels are utilized for data transmission in the data transmission interval. Therefore, the maximum achievable throughput is defined as the product of the average number of sensed idle channels $E[N]$, the amount of time available for the data transmission per cycle interval (T_{tr}/T_{cycle}), and data rate per sensed idle channels R. Hence, the maximum achievable throughput is given as [8]:

$$Th_{max} = \frac{E[N] \times T_{tr} \times R}{T_{cycle}} \quad (3.15)$$

where $E[N]$ is achieved from Eq. (3.9). However, the throughput of successful users in the SMC-MAC protocol is the minimum of the $(Ch_{idle}T)$ and the average number of sensed idle channels from Eq. (3.9) where Ch_{idle} is the number of idle channels that a cognitive user is allowed to use. Therefore, the throughput of cognitive users in SMC-MAC is given as [8]:

$$Th_{SMC-MAC} = \frac{E[\min(Ch_{idle}T, N)] \times T_{tr} \times R}{T_{cycle}} \quad (3.16)$$

where T is the number of successful cognitive users during the contention interval. Therefore, $(Ch_{idle}T)$ defines the total number of idle channels on which all T successful cognitive users can transmit. Moreover, the throughput of successful cognitive users in the proposed scheme is given as:

$$Th_{prop.} = \frac{E[\min(\mathrm{Ch}_{idle} \times Z, \mathrm{N})] \times T_{tr} \times R}{\mathrm{T}_{cycle}} \quad (3.17)$$

where Z is the number of successful users after the backoff algorithm in the contention interval.

3.5 Simulation Results

The proposed distributed MAC protocol parameters for the cognitive user network are employed from IEEE 802.11a [16]. The simulation parameters are as follows: idle interval (T_{idle}) is 34 μs, single slot time is 9 μs, CR-RTS, CR-CTS and CR-SIFS frame time are 24 μs, 24 μs, and 16 μs, respectively. The data rate of each channel is 54 Mbps.

$$T_{idle} = CR - SIFS + 2 \times singleslottime,$$

$T_{ss} = 3 \times N_{ch} \times singleslottime$, and
$T_{ct} = number\ of\ contention\ slots \times ((CR - RTS) + (CR - SIFS) + (CR - CTS))$.

The simulation results of the sensing-sharing analysis discussed in Sect. 3.4.1 are presented in Fig. 3.4 and Fig. 3.5. The total number of licensed channels are assumed to be $N_{ch} = 20$ and $Ch_{idle} = 1$. In, Fig. 3.4 the numerical results are presented from Eq. (3.9) for the case when the total number of cognitive users are $N_{CU} = 5$, $N_{CU} = 10$ and the traffic load α is assumed to be 0.5. Since a cognitive user is able to sense only the fixed number of channels given by Ch_{max}, Fig. 3.4 shows that as the number of channels sensed by each cognitive user increases, the number of idle channels detected by N_{CU}(number of cognitive users) also increases. However, for higher values of Ch_{max}, more mathematical computations are required, making the cognitive radio terminal less energy efficient. Further, Fig. 3.5 demonstrates the actual number of idle channels and the number of idle channels sensed by 10 cognitive users for different values of traffic load α and Ch_{max}. Moreover, Fig. 3.5 reveals that there is gap between the actual number of idle channels present and the number of idle channels detected for different Ch_{max} values, which is due to the lower number of channels sensed by the individual cognitive user in particular defined Ch_{max}. However, Fig. 3.5 demonstrates that as the cognitive user's ability to sense licensed channels increases, i.e., as the value of the parameter Ch_{max} increases, the total number of idle channels sensed by all cognitive users approaches the total number of available idle channels. In addition, various limitations of the SMC-MAC protocol [8, 35, 36], as revealed in the numerical simulation results, are avoided with the use of the proposed scheme, as demonstrated in Figs. 3.6 and 3.7.

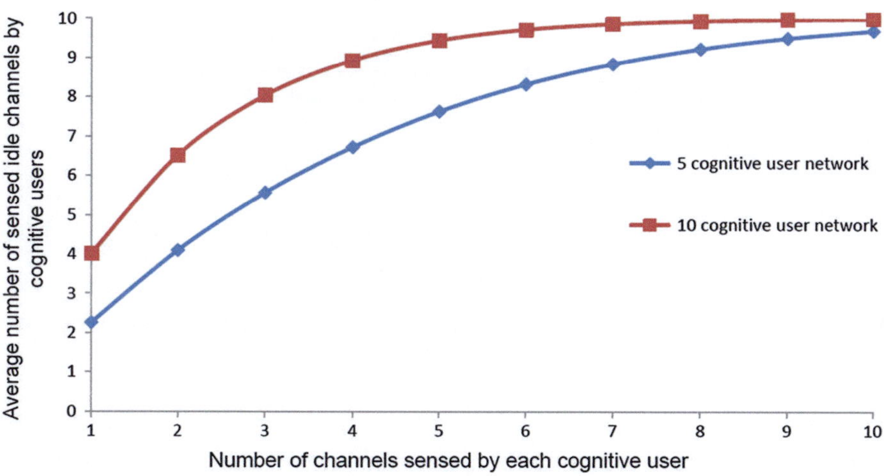

Fig. 3.4 The number of channels sensed by each cognitive user in relation to the average number of sensed idle channels by a 5- and 10-cognitive user network

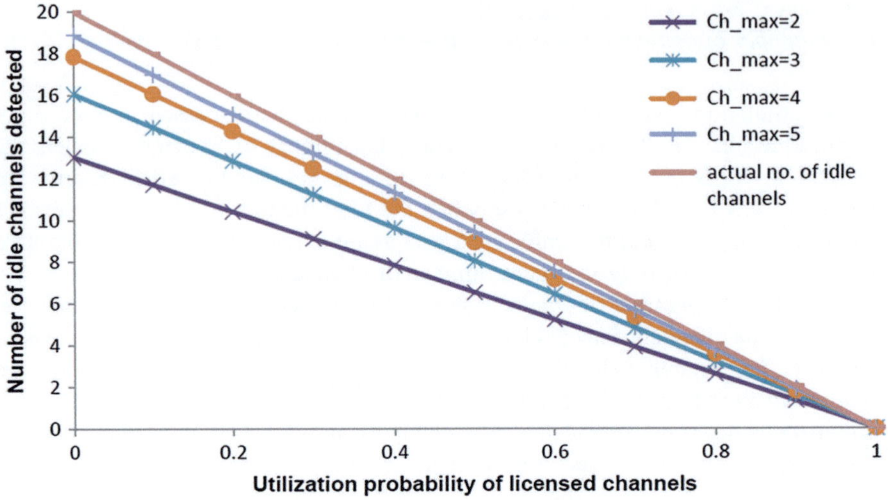

Fig. 3.5 The number of detected idle channels in relation to the utilization probability of licensed channels for a 10-cognitive user network

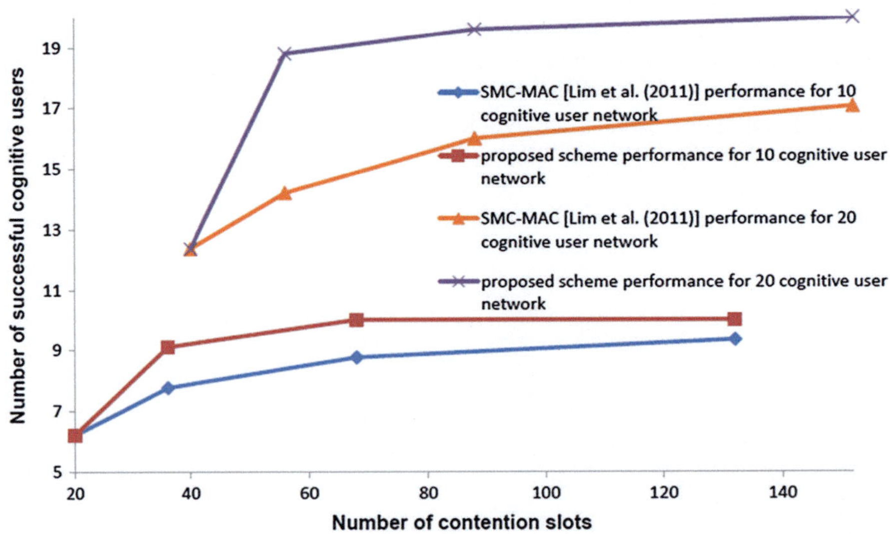

Fig. 3.6 The distribution of successful cognitive users in relation to the number of contention slots for the proposed and SMC-MAC [8] protocol averaged over 10 runs

In the proposed method, applying the binary exponential backoff mechanism to resolve the contention among collided users significantly increases success among users, as illustrated in Fig. 3.6, compared with the SMC-MAC protocol for the same number of contention slots. The SMC-MAC protocol has no contention

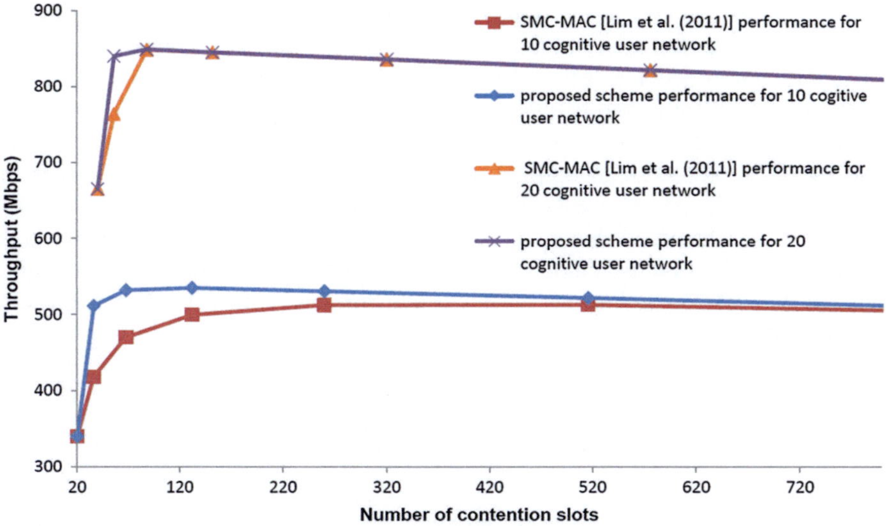

Fig. 3.7 The throughput comparison between the proposed and SMC-MAC protocols [8] with varying contention slots for an average licensed channel utilization probability of (α) 0.1 and data rate of 54 Mbps per channel

resolution algorithm, as discussed in [8, 35, 36], and Fig. 3.6 clearly illustrates that there are appreciably more successful cognitive users when the backoff algorithm is applied under the proposed scheme. In the SMC-MAC scenario, because it is not possible for the collided cognitive users in the contention interval to select a different contention slot during that cycle—resulting in fewer successful cognitive users, as shown in Fig. 3.6—data transmission is thus not possible within the same cycle for the collided cognitive users in the SMC-MAC protocol.

In Fig. 3.7, the throughput is plotted with the number of contention slots in networks with 10 and 20 cognitive users, when channel utilization probability is 0.1. Figures 3.6 and 3.7 illustrate that there is some optimal value of contention slots, depending on the number of cognitive users, for which a maximum number of successful users and throughput is obtained, and if we further increase the contention slots from this value, the throughput decreases due to the decrease in the data transmission interval. Since the number of transmitting cognitive users in a wireless communication system is changing at random, having a fixed number of contention slots, as in the SMC-MAC protocol [8], is not practical. In the proposed scheme, the optimal number of contention slots varies according to the number of cognitive users, with the aim of enhancing performance, as shown in Fig. 3.7. Further, the throughput is greater in the 20-user versus 10-user network, because more users have successfully secured idle channels. In addition, the throughput of the proposed scheme and the SMC-MAC protocol in the 20-cognitive user network scenario is the same at optimal contention slots, as shown in Fig. 3.7. This is because, although the number of successful cognitive users in the proposed scheme is higher at optimal contention slots than in the SMC-MAC protocol (Fig. 3.6), the number of successful users obtaining idle channels for data transmission in the 20-cognitive user network is the same as that in the SMC-MAC protocol, due to the selected $Ch_{\max}$ parameter in this case. Hence, the throughput of the proposed scheme in the 20-cognitive user network could be greater than that of the SMC-MAC protocol at an optimal number of contention slots if all successful cognitive users in the proposed scheme have secured idle channels, as demonstrated in the 10-user network shown in Fig. 3.7. However, the computational results presented for the optimal number of contention slots are simulated results, and a detailed analysis, including analytical results for the proposed scheme, are discussed in the next chapter.

3.6 Summary and Future Direction

In this chapter, a cooperative MAC protocol has been proposed for the distributed cognitive radio communication system, with a backoff algorithm for contention solving. The proposed method significantly enhances the performance of cognitive radio communication systems by increasing the number of successful cognitive users for data transmission. Hence, as the numerical simulation results demonstrate, the proposed method enhances throughput in comparison to that of the existing SMC-MAC protocol reported in [8] for distributed cognitive networks. The proposed MAC protocol optimizes the number of contention slots depending on the number of cognitive users, in contrast to the fixed number of slots in the SMC-MAC protocol. However, additional development is needed and potential technical issues and future research directions for the CR-MAC protocol are summarized below.

- A common control channel is the backbone of a distributed cognitive radio network, as it facilitates communication and coordination among SUs. However, this channel may become saturated when the number of cognitive users or traffic load increases [25]. Therefore, the independent nodes may not observe the same spectrum opportunities, and thus may be unable to share the same channel with other users. Additional dynamic strategies should be developed to achieve a reliable exchange of control and signaling information, to permit synchronization within the cognitive radio network [25].
- Increasing the sensing time enables an increase in the number and quality of detected spectrum opportunities [25]. However, in order to limit the sensing overhead, a cognitive user can observe only a limited part of the radio resource and only a few MAC protocols implement such a criterion (e.g., in [15], probing channel selection is performed based on the state of the Markov process). Therefore, this problem of limiting sensing time and maximizing spectrum opportunities requires further investigation for improving spectrum sensing effectiveness.

- In cognitive radio communication systems, detecting the primary transmitter signal does not always correspond to the discovery of spectrum opportunities [37, 38]. Indeed, even when primary signals can be perfectly detected, spectrum opportunity discovery is affected by three major issues: the hidden transmitter, the exposed transmitter, and the hidden receiver [25]. A hidden transmitter is outside the sensing range of the cognitive sender but is located close to the cognitive receiver. An exposed transmitter is a primary sender that is located in the proximity of the cognitive transmitter, while the licensed receiver is outside the secondary transmitter's interference range. A hidden receiver is a primary receiver that is located in the interference range of a cognitive transmitter, while the primary transmitter is outside the detection range of the cognitive users. The hidden transmitter problem is resolved by performing spectrum sensing at both the transmitter and receiver side, but no adequate solutions are yet available for the other problems. In order to effectively resolve these issues, a cognitive user should be able to detect the presence of a neighbor primary receiver [25]. In [39], the authors describe a sensor that is able to locate a primary receiver by measuring its local oscillator leakage power from its front end. However, as local oscillator leakage power is very low, this approach is only suitable for the detection of TV receivers [40].
- Cooperative sensing has emerged as a means of greatly enhancing the effectiveness of primary user detection in a cognitive radio wireless fading channel. However, as stated in [41, 42], collaborative detection is limited by the effects of spatially correlated shadowing. For a given SNR, a larger number of correlated sensing nodes are needed to achieve the same detection probability of a few independent users [25]. Therefore, developing efficient MAC protocols should consider the correlation effect in cooperative sensing schemes.
- When an incumbent user is detected, cognitive users interrupt transmission and move to a new available channel to continue data transfer. However, the time spent searching for a new channel limits cognitive user performance, and limiting packet loss and delay during the spectrum mobility process is a challenge. Researchers [43, 44] have introduced the concept of a backup channels list during spectrum handoff that reduces latency and avoids performance degradation. Further investigation is needed, however, to increase the number of available channels and to introduce QoS criteria for protecting priority users [25].
- Most cognitive radio MAC protocols have been designed for a spectrum interweave access model that is transmitting only in white space. Due to the increasing demand for spectrum opportunities, however, the underlay model is emerging as an issue in cognitive radio networks. In order to successfully implement the underlay transmission access model, new metrics that represent the performance degradation experienced by the primary system should be investigated [25].
- In addition to spectrum interweave and underlay transmission models, exploiting adaptive modulation and coding (AMC) and power control techniques for spectrum overlay can improve the overall system capacity and the

efficiency of CR-MAC protocols [25]. Further benefits could be obtained by investigating a hybrid transmission scheme for the CR-MAC protocol.

- Historically, researchers engaged in the development of spectrum-efficient systems have attempted to address bandwidth scarcity without considering energy efficiency. Green communication approaches for cognitive radios must now be investigated in order to conserve power consumption, reduce interference, and improve the battery life of consumer devices. The concept of the coexistence of femtocells and microcells with macrocell users in the cellular network can also be investigated for CR networks to enhance energy efficiency and performance [25].

References

1. J. Jia, Q. Zhang, X. Shen, HC-MAC: A hardware-constrained cognitive MAC for efficient spectrum management. IEEE J. Sel. Areas Commun. **26**(1), 106–117 (2008)
2. C. Cordeiro, K. Challapali, C-MAC: A cognitive MAC protocol for multi-channel wireless networks, in *Proceedings of 2nd IEEE International Symposium on New Frontiers in Dynamic Spectrum Access Networks* (DySPAN), Dublin, 17–20 April 2007, pp. 147–157
3. H. Su, X. Zhang, CREAM-MAC: An efficient cognitive radio-enabled multi-channel MAC protocol for wireless networks, in *Proceedings of IEEE International Symposium on World of Wireless, Mobile and Multimedia Networks* (WoWMoM), Newport Beach, CA, June 2008, pp. 1–8
4. L. B. Le, E. Hossain, OSA-MAC: A multi-channel MAC protocol for opportunistic spectrum access in cognitive radio networks, in *Proceedings of IEEE Wireless Communications and Networking Conference* (WCNC), Las Vegas, NV, 31 March–3 April 2008, pp. 1426–1430
5. S.- H. Lee, B.-J. Lee, S.-H. Rhee, Efficient utilization of available channels in dynamic spectrum access networks. Wire. Pers. Comm. **67**(1):95–104 (2012)
6. P. Papadimitratos, S. Sankaranarayanan, A. Mishra, A bandwidth sharing approach to improve licensed spectrum utilization. IEEE Commun. Mag. **43**(12), S10–S14 (2005)
7. A. Sabharwal, A. Khoshnevis, E. Knightly, Opportunistic spectral usage: Bounds and a multi-band CSMA/CA protocol. IEEE/ACM Trans. on Networking **15**(3), 533–545 (2007)
8. S. Lim, T.-J. Lee, A self-scheduling multi-channel cognitive radio MAC protocol based on cooperative communications. *IEICE Trans. Comm.* **E94-B**(6):1657–1668 2011
9. K. Thilina, E. Hossain, and D. Kim, DCCC-MAC: A dynamic common control channel-based MAC protocol for cellular cognitive radio networks. IEEE Trans. Veh. Technol, **PP(** 99), (2015). doi:10.1109/TVT.2015.2438058
10. T.-Z. Oo, N. Tran, D. Dang, Z. Han, L. Le, C.-S. Hong, OMF-MAC: An opportunistic matched filter-based MAC in cognitive radio networks. IEEE Trans. Veh. Tech. **PP**(99), (2015). doi:10.1109/TVT.2015.2415033
11. R. Hossain, R.-H. Rijul, M.-A. Razzaque, A.-M.-J. Sarkar, Prioritized medium access control in cognitive radio ad hoc networks: protocol and analysis. Wire. Pers. Comm. **79**(3), 2383–2408 (2014)
12. P. Xie, L. Li, J. Zhu, R. Zheng, M. Zhang, A cooperation and access spectrum sharing protocol with cooperative interference management. Wire. Pers. Comm. **81**(3), 997–1015 (2015)
13. K. Lee, C. Kim, Distributed sequential access MAC protocol for single hop wireless networks. Wire Pers. Comm. **72**(4), 2177–2184 (2013)

14. H. Kim, K.-G. Shin, Efficient discovery of spectrum opportunities with MAC-layer sensing in cognitive radio networks. IEEE Trans. Mob. Comput. **7**(5), 533–545 (2008)
15. Q. Zhao, L. Tong, A. Swami, Y. Chen, Decentralized cognitive MAC for opportunistic spectrum access in ad hoc networks: A POMDP framework. IEEE J. Sel. Areas Comm. **25**(3), 589–600 (2007)
16. IEEE 802.11, Wireless LAN Medium access control (MAC) and Physical layer (PHY) Specifications., IEEE, (June 2007)
17. S. Pandit, G. Singh, Throughput enhancement of distributed cognitive radio network using bandwidth wastage in medium access control protocol. Wire. Pers. Comm. (under review), (March 2016)
18. G.-A. Shah and O.-B. Akan, Cognitive adaptive medium access control in cognitive radio sensor networks. IEEE Trans. Veh Technol, 64(**2**), 757–767 (2015)
19. K. Bian, J.-M. Park, MAC-layer misbehaviors in multi-Hop cognitive radio networks, in *Proceedings of US-Korea Conference on Science, Technology and Entrepreneurship* (UKC2006), Teaneck, New Jersey, 65–73 2006
20. R. Jain, *The Art of Computer System Performance Analysis: Techniques for Experimental Design, Measurement Simulation and Modeling* (New York, NY, Wiley- Interscience, 1991)
21. M. Timmers, S. Pollin, A. Dejonghe, L.-V. Perre, F. Catthoor, A distributed multichannel MAC protocol for multihop cognitive radio networks. IEEE Trans. Veh. Technol. **59**(1), 446–459 (2010)
22. S.C. Jha, M.M. Rashid, V.K. Bhargava, C. Despins, Medium access control in distributed cognitive radio networks. IEEE Wirel. Commun. **18**(4), 41–51 (2011)
23. J. Xiang, Y. Zhang, T. Skeie, Medium access control protocols in cognitive radio networks. Wirel. Comm. Mobile Comput. **10**(1), 31–49 (2010)
24. C. Cormioand, K.R. Chowdhury, A survey on MAC protocols for cognitive radio networks. Ad. Hoc. Netw. **7**(7), 1315–1329 (2009)
25. A. De Domenico, E.C. Strinati, M.G. Di Benedetto, A survey on MAC strategies for cognitive radio networks. IEEE Comm. Surv. Tutorials **14**(1), 21–44 (2012)
26. B.F. Lo, A survey of common control channel design in cognitive radio networks. Phys. Comm. **4**(1), 26–39 (2011)
27. T.V. Krishna, A. Das, A survey on MAC protocols in OSA networks. Comput. Netw. **53**(9), 1377–1394 (2009)
28. P. Pawelczak, S. Pollin, H.S.W. So, A.R. Bahai, R.V. Prasad, R. Hekmat, Performance analysis of multichannel medium access control algorithms for opportunistic spectrum access. IEEE Trans. Veh. Technol. **58**(6), 3014–3031 (2009)
29. Q. Chen, Y.C. Liang, M. Motani, W.C. Wong, A two-level MAC protocol strategy for opportunistic spectrum access in cognitive radio networks. IEEE Trans. Veh. Technol. **60**(5), 2164–2180 (2011)
30. J. Park, P. Pawelczak, D. Cabric, Performance of joint spectrum sensing and MAC algorithms for multichannel opportunistic spectrum access ad hoc networks. IEEE Trans. Mob. Comput. **10**(7), 1011–1027 (2011)
31. W.S. Jeon, J.A. Han, D.G. Jeong, A novel MAC scheme for multichannel cognitive radio ad hoc networks. IEEE Trans. Mob. Comput. **11**(6), 922–934 (2012)
32. D. Willkomm, S. Machiraju, J. Bolot, A. Wolisz, Primary user behaviour in cellular networks and implications for dynamic spectrum access. IEEE Commun. Mag. **47**(3), 88–95 (2009)
33. I.-F. Akyildiz, W.-Y. Lee, M.-C. Vuran, S. Mohanty, Next generation/dynamic spectrum access/cognitive radio wireless networks: A survey. Comput. Netw. 50(**13**), 2127–2159 (2006)
34. Y.-C. Liang, K.-C. Chen, G.-Y. Li, P. Mahonen, Cognitive radio networking and communications: An overview. IEEE Trans. Veh. Technol. **60**(7), 3386–3407 (2011)
35. S. Pandit, G. Singh, Self-scheduled MAC-layer protocol for spectrum sharing in cognitive radio communication, in *Proceedings of 6th IEEE International Conference on Contemporary Computing* (IC3), India, Aug. 8–10, 2013, pp. 250–255

36. Shweta Pandit and G Singh, Backofff algorithm in cognitive radio MAC-protocol for throughput enhancement, IEEE Trans. Veh. Technol. **64**(5), 1991–2000 (2015)
37. A. Ghasemi, E. S. Sousa, Impact of user collaboration on the performance of sensing-based opportunistic spectrum access, in *Proceedings of IEEE Vehicular Technology Conference* (VTC Fall-06*)*, Montreal, Canada, 1–6 September 2006
38. W. Ren, Q. Zhao, A. Swami, Power control in cognitive radio networks: how to cross a multi-lane highway. IEEE J. Sel. Areas Comm. **27**(7), 1283–1296 (2009)
39. B. Wild, K. Ramchandran, Detecting primary receivers for cognitive radio applications, in *Proceedings of IEEE International Symposium on New Frontiers in Dynamic Spectrum Access Networks*, (DySPAN'05), November 2005, pp. 124–130
40. I.F. Akyildiz, W.Y. Lee, M.C. Vuran, S. Mohanty, A survey on spectrum management in cognitive radio networks. IEEE Comm. Mag. **46**(4), 40–48 (2008)
41. S. Mishra, A. Sahai, and R. Brodersen, Cooperative sensing among cognitive radios, in *Proceedings of IEEE International Conference on Communications* (ICC'06), vol. 4, June 2006, pp. 1658–1663
42. A. Ghasemi and E. S. Sousa, Collaborative spectrum sensing for opportunistic access in fading environments, in *Proceedings of Symposium on Dynamic Spectrum Access Networks* (DySPAN'05), Baltimore, MD, USA, November 2005, pp. 131–136
43. C. M. Cordeiro, K. Challapali, D. Birru, IEEE 802.22: An introduction to the first Wireless Standard based on cognitive radios, J. Comm., Special Issue from selected papers from *DySPAN* 2005, **1**(1):328–337 (April 2006) (Invited Paper)
44. C. Cordeiro, K. Challapali, M. Ghosh, Cognitive PHY and MAC layers for dynamic spectrum access and sharing of TV bands, in *Proceedings of International Workshop on Technology and Policy for Accessing Spectrum* (TAPAS'06), Boston, MA, USA, August 2006

Chapter 4
Distributed Cognitive Radio Medium Access Control Protocol in Perfect and Imperfect Channel Sensing Scenarios

4.1 Introduction

In the previous chapter, it is assumed that sensing licensed channels by cognitive users is perfect, which is very difficult to yield in practice. Therefore, in this chapter, the practical scenario of imperfect sensing/sensing errors is considered in the proposed distributed cognitive radio MAC protocol. The idle channel detection in the cognitive radio MAC protocol is affected by the false alarm probability occurring due to imperfect sensing. The false alarm [1, 2] occurs when the cognitive user falsely (imperfectly) detects a licensed channel as busy, when it is actually idle, and in this situation the cognitive user cannot transmit data. Missed detection also results in imperfect sensing of a licensed channel, where a cognitive user transmits their data on a licensed channel that is already occupied by the primary user and hence causes interference to the primary user. In this chapter, a potential scheme has been proposed to depict the effect of perfect and imperfect sensing on the performance of the proposed distributed cognitive radio MAC protocol. The simulation results are presented for different false alarm probabilities and the throughput is computed in this environment. Moreover, the amount of interference occurring on the primary user network due to missed detection probability is also seen. Further, as we have discussed in the previous chapter and [3], the number of collisions are significantly high if the number of contention slots are limited and cognitive users are significantly greater. However, the large number of contention slots increases the success rate of cognitive users in the cognitive network, but simultaneously decreases the data transmission interval and hence throughput of the cognitive radio network. Therefore, mathematical formulation of the optimum number of contention slots is obtained for the proposed MAC protocol so that the throughput of cognitive radio network is enhanced with the minimum number of contention slots, as discussed in Sect. 4.3.2 of this chapter. In the results and discussion section of this chapter, we have obtained an optimized number of contention slots using the

S. Pandit and G. Singh, *Spectrum Sharing in Cognitive Radio Networks*,
DOI 10.1007/978-3-319-53147-2_4

proposed MAC protocol with the backoff algorithm at which all the users become successful.

Further, an important parameter to observe the performance of the MAC protocol is energy consumption [4] of the proposed system. Since a mobile terminal, generally, has limited battery power, the proposed system should have high energy efficiency. Recently, several researchers/scientists presented significant work in the field of energy consumption and energy efficiency of the cognitive radio system [4–6]. An energy efficient multichannel MAC protocol has been proposed in [7], and four rendezvous algorithms have been provided for cognitive radio MAC protocol when there is no control channel or centralized system. Moreover, a wireless sensor network and cognitive radio technology has been integrated into the cognitive receiver based MAC (CRB-MAC) protocol [8] to provide energy efficiency and better delay performance. Further, Wang et al. [4] have optimized the spectrum sensing and access time to reduce the energy consumption of the cognitive radio user. However, the tradeoff between energy consumption in data transmission and energy overhead is discussed in [5]. Therefore, we have numerically computed the energy efficiency [6] of the proposed distributed multichannel cognitive MAC protocol for different false-alarm probabilities. The energy consumed for sensing the licensed channels, sharing the sensing information, reserving idle channels, and for data transmission is computed. Moreover, the throughput and energy efficiency of the proposed MAC protocol are also compared with that of the perfect sensing scenario.

The remainder of the chapter is organized as follows. In Sect. 4.2, the related work and problem formulation is explained in detail. Mathematical modeling for perfect and imperfect sensing along with the contention interval analysis is performed in Sect. 4.3. In addition, the throughput for perfect and imperfect sensed environments is also computed. Further, in Sect. 4.4, energy efficiency of the proposed MAC protocol is numerically computed, and Sect. 4.5 explores the numerical simulation results. Finally, Sect. 4.6 concludes the work.

4.2 Related Work and Problem Formulation

Wireless networks have become an essential part of modern life. However, currently, 3% of worldwide energy is consumed by ICT infrastructures, which cause about 2% of worldwide CO_2 emissions. The transmitted data-volume increases rapidly, and wireless communications are used broadly; however, network design rules have practically ignored the energy efficient network design approach to limit CO_2 emissions. This design approach is currently named Green Communications. Significant energy savings in mobile networks can be expected by defining and standardizing energy efficiency metrics and combining energy aware flexible radios and networks. There have been several discussions on energy efficiency and power consumption issues in the wireless network, and they are summarized in this chapter. In [9], several techniques such as cross layer approach, multiple antennas, cell size reduction, and cognitive radio, from the system-wide energy efficiency point of view, outlining challenges and open issues have been detailed. Nobar et al. [10] have proposed a green cognitive radio network (RF-GCRN), where a central node, called a power beacon (PB), harvests green energy from ambient sources and wirelessly delivers randomly harvested energy to cognitive users. The proposed RF-GCRN differs from the conventional cognitive radio network because it does not need any battery source at the cognitive user and energy is supplied from the central node only. The simulation results in [10] have shown that the proposed model is better suited for low data rate applications without degrading the primary user's performance.

Further, power loading algorithms have been proposed for orthogonal frequency-division multiplexing (OFDM) systems based on average and outage capacity criteria [11]. Rate-power allocation algorithms have been developed for ergodic and quasi-static channel models in [11]. The throughput achieved by these algorithms and the effects of channel multipath are investigated both analytically and with simulations. Further, in [12] the authors have applied the broadcast approach for maximizing the throughput over a fading channel. They have investigated the performance of the broadcast approach for various fading distributions, which corresponds to different models of partially transmitted channel state information (CSI) and have derived the optimal power allocation for these models. Simulation results have shown that in the MISO channel, uniform power allocation is preferable over beamforming power allocation in the region where broadcasting gain over single level coding is non-negligible [12]. Moreover, Sun and Honig in [13] considered a multicarrier system with partial channel state information and found that uniform power distribution over an optimized subset of subchannels or on-off power allocation in fading channels gives the same asymptotic growth in capacity as the optimal water filling approach. Further, they investigated the correlated subchannels and cellular uplink and downlink channel models [13]. Recent advances and research challenges for a green cellular network is also discussed in [14]. In addition to this, a multicarrier system is again considered in [15] and closed form expression of throughput has been derived by exploiting frequency diversity, which does not require any channel quality information at the transmitter. The authors in [16] have assumed systems with various combinations of single antenna, multiple antenna, narrowband, broadband, single-user, and multiuser technology and have provided the performance. They have shown the role of limited feedback in the standardization of next generation wireless systems and have applied certain power constraints [16]. Another interesting uplink power allocation algorithm has been developed by Parsaeefard and Sharafat in [17] for underlay cognitive radio networks (CRNs) with a view to maximizing the social utility of secondary users (SUs) when channel gains from SUs to primary base stations, and interference caused by primary users (PUs) to the SU's base station are uncertain. They have converted the power allocation problem into a geometric programming problem and solved by using Lagrange dual decomposition [17].

Most of the papers discussed sensing-throughput tradeoff [18] in cognitive radio network; however, the tradeoff between sensing performance and energy consumption has been investigated in cognitive radio (CR) networks with amplify-and-forward relays by Huang et al. in [19]. Their main objective is to minimize the energy consumption during sensing under given false alarm and detection probabilities constraints and for this an optimization problem has been developed. However, in [20] a power-allocation scheme for a decode-and-forward (DF) relaying-enhanced cooperative wireless system has been proposed. It is concluded in [20] that instead of demanding high on-grid power supply or high green energy availability, the proposed system can achieve compatible or higher throughput by utilizing the harvested energy. Traditional power grids are currently being transformed into smart grids (SGs), featuring multi-way communication among energy generation, transmission, distribution, and usage facilities. This smart grid concept has been applied for cognitive radio networks in [21] for efficiently utilizing all available spectrum resources. There is another tradeoff between energy consumption and throughput in the cognitive radio system under local, as well as cooperative sensing, characterized in [22]. In addition to this, improving spectrum efficiency conflicts with increasing energy efficiency; therefore, recently, in [23] the spectrum and energy efficiency tradeoff problem is considered for cognitive radio networks, and an algorithm has been proposed to solve this tradeoff problem. Also, sensing overhead-throughput tradeoff issues are presented in [24].

It is known that cooperative sensing decreases the sensing error; however, it also increases the energy spent for sensing in comparison to non-cooperative sensing method. Therefore, in [25] the problem of energy-efficient (EE) spectrum sensing scheduling with satisfactory PU protection has been considered. They have exploited the diversity of cognitive users in their received signal-to-noise ratio (SNR) of the primary signal to determine the sensing duration for each user/channel pair for higher energy efficiency. The problem has been transformed as an optimization problem with two different objectives. The first objective is to minimize the energy consumption, and the second objective is to minimize the spectrum sensing duration to maximize the remaining time for data transmission to get higher throughput [25]. Further, in order to improve energy consumption from the conventional cooperative sensing method, the authors in [26] have proposed a reliable data combining method for cooperative spectrum sensing, according to which the fusion center uses two threshold values to make the final decision only if it is confident enough in the validity of received local data. Otherwise, an additional sensing will be performed. The simulations have validated that for all SNRs, higher absolute throughput and also higher throughput per energy consumption are accessible, rather than conventional cooperative sensing [26]. Moreover, for a large range of SNRs less energy is consumed [26].

Periodic sensing is generally employed in the cognitive radio network to facilitate spectrum handoff in order to avoid transmission collision via switching to a truly idle channel when the current channel is re-occupied by the primary user. However, byproducts of spectrum handoff, such as extra energy cost on channel switch, switch delay, and collisions due to excessive switching, have been largely underestimated [27]. Therefore, questions on how often should we switch and different periodic sensing and handoff mechanisms have been discussed in [27].

Further, the tradeoff between the secondary user's energy efficiency and transmission reliability has also been elaborated. They proposed a switch-stay model and studied to balance the aforementioned tradeoff by considering sensing accuracy, the probability of collision, throughput, and delay constraints [27]. In [28], instead of periodic sensing, a random sensing order policy is discussed and based on this the performance analysis and optimization of a distributed secondary network is done. With the help of Markov model, the average throughputs of SUs and the average interference level among SUs and PUs are investigated. Fully distributed sensing order algorithms can lead to substantial performance improvements in cognitive radio networks without the need for centralized management or message passing among the users [28]. For random or unknown arrivals of primary-user signals, the authors in [29] have proposed an energy detection-based generalized likelihood ratio test (GLRT) algorithm for spectrum sensing in CR femtocell networks. The simulation results for the proposed detector have shown satisfactory performance and features low complexity. In addition to this, the authors in [30] focused on energy efficiency over a frame. By exploring a parametric problem, they established the optimal threshold structure of the strategy, according to which the SU decides the sensing order, as well as when and which channel to access. Furthermore, they have designed both optimal and approximate algorithms accordingly, and results show that there is an increase in the energy efficiency compared to full sensing [30].

The energy efficiency issue has also been examined in [31] for fast-growing wireless communications in order to enhance spectrum efficiency and energy efficiency simultaneously. They have considered an energy efficient non-cooperative cognitive radio networks from the micro, meso, and macro perspectives, where the micro view means how to design energy-efficient spectrum sensing algorithms for each individual secondary user, the meso view means how to coordinate non-cooperative secondary users to share spectrum efficiently, and the macro view means how to deploy cognitive radio networks in an energy-efficient approach [31]. In addition, the resource allocation schemes for orthogonal frequency-division multiple access (OFDMA)-based cognitive femtocells have been proposed in [32].

The authors in [33] have provided energy efficiency analysis for spectrum sensing, spectrum sharing, and allocation and for spectrum handoff and management. A green energy-powered cognitive radio (CR) network has been proposed, which implements energy harvesters providing renewable energy to cognitive users [33]. Green energy-powered CR network designing is a challenging task; however, this increases the network availability and thus extends emerging network applications. The work in [34] presents a novel scheme that investigates the usage history for energy efficient spectrum sensing in infrastructure-based cognitive radio networks. In the proposed scheme, the cognitive users share their sensing information with the central base station, and the base station provides usage history of the licensed bands, based on which the cognitive user decides the channels to be scanned in order to reduce the sensing and improve energy consumption [34]. Moreover, Son et al. [35] have investigated several power allocation policies in orthogonal frequency division multiplexing-based cognitive radio networks under different availabilities of inter-system channel state information (CSI) and licensed primary users (PUs) with two different capabilities, namely for peak interference-power tolerable and average interference-power tolerable primary users. They have proposed optimal and efficient suboptimal power allocation policies for a cognitive network [35]. Further, the joint bandwidth and power allocations for cognitive radio networks (CRNs) is studied in [36], which opportunistically operate on a set of channels unused by multiple primary user (PU) networks. The main objective is to minimize the total power allocation of all coexisting cognitive users and guarantee their quality of service (QoS) requirements [36]. Further, in a relay network an energy-aware mechanism is implemented in the selected relay's transmission, opting for power reduction, as the channel state information is acquired prior to the signal's forwarding to the user terminal and an energy-aware multi-mode relaying (EA-MMR) scheme has also been proposed [37, 38]. Recently, two approaches for quick spectrum sensing in cognitive radio networks have been proposed in [39].

Because of the false alarm probability, the number of idle channels detected by cognitive users in the sensing-sharing interval of the cognitive radio MAC protocol is less than the actual number of idle channels detected in perfect sensing. Since in the contention interval, cognitive users compete for reserving the idle licensed channels detected in the sensing-sharing interval; therefore, fewer data will be transmitted over the licensed channels in case of a false alarm due to fewer detected idle channels, which results in less throughput compared to that of the perfectly sensed environment. In addition to this, miss-detection can also happen, in which the busy licensed channels will be detected as being idle, and although cognitive users transmit their data on the miss-detected licensed channels, they will not increase the throughput when compared with the perfect sensing environment. This is because the data of cognitive users transmitted on the miss-detected licensed channels undergo collision with the primary user's data and hence does not contribute to the cognitive user's throughput. However, the miss-detection causes interference to the primary user. Hence, we have seen the false alarm effect on the throughput and energy efficiency of the proposed MAC protocol and miss-detection effect on the interference to the primary network.

Moreover, once the channel is detected as busy, either due to perfect or imperfect sensing (false alarm), in the sensing-sharing interval by a cognitive user, this channel will not be utilized or sensed again in the current cycle interval. Hence, only the false alarm has affected the throughput of the proposed MAC protocol due to the detection of fewer idle channels compared to the actual idle channels present. Moreover, in the MAC protocol, the cognitive user's data is only transmitted in the data transmission interval; therefore, the cognitive user can easily know about the primary user's signal in sensing-sharing and contention intervals, thus the situation where both the primary and cognitive users are transmitting simultaneously will never occur, hence there is no need to differentiate between the primary and secondary user's signal. However, in case the primary user becomes active during the data transmission interval, the signal presence is detected immediately by the cognitive user, who is currently using this channel, and they will, therefore, stop data transmission to protect the primary user.

4.3 Mathematical Modeling

In this section, mathematical modeling of perfect and imperfect channel sensing for the distributed cognitive radio MAC protocol is performed and different parameters of the cognitive radio network are analyzed.

4.3.1 *Sensing-Sharing Interval Analysis*

Since it is obvious that false alarm results in detection of fewer idle channels by the cognitive users, it has affected the system performance. Therefore, this subsection computes the total number of idle channels detected by the cognitive users for both perfect and imperfect sensing scenarios and interference probability to the primary network due to missed detection as follows:

4.3.1.1 Perfect Sensing

Firstly, we find out the number of cognitive users needed for a particular number of licensed channel sensing at a given $Ch_{\max}$. The probability distribution that x number of slots out of N_{ch} slots in the sensing-sharing interval is not selected by any cognitive user is given by:

$$p(x) = \binom{N_{\mathrm{ch}}}{x} p_{\mathrm{nosensed}}^{x} (1 - p_{\mathrm{nosensed}})^{N_{\mathrm{ch}}-x}, 0 \leq x \leq N_{\mathrm{ch}} \tag{4.1}$$

where p_{nosensed} is achieved from Eq. (3.6). The average number of sensing-sharing slots not selected by any cognitive user is:

$$E[X] = \sum\nolimits_{x=0}^{N_{\mathrm{ch}}} xp(x) \tag{4.2}$$

Therefore, the average number of sensing-sharing slots selected or number of licensed channels sensed by N_{CU} cognitive users is:

$$E[Y] = N_{\mathrm{ch}} - E[X] \tag{4.3}$$

Equation (4.3) provides the total number of channels selected for sensing from the total licensed channels by all the cognitive users for the given $Ch_{\max}$ value. The number of idle channels detected among the selected licensed channels in Eq. (4.3) by N_{CU} cognitive users for the given utilization probability α of each channel is:

$$p(u) = \binom{E[Y]}{u}(1-\alpha)^u \alpha^{E[Y]-u}, \quad 0 \le u \le E[Y] \tag{4.4}$$

From Eq. (4.4), the average number of idle channels detected by N_{CU} cognitive users is computed as:

$$E[U] = \sum\nolimits_{u=0}^{E[Y]} up(u) \tag{4.5}$$

4.3.1.2 Imperfect Sensing

As discussed earlier that false alarm and missed detection are the two parameters to be considered in imperfect sensing, in this sub-section we show how these parameters affect the proposed MAC protocol.

(a) *False alarm*

For the given probability of the false alarm and idle channels detected by N_{CU} cognitive users, the probability of g channels that are falsely detected busy out of $E[U]$ licensed idle channels by N_{CU} cognitive users is:

$$p(g) = \binom{E[U]}{g} p_{\mathrm{f}}^g (1-p_{\mathrm{f}})^{E[U]-g}, \quad 0 \le g \le E[U] \tag{4.6}$$

Therefore, the average number of falsely detected licensed channels that are the number of channels detected as busy contrary to being idle is:

$$E[G] = \sum\nolimits_{g=0}^{E[U]} gp(g) \tag{4.7}$$

The average number of idle channels detected after certain false alarm probability by N_{CU} cognitive users is:

$$E[H] = E[U] - E[G] \tag{4.8}$$

(b) *Misdetection*

Moreover, the average number of idle channels detected after certain missed detection probability by N_{CU} cognitive users will be more than $E[U]$; however, it does not contribute to the cognitive user's throughput, as discussed earlier. In addition, due to missed detection, the primary user's presence will not be detected on the licensed channel by the cognitive users and, therefore, the interference to the primary user will occur if the miss-detected licensed channel has also been utilized by the cognitive user along with the primary user. Therefore, the probability of interference to the primary user due to missed detection is computed as follows [40]:

$$P_{\mathrm{int}} = p_{\mathrm{m}} \times Prob(p \leq T_{\mathrm{tr}}) \times P_{\mathrm{CU}} \tag{4.9}$$

where p_m is the probability of missed detection. $Prob(p \leq T_{\mathrm{tr}})$ defines the probability that the primary user transmits in the data transmission interval, and P_{CU} gives the probability of cognitive user grabbing a channel after successful contention slot.

$$Prob(p \leq T_{\mathrm{tr}}) = 1 - exp(-\lambda_{\mathrm{p}} T_{\mathrm{tr}})$$

λ_{p} is the average primary user ON rate as is discussed in [40] and

$$P_{\mathrm{CU}} = \begin{cases} \dfrac{\binom{N_{\mathrm{CU}}-1}{E[I]-1}}{\binom{N_{\mathrm{CU}}}{E[I]}}, & E[I] \leq N_{\mathrm{CU}} \\ 1, & otherwise \end{cases} \tag{4.10}$$

4.3.2 Contention Interval Analysis

The cognitive users compete with each other to reserve the idle licensed channels during the contention interval after the sensing-sharing interval, as is described in the previous chapter. However, each cognitive user, which has data to send to its intended receiver, randomly selects a contention slot among the total number of contention slots. As discussed in Chap. 3, the comparison has revealed that the application of backoff algorithm in the contention interval has enhanced performance of the cognitive radio network.

Analysis of the contention interval with backoff algorithm is described in detail in this section. Let the number of contention slots initially be CW_1, and each cognitive user randomly selects a contention slot with probability r_1. CW_1 is given as: $CW_1 = 2 \times N_{\mathrm{CU}}$. Therefore, the relation between the contention slots CW_1 and r_1 is given as:

$$r_1 = \frac{1}{CW_1}$$

Let s_1 be the number of cognitive users, who select a contention slot with probability r_1, and the probability distribution is given as:

$$p(s_1) = \binom{N_{\mathrm{CU}}}{s_1} (r_1)^{s_1} (1 - r_1)^{N_{\mathrm{CU}} - s_1}, \quad 0 \leq s_1 \leq N_{\mathrm{CU}} \tag{4.11}$$

Further, the probability that a contention slot is selected by only a single cognitive user is:

$$\begin{aligned} p_{\text{success}(1)} &= p(1) = \binom{N_{\text{CU}}}{1} (r_1)^1 (1 - r_1)^{N_{\text{CU}}-1} \\ &= N_{\text{CU}} r_1 (1 - r_1)^{N_{\text{CU}}-1} \end{aligned} \tag{4.12}$$

Let t_1 be the random variable, which denotes the number of successful cognitive users and the probability of t_1 cognitive users being successful is:

$$p(t_1) = \binom{\text{CW}_1}{t_1} \left(p_{\text{success}(1)}\right)^{t_1} \left(1 - p_{\text{success}(1)}\right)^{CW_1 - t_1}, \quad 0 \le t_1 \le \text{CW}_1 \tag{4.13}$$

The average number of successful cognitive users is numerically computed from Eq. (4.13) and is defined as:

$$E[T_1] = \sum\nolimits_{t=0}^{CW_1} t_1 p(t_1) \tag{4.14}$$

From Eq. (4.14), the average number of collided cognitive users is:

$$E[C_1] = N_{\text{CU}} - \sum\nolimits_{t=0}^{CW_1} t_1 p(t_1) \tag{4.15}$$

Further, to increase the contention interval size in order to make all the cognitive users successful, we follow the procedure as:

$r_{\text{i}} = \frac{1}{CW_{\text{i}}}$, where $i = 2,3,4,\ldots$, and $CW_2 = 2^4$, $CW_3 = 2 \times CW_2$, $CW_4 = 2 \times CW_3$, …….

Therefore, the contention interval is increased according to the binary exponential backoff algorithm. The number of cognitive users who have collided in the former contention interval are competing for the individual contention slot during the increased contention interval, which is described as:

$$p(s_{\text{i}}) = \binom{E[C_{\text{i}-1}]}{s_{\text{i}}} (r_{\text{i}})^{s_{\text{i}}} (1 - r_{\text{i}})^{N_{\text{CU}} - s_{\text{i}}}, \quad 0 \le s_{\text{i}} \le E[C_{\text{i}-1}],\ i = 2, 3, 4, \ldots \tag{4.16}$$

$$p_{\text{success(i)}} = E[C_{\text{i}-1}] \times r_{\text{i}} \times (1 - r_{\text{i}})^{E[C_{\text{i}-1}]-1} \tag{4.17}$$

$$p(t_i) = \binom{\text{CW}_{\text{i}}}{t_i} \left(p_{\text{success(i)}}\right)^{t_i} \left(1 - p_{\text{success(i)}}\right)^{CW_{\text{i}} - t_{\text{i}}}, \quad 0 \le t_{\text{i}} \le \text{CW}_{\text{i}},\ i = 2, 3, 4, \ldots. \tag{4.18}$$

The average number of successful cognitive users is computed from Eq. (4.18) and is defined as:

$$E[T_{\mathrm{i}}] = \sum\nolimits_{t_i=0}^{CW_i} t_{\mathrm{i}} p(t_{\mathrm{i}}) \tag{4.19}$$

and the average number of collided cognitive users are:

$$E[C_i] = E[C_{\mathrm{i}-1}] - E[T_{\mathrm{i}}] \tag{4.20}$$

Further, the total number of contention slots,CW_{total}, are:

$$CW_{\mathrm{total}} = \sum\nolimits_{i=1}^{i} CW_{\mathrm{i}} \tag{4.21}$$

Hence, the total number of successful cognitive users until CW_{total} contention slots are:

$$E[T_{\mathrm{total}}] = E[T_{\mathrm{i}-1}] + E[T_{\mathrm{i}}] \tag{4.22}$$

We have assumed a maximum contention window size CW_{max} of 1024. However, in case the maximum contention window is reached, that is $CW_{\mathrm{total}} = CW_{\mathrm{max}}$ and all the cognitive users in the network have not become successful, then the contention interval will not increase further and the cognitive users become successful until the maximum contention interval enters into the data transmission period.

4.3.3 Data Transmission Interval Analysis

The data transmission interval T_{tr} is defined as:

$$\begin{aligned} T_{\mathrm{tr}} &= T_{\mathrm{cycle}} - (T_{\mathrm{idle}} + T_{\mathrm{ss}} + T_{\mathrm{ct}}) \\ &= T_{cycle} - (\mathrm{T}_{\mathrm{idle}} + 3 \times \mathrm{T}_{\mathrm{slot}} \times \mathrm{N}_{\mathrm{ch}} + \mathrm{CW}_{\mathrm{total}} \times (\mathrm{CR} - \mathrm{RTS} + \mathrm{CR} - \mathrm{SIFS} + \mathrm{CR} - \mathrm{CTS})) \end{aligned} \tag{4.23}$$

where T_{cycle} is the total cycle time, $T_{\mathrm{idle}}, T_{\mathrm{ss}}$, and T_{ct} are idle interval, sensing-sharing interval, and contention interval duration, respectively. Since the sensing-sharing interval contains N_{ch} number of slots and each sensing-sharing slot have three sub-slots, $3 \times T_{\mathrm{slot}} \times N_{\mathrm{ch}}$ denotes the whole sensing-sharing interval duration. Similarly, $CW_{\mathrm{total}} \times (CR - RTS + CR - SIFS + CR - CTS)$ is the whole contention interval duration.

As discussed in the previous chapter, only those successful cognitive users transmit their data in the data transmission interval that have the idle licensed channels. Further, the throughputs for the following two cases are considered: (1) for the perfectly sensed licensed channels and (2) for the licensed channels imperfectly detected busy or for false alarm case. These two cases are discussed below:

4.3.3.1 Throughput for Perfect Sensing

The throughput T is the product of the minimum of the $E(\mathrm{Ch}_{\mathrm{idle}} \times T_{\mathrm{total}})$ and the average number of sensed idle channels from (4.5), the amount of time available for the data transmission per cycle interval $(T_{\mathrm{tr}}/T_{\mathrm{cycle}})$, and the data rate per sensed idle channels R. Further, the throughput T for the proposed MAC protocol is given as:

$$T = \frac{E[\min(\mathrm{Ch}_{\mathrm{idle}} \times T_{\mathrm{total}}, U)] \times T_{\mathrm{tr}} \times R}{T_{\mathrm{cycle}}} \tag{4.24}$$

where Ch_{idle} is the number of idle channels that a cognitive user is allowed to use simultaneously. $E[T_{\mathrm{total}}]$ is the number of successful users after using the backoff algorithm in the contention interval which is obtained from Eq. (4.22), and the number of idle channels detected $E[U]$ is obtained from Eq. (4.5).

4.3.3.2 Throughput for Imperfect Sensing

The throughput for imperfect sensing scenario (false alarm), $\mathrm{T_I}$, is computed from Eq. (4.8), since fewer idle channels are detected in the false detection, and is given as:

$$T_I = \frac{E[\min(\mathrm{Ch}_{\mathrm{idle}} \times T_{\mathrm{total}}, \mathrm{H})] \times T_{\mathrm{tr}} \times R}{T_{\mathrm{cycle}}} \tag{4.25}$$

$E[H]$ is obtained from Eq. (4.8), which is the total number of idle channels detected in the false alarm scenario. However, the throughput for the missed detection scenario is same as that for the perfectly sensed scenario, as discussed earlier in this chapter, because data from cognitive users transmitted over the undetected channels undergo collision with the primary user's data and hence does not contribute to the cognitive radio user throughput.

4.4 Energy Efficiency

It is known that before accessing a licensed channel the cognitive radio performs spectrum sensing on the channel, which consumes energy due to the radio frequency (RF) circuit operation and baseband signal processing, as discussed in [5, 41]. In addition, in the proposed MAC protocol, there are energy overheads due to sensing, competing, and idling [5] before data transmission. Therefore, it is clear that the energy consumption is not only in the data transmission interval for information transfer, but also in the sensing-sharing and contention interval in which even idle users also consume energy. Performance of the proposed MAC protocol in terms of energy consumption is further computed in this section, and the energy efficiency parameter is defined for this purpose as:

$$\mathrm{EE} = \frac{\text{Total amount of useful data delivered (bits)}}{\text{Total energy consumed (Joules)}},$$

where EE is the energy efficiency and the total amount of useful data delivered is given by the throughput per cycle time. The total energy consumed is computed by the data transmitted during each interval of total cycle time. We have used three parameters, namely, (1) the transmission power (P_T) that is required by a cognitive node for transmitting data, (2) reception power (P_R) that is consumed by a cognitive user terminal while receiving data, and (3) idle mode power (P_I) is the power consumed by the cognitive terminal when it is neither transmitting nor receiving data and is only tuned to a particular channel [42]. Therefore, the energy consumption in different intervals is as follows.

4.4.1 Energy Consumed in the Sensing-Sharing Interval

In the sensing-sharing interval, each cognitive user senses $Ch_{\max}$ number of channels by randomly selecting the sensing-sharing slot, and in the first sub-slot of the selected sensing-sharing slot, the licensed channel is sensed, and in the second and third sub-slots, sensing results are broadcasted for sharing with other cognitive users. Therefore, the total energy consumed by N_{CU} cognitive users for sensing and broadcasting sensing results is:

$$(P_\mathrm{R} \times T_\mathrm{slot} + P_\mathrm{T} \times 2 \times T_\mathrm{slot}) \times N_\mathrm{CU} \times Ch_{\max},$$

where T_slot is the single slot duration. The cognitive users remain idle for the number of slots that are not selected by any cognitive user, and the energy consumption for these slots is:

$$E[X] \times P_\mathrm{I} \times 3 \times T_\mathrm{slot},$$

where $E[X]$ is from (4.2). Therefore, the total energy consumed in the sensing-sharing interval is:

$$\mathrm{E}_{\mathrm{T}_{ss}} = (\mathrm{P_R} \times \mathrm{T_{slot}} + \mathrm{P_T} \times 2 \times \mathrm{T_{slot}}) \times \mathrm{N_{CU}} \times \mathrm{Ch_{max}} + \mathrm{E[X]} \times \mathrm{P_I} \times 3 \times \mathrm{T_{slot}} \tag{4.26}$$

4.4.2 Energy Consumed in the Contention Interval

In the contention interval, the collision by a cognitive user is detected by hearing the cognitive radio clear-to-send (CR-CTS) frame. The CR- CTS frame has been sent by the destination cognitive user in response to the cognitive radio ready-to-send (CR-RTS) frame transmitted by the source cognitive user on the selected contention slot in the control channel, and it is well understood that if more than one source cognitive user has selected the same contention slot they will not receive the CR-CTS frame correctly, hence detecting collision. The time interval of the CR-RTS and CR-CTS frames is T_{RTS} and T_{CTS}, respectively and the interval of the CR-SIFS (cognitive radio short-inter-frame spacing) between the CR-RTS and CR-CTS frames is T_{SIFS}. Therefore, in the contention interval, the cognitive user's energy consumption due to collisions, successes, and for being in an idle state in the non-selected contention slots is given as:

$$\begin{aligned} E_{\mathrm{T_{ct}}} = {} & P_{\mathrm{T}} \times T_{\mathrm{RTS}} \times \text{total number of collided users} + P_{\mathrm{I}} \times T_{\mathrm{SIFS}} \\ & \times \text{total number of collided users} + P_{\mathrm{I}} \times T_{\mathrm{CTS}} \times \text{total number of collided users} \\ & + P_{\mathrm{T}} \times T_{\mathrm{RTS}} \times E[T_{\mathrm{total}}] + P_{\mathrm{I}} \times T_{\mathrm{SIFS}} \times E[T_{\mathrm{total}}] + P_{\mathrm{R}} \times T_{\mathrm{CTS}} \times E[T_{\mathrm{total}}] \\ & + [CW_{\mathrm{total}} - (\text{total number of collided users} + E[T_{\mathrm{total}}])] \times P_{\mathrm{I}} \times T_{\mathrm{slot}} \end{aligned} \tag{4.27}$$

where the total number of collided users is taken from Eqs. (4.15) and (4.20) and $E[T_{total}]$ is from Eq. (4.22).

4.4.3 Energy Consumed in the Data Transmission Interval

The information/data is transmitted by the cognitive users over the detected idle licensed channels. The number of channels utilized for the data transmission is a minimum of $(Ch_{\mathrm{idle}} \times E(T_1), E(U))$ and $(Ch_{\mathrm{idle}} \times E(T_{\mathrm{total}}), E(H))$ for perfect and imperfect sensing, respectively. Therefore, the energy consumption over the information/data transmission interval for perfect and imperfect sensing is:

$$E_{\mathrm{T_{tr}}} = P_{\mathrm{T}} \times T_{\mathrm{tr}} \times E[min(Ch_{\mathrm{idle}} \times T_1, U)] \tag{4.28}$$

and

$$E_{\mathrm{T_{tr}I}} = P_{\mathrm{T}} \times T_{\mathrm{tr}} \times E[min(Ch_{\mathrm{idle}} \times T_{\mathrm{total}}, H)] \tag{4.29}$$

respectively, $E_{\mathrm{T_{tr}}}$ and $E_{\mathrm{T_{tr}I}}$ are the consumed energy for perfect and imperfect sensing, respectively, in the transmission time and $E[U]$ and $E[H]$, which are obtained from Eqs. (4.5) and (4.8). With the above defined energy consumption in different intervals, the energy efficiency of the proposed cognitive MAC protocol is:

$$EE = \frac{T}{E_{\text{total}}} \tag{4.30}$$

$$EE_I = \frac{T_{\text{I}}}{E_{\text{I\_total}}} \tag{4.31}$$

where EE and EE_I are the energy efficiency in the perfect and imperfect sensing, respectively. Moreover,

$$E_{\text{total}} = E_{\text{T}_{\text{ss}}} + E_{\text{T}_{\text{ct}}} + E_{\text{T}_{\text{tr}}}, \text{ and}$$

$E_{\text{I\_total}} = E_{\text{T}_{\text{ss}}} + E_{\text{T}_{\text{ct}}} + E_{\text{T}_{\text{tr}}\text{I}}$, are the total energy consumption over a cycle-time for the perfect and imperfect sensing, respectively.

4.5 Results and Discussion

For the proposed MAC protocol, the simulation parameters are shown in Table 4.1 and are also employed for IEEE 802.11a [43]. The numerically simulated results of the cognitive MAC protocol for energy efficiency, as well as the perfectly and imperfectly sensed licensed channels are presented in this section. Figure 4.1 shows

Table 4.1 The simulation parameters of the proposed MAC protocol for the distributed cognitive radio network

Simulation parameters	Numerical values
Number of licensed channels (N_{ch})	20
Utilization probability of licensed channels (α)	0–1
Number of sensed channel by each cognitive user ($Ch_{\max}$)	2–5
Number of cognitive users (N_{CU})	10–30
Probability of false detection (P_{m})	0–1
Cycle time (T_{cycle})	1 s
Single slot time (T_{slot})	9 μs
CR-RTS frame duration	24 μs
CR-CTS frame duration	24 μs
CR-SIFS frame duration	16 μs
Transmit power	916 mW
Reception power	550 mW
Idle mode power	550 mW
Channel bandwidth	20, 6, 5, 1.25 MHz
Data rate	54, 16.197, 13.49, 3.37 Mbps
Modulation	64 QAM
Ch_{idle}	1

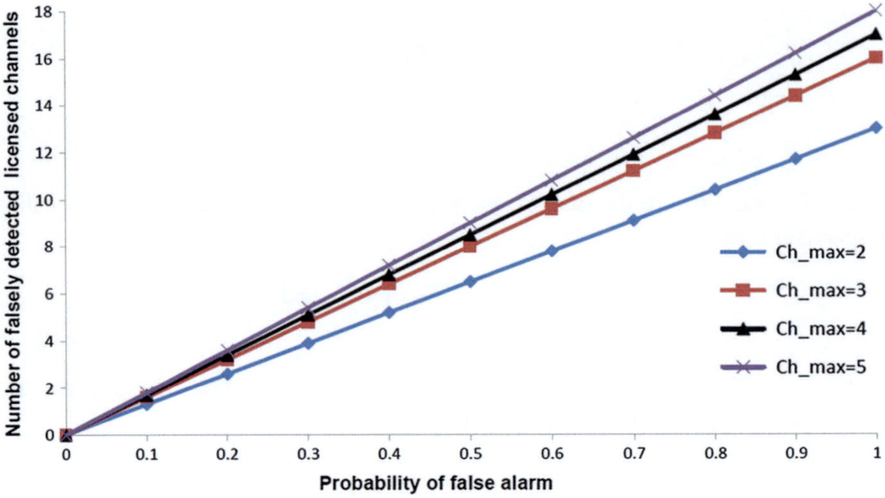

Fig. 4.1 The number of imperfect/falsely detected licensed channels for different probabilities of false alarm and $\mathrm{Ch}_{\max} = 2, 3, 4,$ and 5 in 20 licensed channels and the 10 cognitive users network when it is assumed that all sensed channels actually are idle

the number of imperfectly (falsely) detected licensed channels, which is the number of channels detected as busy; however, those are idle with 10 cognitive users for different probabilities of false alarm, and is computed from Eq. (4.7). It is also illustrated from Fig. 4.1 that as the false alarm probability increases for an arbitrarily chosen value of $Ch_{\max}$, the number of imperfect/falsely detected licensed channels increases linearly. It should be noted that we have simulated the results when it is assumed that all sensed channels actually are idle for different $Ch_{\max}$. Moreover, with the increase of $Ch_{\max}$ for the chosen value of false alarm probability, the number of imperfectly detected licensed channels is more for the higher value of Ch_{max} due to the greater number of sensed licensed channels. Further, the simulation results of the sensing-sharing analysis, which is discussed in Sect. 4.3.1, have been presented in Fig. 4.2. The utilization probability of licensed channels with the number of idle channels detected for different $Ch_{\max}$ value is shown in Fig. 4.2a, and it reveals that for perfect sensing, the number of sensed idle channels is significantly greater in comparison to that of the false alarm scenario for a particular value of $Ch_{\max}$. This behavior is well understood from Eq. (4.5), which computed the idle channels detected for a chosen α in the perfectly sensed environment and from Eqs. (4.7) and (4.8) that reveals the effect of false alarm on the detection of idle channels. Moreover, as the $Ch_{\max}$ value increases, significantly more licensed channels are sensed and hence detected as idle depending on α, which is illustrated from Fig. 4.2a. Since each cognitive user can utilize only a single idle channel, for the network with 10 cognitive users, the maximum number of idle channels utilized for data transmission is 10. However, Fig. 4.2a has illustrated that for some values of $Ch_{\max}$ and α, the number of idle channels detected

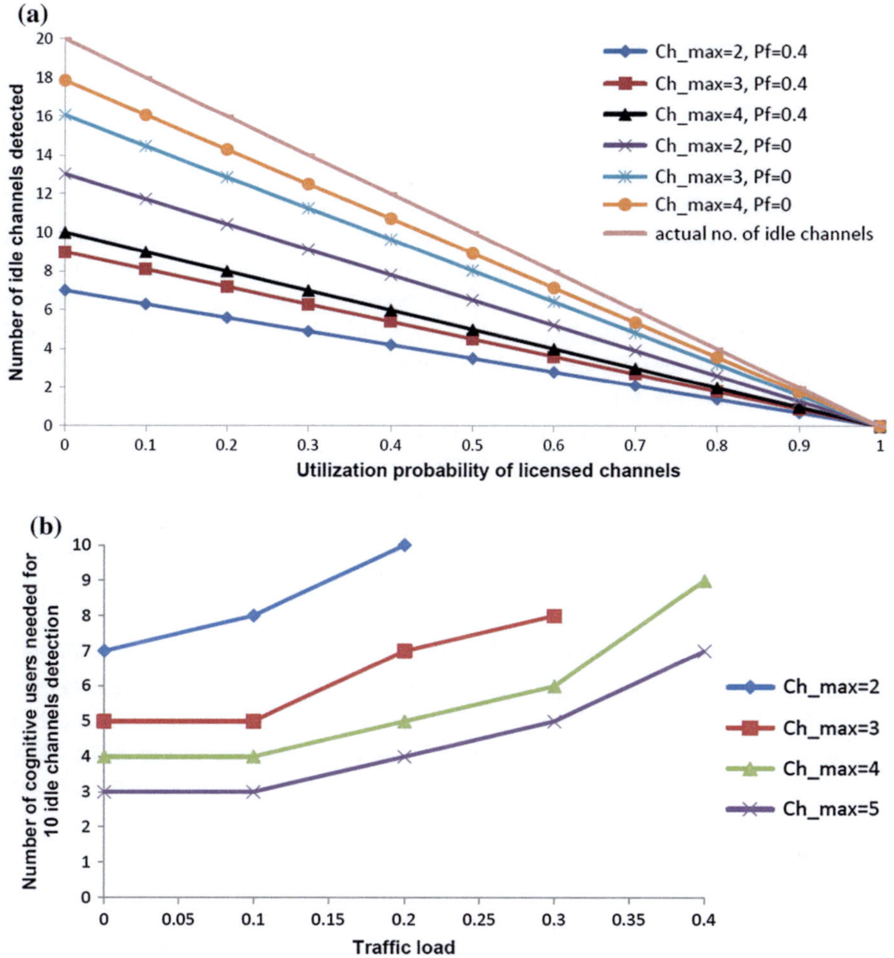

Fig. 4.2 The effects of variation of the utilization probability/traffic load of the licensed channels on the (a) number of idle channels detected for perfect $(P_f = 0)$ and imperfect sensing/false alarm $(P_f = 0.4)$ with $Ch_{\max} = 2, 3, 4$, and (b) number of cognitive users required to detect all needed idle channels with different $Ch_{\max} = 2, 3, 4, 5$ in 10 cognitive users and 20 licensed channel networks

is more than 10 for the network with 10 cognitive users. Therefore, it is proposed that after detecting the required number of idle channels by particular cognitive users in the sensing-sharing interval's slots, further licensed channels are not sensed by the assigned cognitive users, which has resulted in the adaptation of the number of channels sensed and also adaptation in the number of cognitive users used for sensing.

Thus, Fig. 4.2b depicts the number of cognitive users required to detect the 10 idle channels for different utilization probabilities and for different values of $Ch_{\max}$

in the perfect sensing environment. As the utilization probability of a licensed channel increases for particular $Ch_{\max}$, as shown in Fig. 4.2b, even 10 users cannot sense 10 idle channels. For example, with $Ch_{\max} = 2$ and for $\alpha \geq 0.3$, all 10 cognitive users cannot find required 10 idle channels, and this is also verified from Fig. 4.2a, where the number of idle channels detected by 10 cognitive users is fewer than 10 for $\alpha \geq 0.3$. In addition, the number of cognitive users needed is fewer for higher values of $Ch_{\max}$ at a particular value of α. Thus, after detecting the required number of idle channels, further users do not have need to sense any other licensed channels and hence can minimize the energy consumed in sensing and broadcasting the sensed information. Moreover, all the cognitive users cannot detect 10 idle channels for $\alpha \geq 0.4$ with $Ch_{\max} = 2, 3, 4, 5$, which is shown in Fig. 4.2a and, therefore, these values of α are not plotted in Fig. 4.2b. All 10 cognitive users sense the licensed channels for these values. Further, the contention interval analysis presented in Sect. 4.3.2 of this chapter is simulated and demonstrated in Fig. 4.3, which shows the average number of successful cognitive users in the various number of contention slots for different numbers of cognitive users in a network. Figure 4.3 also illustrates the comparison between the existing SMC-MAC protocol [44] and the proposed method, which reveals that with fewer contention slots, more users are successful in the proposed scheme in comparison to that of the existing SMC-MAC.

Moreover, it is clear from Fig. 4.3 that the optimum number of contention slots in the proposed scheme is: $\sum_{i=1}^{3} CW$ at which all the cognitive users become successful. For example, with $N_{CU} = 10$ only 68 slots are required to make all cognitive users successful in the proposed scheme; however, in the SMC-MAC, approximately 200 slots are needed for this purpose, which reduces the data

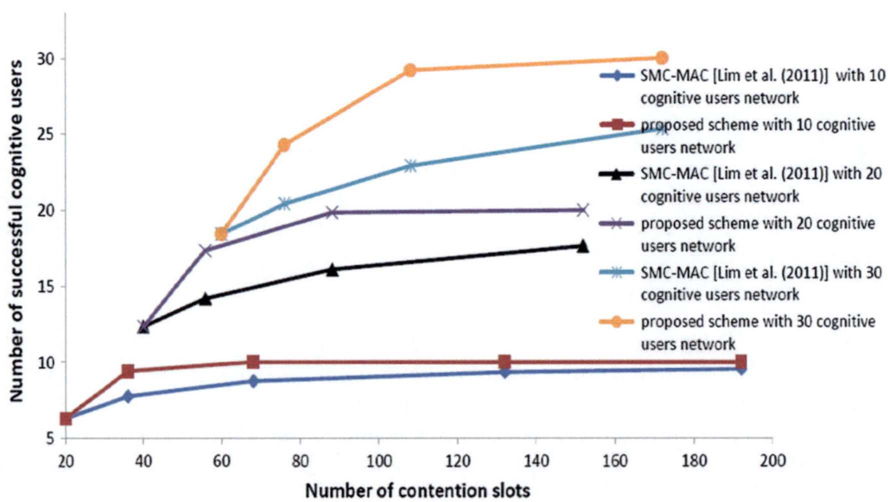

Fig. 4.3 The number of successful cognitive user variation with the number of contention slots for the proposed and SMC-MAC [44] protocol in the networks with 10, 20, and 30 cognitive users

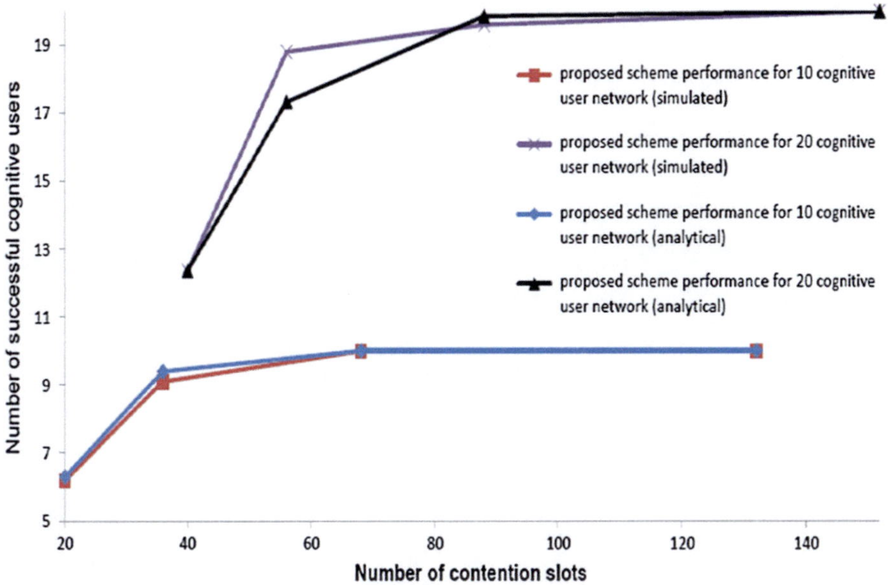

Fig. 4.4 The comparison of the analytical and simulated results of the proposed MAC protocol

transmission time of the cognitive users. Further, the results presented in the previous chapter are simulated results of the proposed scheme; the comparison with the analytical results whose mathematical modeling is discussed in Sect. 4.3.2 of this chapter, is shown in Fig. 4.4. It is illustrated from Fig. 4.4 that there is a small difference among the analytical and simulated results when we have applied the backoff algorithm for contention solving in the contention interval and, therefore, the throughput is assumed to be the same for both cases.

Further, Fig. 4.5 shows the throughput of the MAC protocol for perfect and imperfect sensing due to false alarm with 10 and 20 cognitive users. Because of the limited number of idle channels detected in the false alarm/imperfect sensing scenario, the cognitive users are unable to utilize the other idle channels present and they have limited their throughput when compared to that of the perfectly sensed scenario, as shown in Fig. 4.5. According to Fig. 4.2a, the number idle channels detected for $Ch_{\max} = 2$ and $P_f = 0$ is more than 10 for $\alpha = 0, 0.1, 0.2$. However, since the cognitive radio network can utilize a maximum of 10 idle channels because there are 10 cognitive users in the network, the maximum throughput is for 10 users and not more, which is the reason that for $\alpha = 0, 0.1, 0.2$ the throughput is the same. However, as α is increasing further from 0.2, the number of idle channel detection decreases from 10, and all 10 cognitive users cannot get 10 idle channels. Therefore, some of the cognitive users cannot transmit their data due to the lack of idle channels present, and hence the throughput is linearly decreasing for all other

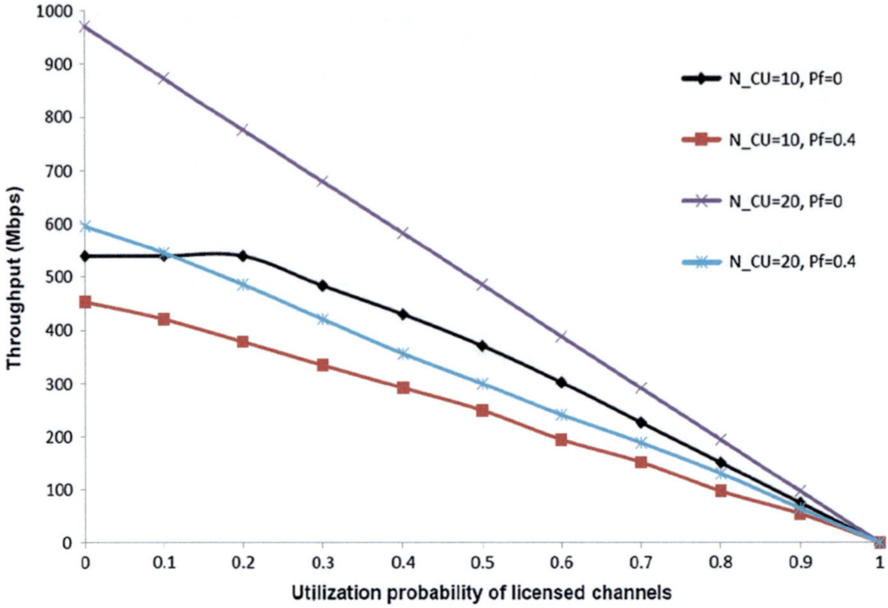

Fig. 4.5 The throughput of cognitive network with different licensed channel utilization probability for $Ch_{\max} = 2$, $N_{ch} = 20$, $N_{\text{CU}} = 10, 20$, data rate of 54 Mbps, and $P_f = 0, 0.4$

values of α as shown in Fig. 4.5. The mathematical description of this simulation is also discussed in the analysis section.

Further, Fig. 4.6 shows the throughput of a cognitive network utilizing varying channel bandwidths of different licensed networks because of the cognitive user terminal's heterogeneous network support, for example TV broadcast network, WCDMA 3G cellular network, and CDMA network of 6, 5, and 1.25 MHz channel bandwidths. Moreover, Fig. 4.7 represents energy efficiency of the MAC protocol as computed using Eq. (4.30) for different values of $Ch_{\max}$ and the perfect sensing scenario in the networks with 10, 20, and 30 cognitive user. The energy efficiency of the 10-users network is higher than those with 20 and 30 users, because the total number of licensed channels is fixed at 20, and more cognitive users have increased the sensing-sharing and contention interval, which results in decreased data transmission time. In addition to this, more cognitive users result in more collisions and successful slots in the contention interval, which causes more energy consumption. Therefore, the combined effect of the above two factors which are less data transmission time and more collisions, has resulted in less useful data transmission

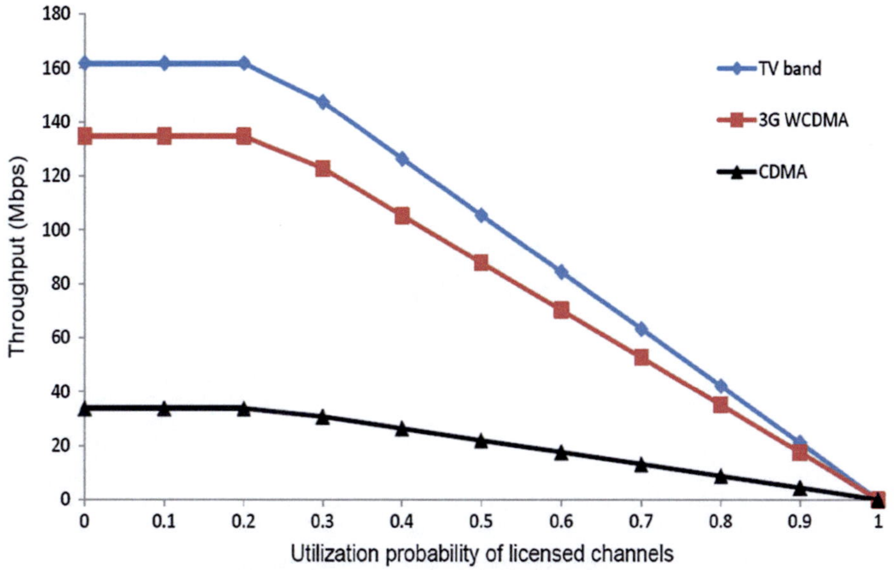

Fig. 4.6 The throughput variation of cognitive network in different primary user network with traffic load of licensed channels for $Ch_{\max} = 2$, $N_{\mathrm{ch}} = 20$, $N_{CU} = 10$ and $R = 16.197$ Mbps (TV band),13.49 Mbps (3G WCDMA), 3.37 Mbps (CDMA)

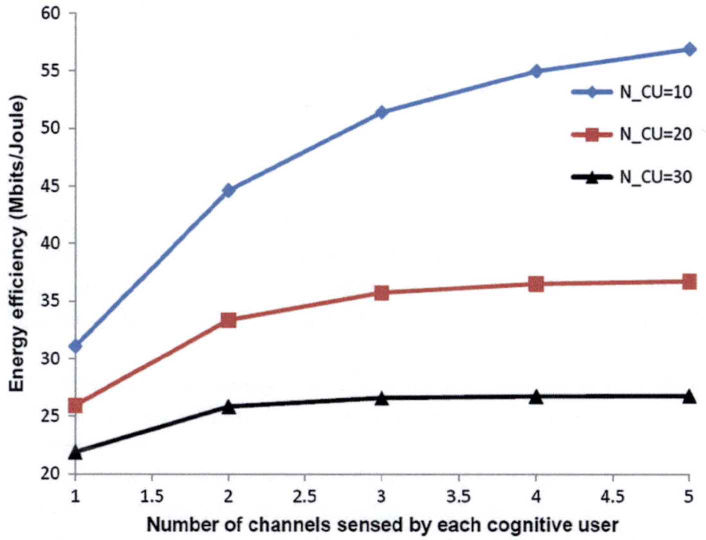

Fig. 4.7 The energy efficiency of the proposed protocol with different values of $Ch_{\max}$ where the simulation parameters are $\alpha = 0.5$, $R = 54$ Mbps, $N_{\mathrm{ch}} = 20$, $N_{\mathrm{CU}} = 10, 20, 30$ and $Ch_{\max} = 2$

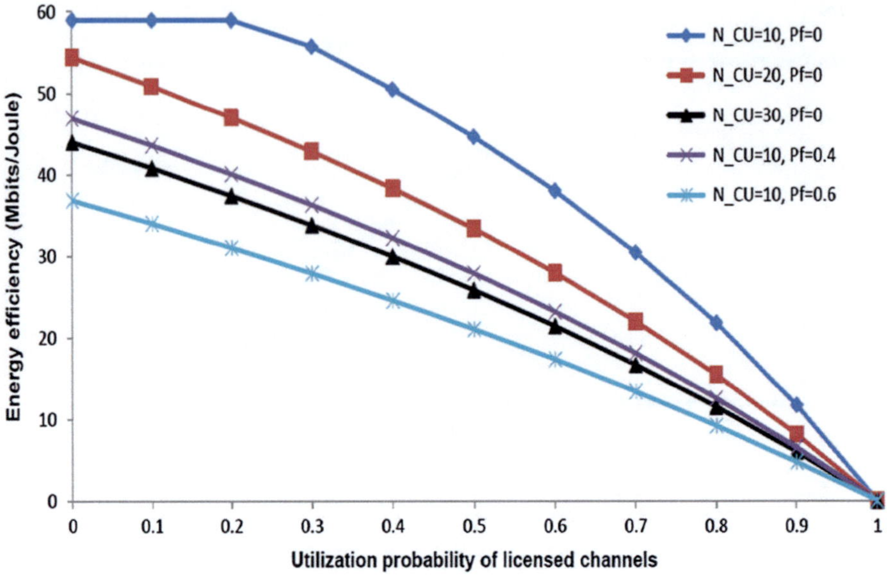

Fig. 4.8 The energy efficiency variation with the traffic load for various number of cognitive users and different false alarm probabilities, where R = 54 Mbps, $N_{\mathrm{ch}} = 20$, and $Ch_{\max} = 2$

with more energy consumption for increased cognitive users in a network and has decreased the energy efficiency of the system.

Further, in Fig. 4.8 the energy efficiency is depicted with the traffic load utilization (α) for 10, 20, and 30 cognitive users networks with perfect and imperfect/falsely sensed licensed channels. Since with greater false alarm probability there are fewer idle channels used for transmitting data, there are also fewer information bits transmitted with less energy efficiency. Moreover, the probability of interference to the primary users due to different miss-detection probabilities for optimized contention slots in the 10 cognitive users network with 20 licensed channels is shown in Fig. 4.9. It is illustrated from Fig. 4.9 that in the proposed scheme, the interference probability is less for the lower values of missed detection probability.

Further, Fig. 4.10 compares the average idle channel utilization with the number of cognitive users in the proposed scheme in this chapter and the one presented in [40]. It is clear from Fig. 4.10 that the idle channel utilization decreases rapidly with the number of cognitive users in the contention-based multichannel protocol presented in [40] due to the fixed number of contention slots; however, in the proposed scheme we have a flexible contention window that varies its size according to the number of cognitive users and hence has resulted in maximum idle channel utilization even for a higher number of cognitive users.

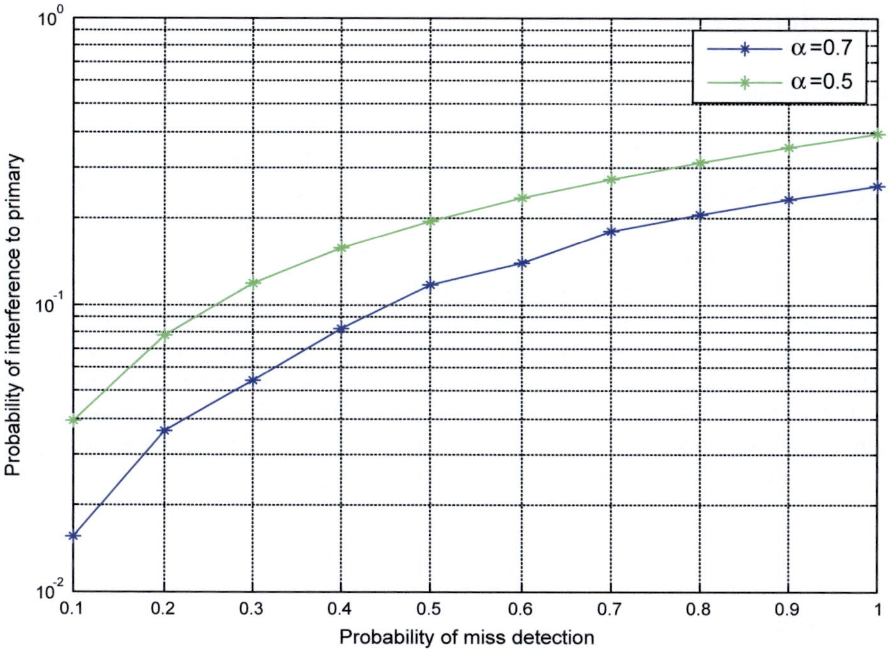

Fig. 4.9 The probability of interference to the primary user due to different missed detection probability for optimized contention slots in the 10 cognitive users network with $N_{ch} = 20$

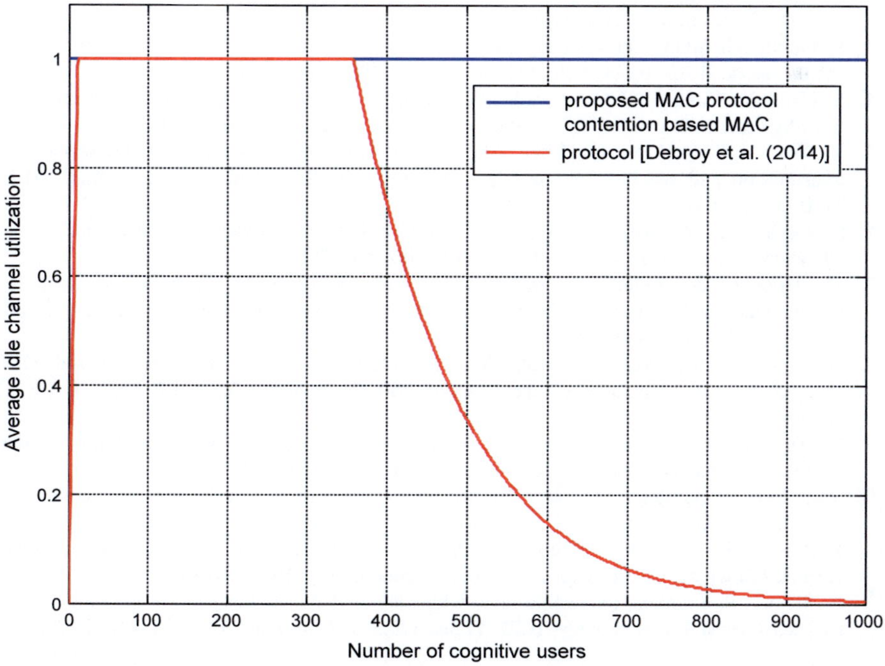

Fig. 4.10 The average idle channel utilization with the number of cognitive users for $N_{ch} = 20$ and $\alpha = 0.5$ for the proposed scheme and contention-based MAC protocol [40]

4.6 Summary

In this chapter, the cognitive radio MAC protocol in a practical scenario is considered, and the perfect and imperfect sensing effect on the performance of throughput and energy efficiency of the cognitive radio network is presented. The imperfect sensing as a result of false alarm has affected the system performance of the cognitive radio network by missing the opportunities of spectrum use in comparison to perfect sensing, as demonstrated in the simulation results. In addition to this, an optimum number of contention slots has been obtained for the proposed MAC protocol, which has avoided the problem of contention slots throughput tradeoff. Moreover, performance of the MAC protocol for different licensed channel utilization probability has been simulated. The simulation results have illustrated that throughput and energy efficiency of the MAC protocol for an imperfectly sensed environment is less as compared to that of the perfect sensing scenario, and the interference to the primary user is less in the proposed protocol for lower values of missed detection probability.

References

1. I.- F. Akyildiz, W.- Y. Lee, M.- C. Vuran, S. Mohanty, Next generation/dynamic spectrum access/cognitive radio wireless networks: A survey. Comput. Netw. **50**(13), 2127–2159 (2006)
2. S. Pandit, G. Singh, Throughput maximization with reduced data loss rate in cognitive radio network. Telecommun. Syst. **57**(2), 209–215 (2014)
3. S. Pandit, G. Singh, Backofff algorithm in cognitive radio MAC-protocol for throughput enhancement. IEEE Trans. Veh. Technol. **64**(5), 1991–2000 (2015)
4. S. Wang, Y. Wang, J.- P. Coon, A. Doufexi, Energy efficient spectrum sensing and access for cognitive radio networks. IEEE Trans. Veh. Technol. **61**(2), 906–912 (2012)
5. Y. Wu, V.-K.-N. Lau, D.-H.-K. Tsang, L.-P. Qian, Energy-efficient delay-constrained transmission and sensing for cognitive radio systems. IEEE Trans. Veh. Technol. **61**(7), 3100–3113 (2012)
6. S. Bayhan, F. Alagoz, Scheduling in centralized cognitive radio networks for energy efficiency. IEEE Trans. Veh. Technol. **62**(3), 582–595 (2013)
7. M. Timmers, S. Pollin, A. Dejonghe, L. Van der Perre, F. Catthoor, A distributed multichannel MAC protocol for multihop cognitive radio networks. IEEE Trans. Veh. Technol. **59**(1), 446–459 (2010)
8. A. Aijaz, S. Ping, M.R. Akhavan, A.H. Aghvami, CRB-MAC: A receiver-based MAC protocol for cognitive radio equipped smart grid sensor networks. IEEE Sens. J. **14**(12), 4325–4333 (2014)
9. M. De Sanctis, E. Cianca, V. Joshi, Energy efficient wireless networks towards green communications. Wireless Pers. Commun. **59**(3), 537–552 (2011)
10. S.K. Nobar, K.A. Mehr, J.M. Niya, RF-powered green cognitive radio networks: architecture and performance analysis. IEEE Commun. Lett. **20**(2), 296–299 (2016)
11. Y. Yao, G.B. Giannakis, Rate-maximizing power allocation in OFDM based on partial channel knowledge. IEEE Trans. Wireless Commun. **4**(3), 1073–1083 (2005)
12. A. Steiner, S. Shamai, Achievable rates with imperfect transmitter side information using a broadcast transmission strategy. IEEE Trans. Wireless Commun. **7**(3), 1043–1051 (2008)

13. Y. Sunand, M.L. Honig, Asymptotic capacity of multicarrier transmission with frequency-selective fading and limited feedback. IEEE Trans. Inf. Theory **54**(7), 2879–2902 (2008)
14. M.H. Alsharif, R. Nordin, M. Ismail, Classification, recent advances and research challenges in energy efficient cellular networks. Wireless Pers. Commun. **77**(2), 1249–1269 (2014)
15. A. Kuhne, A. Klein, Throughput analysis of multi-user OFDMA-systems using imperfect CQI feedback and diversity techniques. IEEE J. Sel. Areas Commun. **26**(8), 1440–1450 (2008)
16. D.J. Love, R.W. Heath, V.K.N. Lau, D. Gesbert, B.D. Rao, M. Andrews, An overview of limited feedback in wireless communication systems. IEEE J. Sel. Areas Commun. **26**(8), 1341–1365 (2008)
17. S. Parsaeefard, A.R. Sharafat, Robust distributed power control in cognitive radio networks. IEEE Trans. Mob. Comput. **12**(4), 609–620 (2013)
18. Y.C. Liang, Y. Zeng, E.C.Y. Peh, A.T. Hoang, Sensing-throughput tradeoff for cognitive radio networks. IEEE Trans. Wireless Commun. **7**(4), 1326–1337 (2008)
19. S. Huang, H. Chen, Y. Zhang, H.H. Chen, Sensing-energy tradeoff in cognitive radio networks with relays. IEEE Syst. J. **7**(1), 68–76 (2013)
20. X. Huang, N. Ansari, Optimal cooperative power allocation for energy-harvesting-enabled relay networks. IEEE Trans. Veh. Technol. **65**(4), 2424–2434 (2016)
21. A.A. Khan, M.H. Rehmani, M. Reisslein, Cognitive radio for smart grids: survey of architectures, spectrum sensing mechanisms, and networking protocols. IEEE Commun. Surv. Tutorials **18**(1), 860–898 (2016)
22. H.S. Ghadikolaei, I. Glaropoulos, V. Fodor, C. Fischione, A. Ephremides, Green sensing and access: Energy-throughput trade-offs in cognitive networking. IEEE Commun. Mag. **53**(11), 199–207 (2015)
23. H. Hu, H. Zhang, Y.-C. Liang, On the spectrum-and energy-efficiency tradeoff in cognitive radio networks. IEEE Trans. Commun. **64**(2), 490–501 (2016)
24. Y. Wang, M. Xu, W. Zhang, C. Wang, Overhead-throughput tradeoff under a novel frame structure in centralized cooperative cognitive networks. Wireless Pers. Commun. **77**(1), 345–363 (2014)
25. S. Eryigit, S. Bayhan, T. Tugcu, Energy-efficient multichannel cooperative sensing scheduling with heterogeneous channel conditions for cognitive radio networks. IEEE Trans. Veh. Technol. **62**(6), 2690–2699 (2013)
26. M. Moradkhani, P. Azmi, M.A. Pourmina, Optimized reliable data combining cooperative spectrum sensing method in cognitive radio networks. Wireless Pers. Commun. **74**(2), 569–583 (2014)
27. S. Wang, F. Tosato, J.P. Coon, Reliable energy-efficient spectrum management and optimization in cognitive radio networks: How often should we switch? IEEE Wirel. Commun. **20**(6), 14–20 (2013)
28. H.S. Ghadikolaei, C. Fischione, Analysis and optimization of random sensing order in cognitive radio networks. IEEE J. Sel. Areas Commun. **33**(5), 803–819 (2015)
29. W.L. Chin, J.M. Li, H.H. Chen, Low-complexity energy detection for spectrum sensing with random arrivals of primary users. IEEE Trans. Veh. Technol. **65**(2), 947–952 (2016)
30. X. Feng, X. Gan, Energy-efficient design of sensing and transmission in cognitive radio networks. Wireless Pers. Commun. **84**(3), 1647–1662 (2015)
31. C. Jiang, H. Zhang, Y. Ren, H.H. Chen, Energy-efficient non-cooperative cognitive radio networks: micro, meso, and macro views. IEEE Commun. Mag. **52**(7), 14–20 (2014)
32. H. Zhang, C. Jiang, X. Mao, H.H. Chen, Interference-limited resource optimization in cognitive femtocells with fairness and imperfect spectrum sensing. IEEE Trans. Veh. Technol. **65**(3), 1761–1771 (2016)
33. X. Huang, T. Han, N. Ansari, On green-energy-powered cognitive radio networks. IEEE Commun. Surv. Tutorials **17**(2), 827–842 (2015)
34. T.S. Syed, G.A. Safdar, On the usage of history for energy efficient spectrum sensing. IEEE Commun. Lett. **19**(3), 407–410 (2015)

35. K. Son, B.C. Jung, S. Chong, D.K. Sung, Power allocation policies with full and partial inter-system channel state information for cognitive radio networks. Wireless Netw. **19**(1), 99–113 (2013)
36. Y. Wu, D.H.K. Tsang, Joint bandwidth and power allocations for cognitive radio networks with imperfect spectrum sensing. Wireless Personal Commun. **57**(1), 19–31 (2011)
37. N. Nomikos, D.N. Skoutas, D. Vouyioukas, C. Verikoukis, C. Skianis, Capacity maximization through energy-aware multi-mode relaying. Wireless Pers. Commun. **74**(1), 83–99 (2014)
38. X. Wang, E. Li, D.-B. Ha, T.M. Hoang, H. Zhang, Secure cognitive relay network: joint the impact of imperfect spectrum sensing and outdated feedback. Wireless Pers. Commun. **87**(1), 165–192 (2016)
39. Q. Zhao, B. Wang, X. Li, Two classes of improved quickest spectrum sensing approaches in cognitive radio. Wireless Pers. Commun. pp 1–15, Published Online: July 2016, pp 1–15, doi:10.1007/s11277-016-3494-3
40. S. Debroy, S. De, M. Chatterjee, Contention based multichannel MAC protocol for distributed cognitive radio networks. IEEE Trans. Mob. Comput. **13**(12), 2749–2762 (2014)
41. A. Zappone, C. Zhijiat Chong, E. Jorswieck, S. Buzzi, Green resource allocation in relay-assisted multicarrier IC networks considering circuit dissipated power. in *Proc. of 3rd Int. Conf. on Future Energy Systems: Where Energy, Computing and Communication Meet (e-Energy)*, Madrid, Spain, 9–11 May 2012, pp. 1–10
42. A. Antonopoulos, C. Verikoukis, C. Skianis, O.-B. Akan, Energy efficient network coding-based MAC for cooperative ARQ wireless networks. Ad Hoc Netw. **11**(1), 190–200 (2013)
43. IEEE 802.11, Wireless LAN Medium Access Control (MAC) and Physical Layer (PHY) Specifications. *IEEE* June 2007
44. S. Lim, T.- J. Lee, A self-scheduling multi-channel cognitive radio MAC protocol based on cooperative communications. IEICE Trans. Commun. **E94-B**(6), 1657–1668 (2011)

Chapter 5
Throughput Enhancement Using Bandwidth Wastage in MAC Protocol of the Distributed Cognitive Radio Network

5.1 Introduction

Various MAC protocols have been compared in [1–9] based on network architecture and spectrum access techniques. For multimedia applications having different data rate requirements, a QoS provisioned MAC layer protocol (MQPP) for IEEE 802.11 based cognitive radio networks was proposed in [10]. In [11], the authors divided the multichannel MAC protocols into four different categories and compared them both analytically and through simulation. In reference [12], the tradeoff between QoS of the primary and secondary user was considered, and the QoS of the secondary network for the given primary network model was found. Further, the authors in [13] have proposed a cognitive radio MAC (COMAC) protocol without assuming a predefined interference power level at the primary receiver. A blind rendezvous algorithm that allows cognitive users to meet at a point without any synchronization was proposed in [14] and the channel access delay was also computed. A generalized MAC signaling protocol was proposed in [15], based on collaborative sensing. Another collaborative spectrum sensing MAC protocol, namely truncated time division multiple access (TTDMA) protocol, was proposed in [16], and has provided analytical throughput of multichannel opportunistic spectrum access MAC (OSA) for ad hoc cognitive radio networks. In [17], the authors have provided alternatives to the control channel requirement in cognitive radio ad hoc networks by introducing a gossip enabled stochastic channel negotiation (GES-CN) framework. A memory enabled MAC protocol for cognitive radio was proposed in [18], where the primary user signals cannot be distinguished from cognitive users. The memory enabled protocols help in adjusting the backoff parameters of cognitive users based on channel information and their transmission. An opportunistic multichannel-MAC (OMC-MAC) protocol was proposed in [19] for distributed cognitive radio networks. This protocol provides QoS guarantees to

S. Pandit and G. Singh, *Spectrum Sharing in Cognitive Radio Networks*,
DOI 10.1007/978-3-319-53147-2_5

delay sensitive applications in distributed cognitive networks, which is a challenging task. A coalition game theory based MAC protocol was proposed in [20]. In order to access the channel, a contention resolution multichannel MAC protocol was proposed in [21], and the impact of sensing error was also considered.

In order to design an efficient MAC protocol, channel modeling is performed to know the primary user's channels idleness distribution [22]. A preemptive opportunistic MAC (PO-MAC) with initialization, reporting, and contention phase was proposed in [23], and made a major contribution towards critical real time information transmission in distributed cognitive radio networks. PO-MAC is suitable for real time data transmission because of low latency provided to the cognitive user's data, due to its efficient design. The letter [24] proposed a contention based distributed cognitive radio MAC protocol that exploits channel diversity at the transmitter with the help of channel state information at the transmitter. A full duplex MAC protocol with simultaneous sensing and transmission is presented in [25] that gives the advantage of self-interference cancellation in comparison to other cognitive radio MAC protocols. In order to avoid the adjacent channel interference which results from imperfect filter design, a guard band aware MAC protocol [26] was proposed for cognitive radio networks which maximizes network throughput. A fairness based collision free MAC protocol with its mathematical modeling was proposed in [27]. Practical imperfections like channel uncertainty, noise uncertainty, signal uncertainty, and synchronization issues in cognitive radio system were discussed in [28] and addressed these problems with the authors' particular solution.

The backoff algorithm in SMC-MAC protocol proposed in the previous chapters has enhanced cognitive radio system performance. However, in the proposed scheme, licensed channels are not utilized by the cognitive users during the sensing-sharing and contention interval. Only the control channel is utilized in the sensing-sharing and contention interval, which is a wastage of bandwidth over these intervals on idle licensed channels, as shown in Fig. 5.1.

It is clear that idle channel detected by the cognitive user in a sensing-sharing slot is utilized only in the data transmission interval. Therefore, all the remaining sensing-sharing slots after idle channel detection and contention interval of that licensed channel remain unutilized, causing waste of bandwidth. Bandwidth is one of the scarce resources of wireless communication, therefore this chapter deals with the potential issue of bandwidth wastage that arises in the proposed distributed MAC protocol for cognitive radio communication systems.

This chapter is organized as follows. Section 5.2 describes the system model and proposed method for the enhancement of throughput using the wasted bandwidth. Section 5.3 presents the numerical analysis for the proposed scheme. Section 5.4 presents numerically simulated results of the analysis. Finally, Sect. 5.5 concludes the chapter.

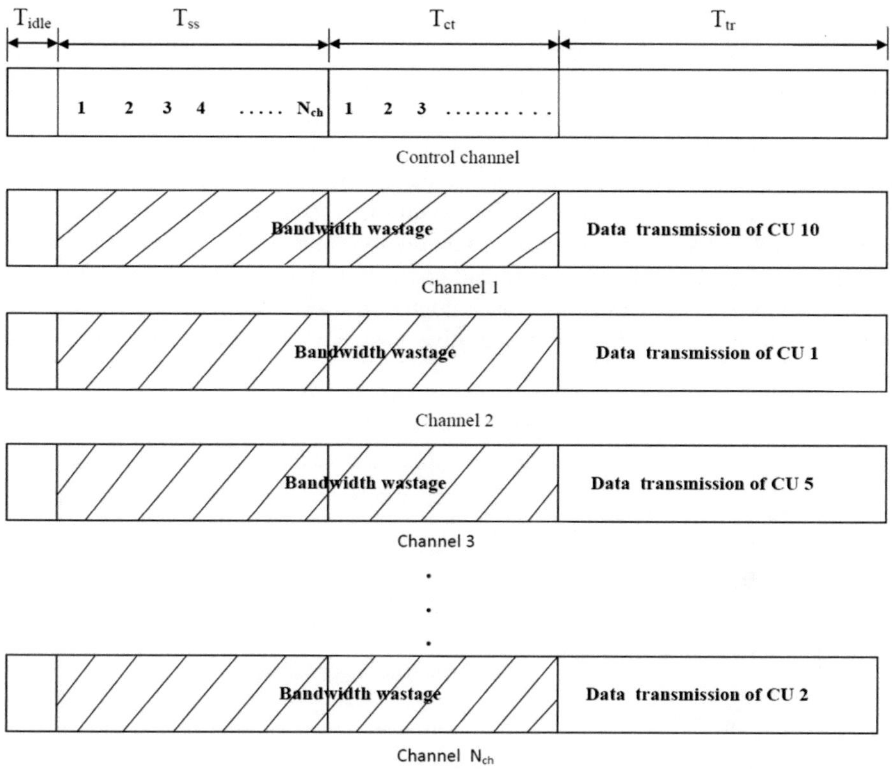

Fig. 5.1 Bandwidth wastage of licensed channels in the cognitive radio medium access control protocol for a cooperative distributed network

5.2 System Model

The system model is similar to the one presented in the previous chapters in which there is one primary user network comprising N_{ch} licensed channels and a cognitive radio network having N_{CU} cognitive users. In this chapter, however, a novel scheme is proposed in which the data is also transmitted over the idle licensed channels during the sensing-sharing and contention interval, which is an improvement over the scheme proposed in the previous chapters.

5.2.1 Proposed Method

In the sensing-sharing interval, cognitive users scan the licensed channels during their assigned slot numbers in the control channel. If the licensed channels are detected to be idle, only then after contention, the cognitive users are allowed to transmit their data in the data transmission interval of the licensed channels. We can see that there is bandwidth wastage during the T_{ss} and T_{ct} intervals in the proposed scheme, as shown in Fig. 5.1. This whole process is performed on the control channel till the data transmission interval, which results in bandwidth wastage due to no information being transmitted during that interval on the idle licensed channels. Examination reveals that before the T_{tr} interval, the licensed channels are not utilized if they are idle, and the bandwidth is wasted. Hence, in order to avoid the bandwidth wastage, we propose a scheme to transmit data on the licensed idle channels during the sensing-sharing and contention interval, as shown in Fig. 5.2.

A channel is sensed randomly by a cognitive user. For example, suppose that channel 1 is sensed by the tenth cognitive user (CU 10) on the control channel

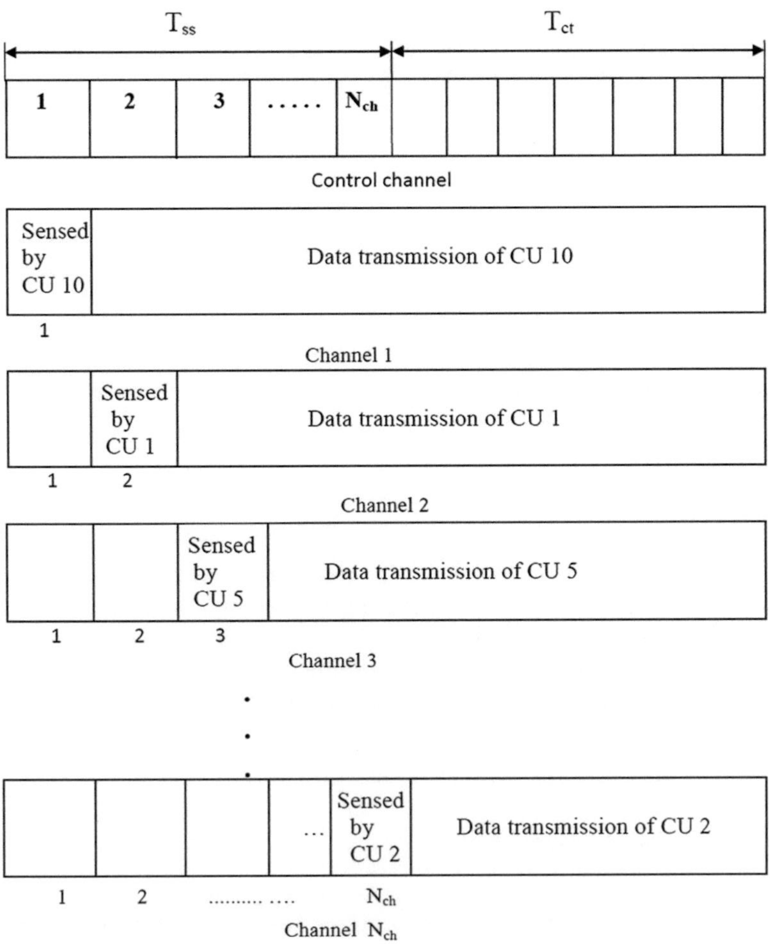

Fig. 5.2 Proposed scheme to avoid bandwidth wastage in the proposed cognitive radio MAC protocol for the cooperative distributed network

during the first slot of the sensing-sharing interval, as shown in Fig. 5.2. In the proposed scheme, if the sensed channel 1 is idle, then the tenth cognitive user starts transmitting its data on channel 1, following the first slot of the sensing-sharing interval till the start of the data transmission interval, after which the idle channels selected by the cognitive users' cooperative communication on the control channel will be utilized. Similarly, the first, fifth, and second cognitive users have sensed channel 2, channel 3, and channel N_{ch} during the second, third, and N_{ch} slots, respectively. If the channels detected are idle, then these cognitive users start transmitting data. It is also assumed that a cognitive user, after detecting first idle channel, will start transmitting on that channel and will continue its transmission on the same channel, even if another idle channel is detected by the same user on successive sensing-sharing slots. This can happen, depending on the parameter defining the number of channels sensed by a cognitive user (Ch_{max}).

In this model, the wasted bandwidth of a licensed channel in the sensing-sharing and contention interval cannot be utilized by the users that have not sensed that channel. In addition, we have assumed that after sensing the licensed channel during the sensing-sharing interval, the status of a licensed channel's availability does not change in that particular cycle. Further, the proposed scheme needs two transceivers, one to transmit data over idle channels detected during sensing-sharing and contention interval, and the other that is tuned to the control channel during these intervals.

5.3 Performance Analysis

In this section, we perform a numerical analysis of the proposed MAC protocol and discuss several performance parameters of the cognitive radio network.

5.3.1 Sensing-Sharing Analysis

The probability distribution of the number of sensed idle channels n by N_{CU} cognitive users is determined by using Eqs. (3.2) and (3.7):

$$p(n) = \binom{E[L]}{n} p_{\text{sensed}}^{n} (1 - p_{\text{sensed}})^{E[L]-n}, \quad 0 \leq n \leq E[L] \tag{5.1}$$

From Eq. (5.1), the average number of idle channels sensed by N_{CU} cognitive users is calculated as:

$$E[N] = \sum\nolimits_{n=0}^{E[L]} np(n) \tag{5.2}$$

Therefore, we can find from Eq. (5.2) the maximum number of cognitive users transmitting their data over the detected idle licensed channels $E[N]$, which yields maximum sensing sharing slots and is given as:

$$E[O] = \min\{E[N], N_{CU}\} \tag{5.3}$$

The maximum number of slots is available for data transmission in the sensing-sharing interval for the case when $E[O]$ cognitive users have sensed different licensed channels in the starting slots and detected that those channels are idle, as shown in the Fig. 5.3. For example, suppose there are 5 cognitive users, 10 licensed channels, $Ch_{\max} = 1$, and 5 idle channels are detected for a particular traffic load value. When 5 cognitive users have selected the first five slots for sensing and the respective channels are idle as shown in Fig. 5.3, the first to fifth cognitive users will have 9, 8, 7, 6, and 5 sensing-sharing slots, respectively, available for data transmission on the respected licensed channels, as numbered in Fig. 5.3. This is the maximum number of slots available for data transmission during the sensing-sharing interval under these conditions; in no other way can we get more sensing-sharing slots than these for data transmission. Therefore, the maximum number of sensing-sharing slots available for data transmission during the sensing-sharing interval is given by:

$$X_{\max} = \sum\nolimits_{i=N_{ch}-E[O]}^{N_{ch}-1} i \tag{5.4}$$

Similarly, the minimum number of cognitive users that can detect $E[N]$ idle channels, yielding the minimum number of sensing-sharing slots for data transmission, is given as:

$$E[P] = min\left\{\frac{E[N]}{Ch_{\max}}, N_{CU}\right\} \tag{5.5}$$

Given a particular value of i, for which $Ch_{\max} \times i = E[N]$, where $1 \le i \le N_{CU}$, we will get a minimum number of slots for data transmission, and in this case the idle channels detected by i cognitive users at ending slots, is shown in Fig. 5.4 and is given by:

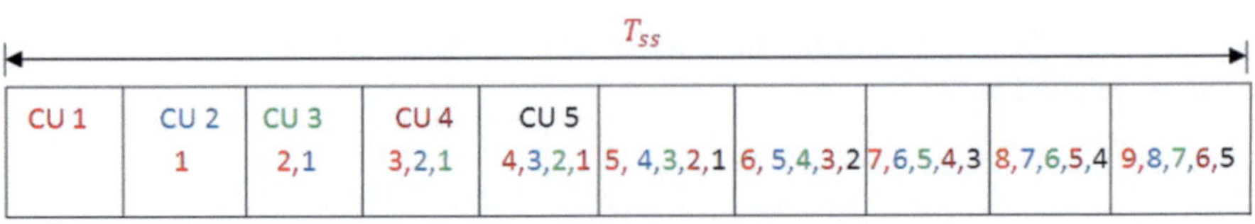

Fig. 5.3 The maximum number of slots for data transmission

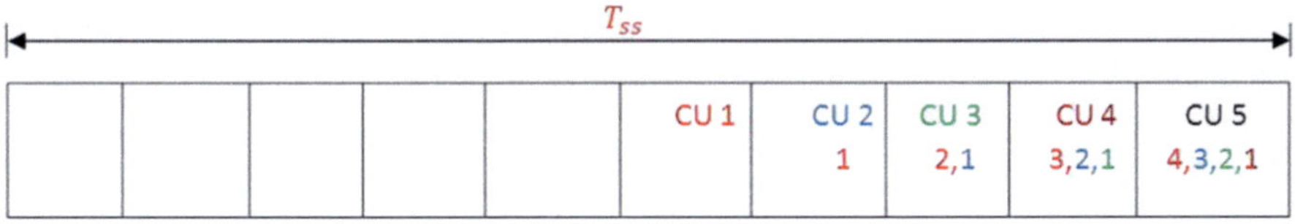

Fig. 5.4 The minimum number of slots for data transmission

$$X_{\min} = \sum_{i=0}^{E[P]-1} (Ch_{\max} \times i + (Ch_{\max} - 1)) \tag{5.6}$$

Figure 5.4 shows the minimum number of slots available for data transmission when $Ch_{max} = 1$, with 5 idle channels detected according to Eq. (5.2) and 10 licensed channels. Therefore, the number of slots for data transmission would be 4, 3, 2, and 1 for the first, second, third, fourth, and fifth cognitive user, respectively.

The minimum number of slots available is constant but the selection of a particular slot may vary among the cognitive users. That is, the first cognitive user (CU 1) can select either the first, second, third, fourth or fifth slot. The other cognitive users can do the same, therefore the maximum number of slots available is also constant. Hence, the number of sensing-sharing slots for data transmission varies between the upper limit and lower limit given by Eqs. (5.4) and (5.6), respectively, and quantifies the number of slots utilized from the wasted bandwidth.

5.3.2 *Data Transmission and Throughput Analysis*

When cognitive users successfully transmit their data in the data transmission interval on the idle channels selected during the contention interval, the data transmission interval T_{tr} is defined by:

$$\begin{aligned} T_{tr} &= T_{\text{cycle}} - (T_{\text{idle}} + T_{\text{ss}} + T_{\text{ct}}) \\ &= T_{\text{cycle}} - (T_{\text{idle}} + T_{\text{ss\_slot}} \times N_{\text{ch}} + \text{CW}_{\text{total}} \times T_{\text{ct\_slot}}) \end{aligned}$$

where $T_{\text{ss\_slot}} = 3 \times T_{\text{slot}}$ is the single sensing-sharing slot duration and T_{slot} is the duration of the sub-slot of the sensing-sharing slot. Similarly, $T_{\text{ct\_slot}}$ is the single contention slot duration. However, as we have already discussed, the throughput of the proposed MAC protocol is given by:

$$Th_{\text{prop.}} = \frac{E[\min(\text{Ch}_{\text{idle}} \times T_{\text{total}}, \text{N})] \times T_{\text{tr}} \times R}{T_{\text{cycle}}} \tag{5.7}$$

where $E[T_{\text{total}}]$ is the number of successful users after the backoff algorithm in the contention interval, and is obtained by using Eq. (4.22). Ch_{idle} is the number of idle channels that a cognitive user is allowed to use simultaneously. The throughput of the SMC-MAC protocol proposed in [29] is given as:

$$Th_{SMC-MAC} = \frac{E[\min(T \times Ch_{\text{idle}}, \text{N})] \times T_{\text{tr}} \times R}{T_{\text{cycle}}}$$

Let us further examine the proposed scheme throughput, in which data is also transmitted over the sensing-sharing and contention intervals. After utilizing the unoccupied bandwidth in the sensing-sharing interval, the cognitive user continues its transmission on the same occupied licensed channels during the contention interval. In this case, the total throughput is the sum of throughput computed in the previous chapter by applying the backoff algorithm and the throughput of the sensing-sharing and contention intervals, which is given as:

$$Th_{\text{total}_{\max}} = Th_{\text{prop.}} + \frac{\sum_{i=N_{ch}-E[O]}^{N_{ch}-1} (R \times (T_{ss\_slot} \times i + T_{ct\_slot} \times CW_{total}))}{T_{cycle}} \tag{5.8}$$

$$Th_{\text{total}_{\min}} = Th_{\text{prop.}} + \frac{\sum_{i=0}^{E[P]-1} (R \times (T_{\text{ss\_slot}} \times ((Ch_{\max} \times i) + (Ch_{\max} - 1)) + T_{\text{ct\_slot}} \times CW_{\text{total}}))}{T_{\text{cycle}}} \tag{5.9}$$

Equation (5.8) and Eq. (5.9) provide the maximum and minimum achievable throughputs, respectively, after utilizing the wasted bandwidth.

5.4 Results and Discussion

The simulation parameters for the proposed scheme are taken as: $N_{\text{ch}} = 20$, $Ch_{\text{idle}} = 1$, $T_{\text{slot}} = 900\mu s$, $T_{\text{idle}} = 1ms$, $T_{\text{ct\_slot}} = 2ms$ and $R = 1Mbps$. The simulation parameters are modified from the previous chapters to observe the prominent effect of bandwidth wastage on throughput and cognitive radio system performance, otherwise the data transmission over small time duration sensing-sharing and contention intervals as considered in the previous chapters does not contribute much to the enhanced throughput. Figure 5.5 shows the effect of the number of cognitive users on the average number of channels sensed for different Ch_{max} values. Figure 5.5 also reveals that as the capability of each cognitive user to sense the channels increases, that is with increase in the value of $Ch_{\max}$, a greater number of licensed channels are sensed, which increases the complexity of the cognitive terminal [30]. Figure 5.6 depicts the probability of collision among cognitive users in the contention interval due to the selection of the same slot by two or more cognitive users. Figure 5.6 also illustrates that as the number of cognitive users increases, the collision probability also increases, which is obvious from the defined

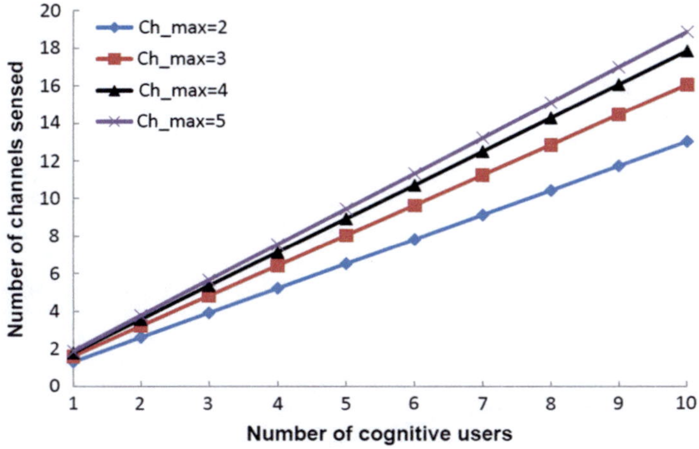

Fig. 5.5 The response of the number of cognitive users on the average number of licensed channels sensed for different values of the parameter defining number of sensed channels by each cognitive user for $Ch_{max} = 2, 3, 4, 5$

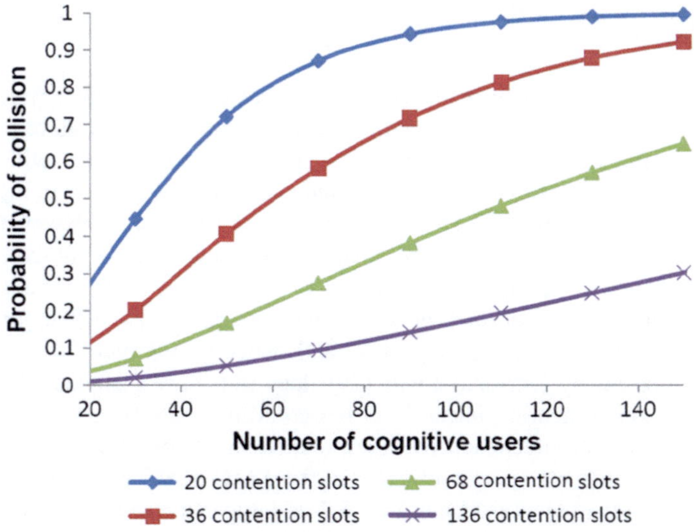

Fig. 5.6 The role of number of cognitive users on the probability of collision for different numbers of contention slots

system model. For lower numbers of contention slots, the collision probability is significantly high in comparison to that of higher numbers of contention slots.

Figure 5.7 depicts the maximum and minimum achievable throughput computed from Eqs. (5.8) and (5.9), which utilize the sensing-sharing and contention intervals (wasted bandwidth) for data transmission, and compares the results with

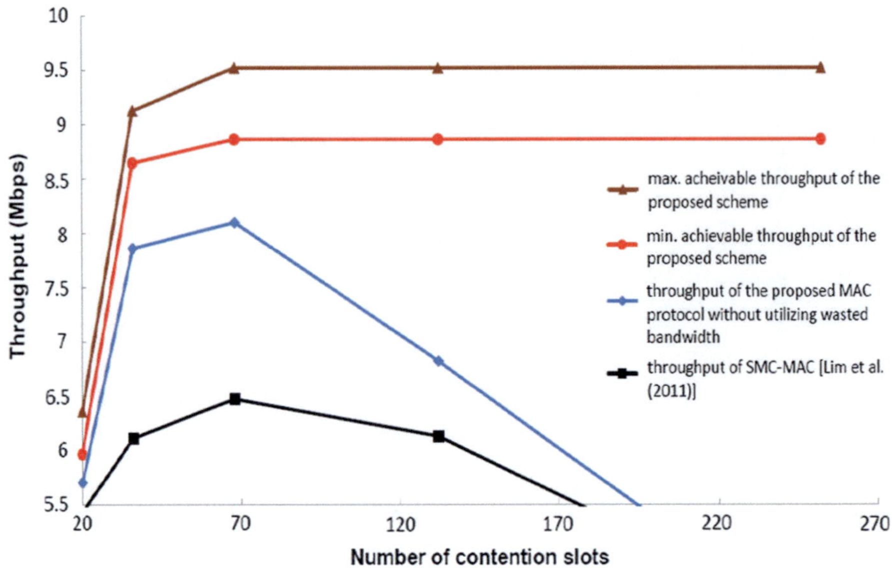

Fig. 5.7 Throughput variation with the number of contention slots for 10 cognitive users, 20 licensed channels. $\alpha = 0.2$ and $T_{\text{cycle}} = 1s$ for schemes that do and do not utilize wasted bandwidth

SMC-MAC and our earlier proposed scheme that does not utilize wasted bandwidth. It is obvious that more contention slots are required if we want to make more users successful, but at the same time, the data transmission time will decrease. Therefore, the throughput of the proposed scheme increases initially in Fig. 5.7 due to more users succeeding, till all the users become successful. However, there is reduction in the throughput of the proposed scheme without utilizing wasted bandwidth beyond the optimum number of contention slots, because further increasing the contention interval keeps the successful users the same while decreasing the data transmission interval and hence throughput. The SMC-MAC protocol proposed by Lim and Li in [29] does not have a contention resolving algorithm in the contention interval, so the colliding users have no provision for succeeding in the current cycle time. Also, wasted bandwidth in the sensing-sharing and contention intervals is not utilized in the SMC-MAC protocol, resulting in throughput degradation as shown in Fig. 5.7 in comparison to the scheme proposed in this chapter. Moreover, for the optimized contention slots, the maximum and minimum achievable throughputs proposed in this chapter are always greater than that of the throughput computed without utilizing wasted bandwidth and the SMC-MAC protocol throughput. However, maximum and minimum achievable throughputs remain constant above the optimum number of contention slots because the decrease in the data transmission interval throughput due to increasing contention slots is balanced with the increasing throughput of the contention

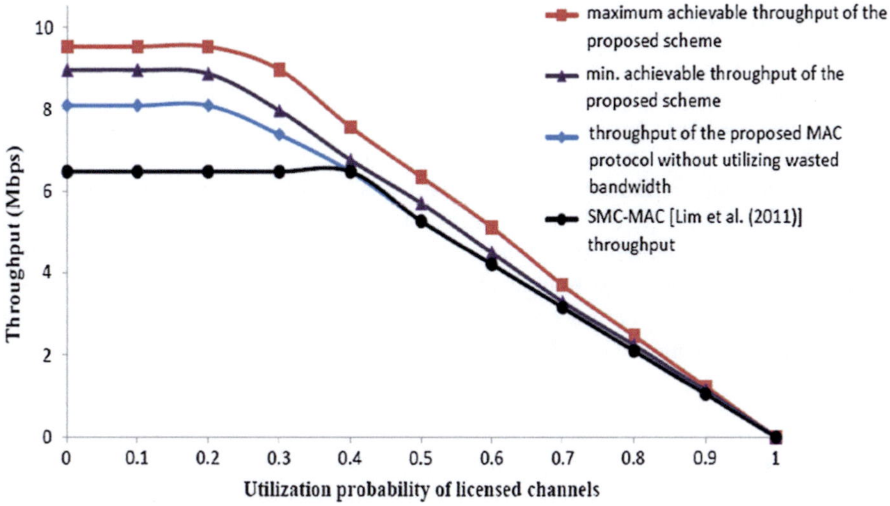

Fig. 5.8 Throughput variation with the utilization probability of licensed channels for 10 cognitive users, $T_{\mathrm{cycle}} = 1s$ and 68 contention slots

interval, since in this case data is also transmitted in the contention interval. In Fig. 5.8, the throughput of a cognitive network with the traffic load of licensed channels is demonstrated for the optimum number of contention slots. The simulation result depicts that there is significant improvement in the throughput when wasted bandwidth is also utilized for data transmission, in comparison to that of the SMC-MAC [29] and the other scheme proposed by us in the previous chapters that did not utilize wasted bandwidth. However, the throughput of the proposed scheme, for which all users are successful, is almost constant for traffic load values from 0 to 0.2 because at these values all 10 cognitive users will transmit their data in the 10 idle channels detected in the sensing-sharing interval. However, in the SMC-MAC protocol, not all users are successful at the selected optimum contention slots, and therefore, for traffic load values from 0 to 0.4, the number of successful users is less than the idle channels detected; hence, throughput is only for those users which are successful, and remains constant at these values. Furthermore, in the SMC-MAC protocol, as traffic load increases above 0.4, the idle channels detected decrease in comparison to the successful users. In this case, throughput is that of the number of successful users getting idle channels, and hence decreases with the increasing traffic load probability. Above a traffic load of 0.4, the SMC-MAC protocol and the proposed scheme that did not utilize wasted bandwidth yield similar throughput, as shown in Fig. 5.8. The increased number of successes in the later scheme does not result in more throughput due to an insufficient number of idle channels.

5.5 Summary

In this chapter, we proposed a scheme for maximizing bandwidth efficiency by utilizing the wasted bandwidth of the licensed channels in the distributed cognitive radio MAC protocol. In addition, we applied the contention resolving algorithm discussed in Chap. 3. Further, the bandwidth wastage in the cooperative distributed MAC protocol was minimized by transmitting cognitive users' data over idle licensed channels, which are unutilized in the sensing-sharing and contention intervals. The proposed technique significantly enhances the throughput of the cooperative distributed network when compared with the SMC-MAC [29] protocol and the scheme proposed in Chap. 3.

References

1. J. Marinho, E. Monteiro, Cognitive radio: survey on communication protocols, spectrum decision issues, and future research directions. Wireless Netw. **18**(2), 147–164 (2012)
2. A. De Domenico, E.-C. Strinati, M.-G.-D. Benedetto, A survey on MAC strategies for cognitive radio networks. IEEE Commun. Surv. Tutorials **14**(1), 21–44 (2012)
3. S.-C. Jha, M.-M. Rashid, V.-K. Bhargava, C. Despins, Medium access control in distributed cognitive radio networks. IEEE Wireless Commun. **18**(4), 41–51 (2011)
4. C. Cormio, K.-R. Chowdhury, A survey on MAC protocols for cognitive radio networks. Ad Hoc Netw. **7**(7), 1315–1329 (2009)
5. A.-C.-V. Gummalla, J.-O. Limb Wireless medium access control protocols. IEEE Commun. Surv. Tutorials **3**(2), 2–15 (2000)
6. S. Kumar, V.-S. Raghavan, J. Deng, Medium access control protocols for ad hoc wireless networks: a survey. Ad Hoc Netw. **4**(3), 326–358 (2006)
7. M. Chowdhury, Asaduzzaman, Md. F. Kader, Cognitive radio MAC protocols: A survey, research issues and challenges. Smart Comput. Rev. **5**(1), 19–29 (2015)
8. N. Mansoor, A.-K.-M. Muzahidul Islam, M. Zareei, S. Baharun, T. Wakabayashi, S. Komaki, Cognitive radio ad-hoc network architectures: a survey. Wireless Person. Commun. **18**(3), 1117–1142 (2015)
9. Le Thanh Tan, Medium access control for dynamic spectrum sharing in cognitive radio networks. Ph.D. Dissertation, INRS-EMT, Canada, Aug 2015
10. K.-C. How, M. Ma, Y. Qin, A MAC-layer QoS provisioning protocol for cognitive radio networks. Wireless Pers. Commun. **65**(1), 203–222 (2012)
11. J. Mo, H.-S. Wilson So, J. Walrand, Comparison of Multichannel MAC Protocols. IEEE Trans. Mobile Comput. **7**(1), 50–65 (2008)
12. P Pawelczak, S. Pollin, H.-S.-W. So, A. Bahai, R.-V. Prasad, R. Hekmat, Quality of service assessment of opportunistic spectrum access: a medium access control approach. IEEE Wireless Commun. **15**(5), 20–29 (2008)
13. H.-B. Salameh, M. Krunz, O. Younis, MAC protocol for opportunistic cognitive radio networks with soft guarantees. IEEE Trans. Mobile Comput. **8**(10), 1339–1352 (2009)
14. Q. Liu, X. Wang, B. Han, X. Wang, X. Zhou, Access delay of cognitive radio networks based on asynchronous channel-hopping rendezvous and CSMA/CA MAC. IEEE Trans. Veh. Technol. **64**(3), 1105–1119 (2015)
15. R.-R. Chen, K.-H. Teo, B. Farhang-Boroujeny, Random access protocols for collaborative spectrum sensing in multi-band cognitive radio networks. IEEE J. Sel. Top. Sig. Process. **5**(1), 124–136 (2011)

16. J. Park, P. Pawelczak, D. Cabric, Performance of joint spectrum sensing and MAC algorithms for multichannel opportunistic spectrum access ad hoc networks. IEEE Trans. Mob. Comput. **10**(7), 1011–1027 (2011)
17. X.-Y. Wang, P.-H. Ho, Gossip-enabled stochastic channel negotiation for cognitive radio ad hoc networks. IEEE Trans. Mobile Comput. **10**(11), 1632–1645 (2011)
18. J. Park, M. van der Schaar, Cognitive MAC protocols using memory for distributed spectrum sharing under limited spectrum sensing. IEEE Trans. Commun. **59**(9), 2627–2637 (2011)
19. S.-C. Jha, U. Phuyal, M.-M. Rashid, V.-K. Bhargava, Design of OMC-MAC: An opportunistic multi-channel MAC with QoS provisioning for distributed cognitive radio networks. IEEE Trans. Wireless Commun. **10**(10), 3414–3425 (2011)
20. M.-H. Cheung, V.-W.-S. Wong, R. Schober, SINR-based random access for cognitive radio: distributed algorithm and coalitional game. IEEE Trans. Wireless Commun. **10**(11), 3887–3897 (2011)
21. L.-T. Tan, L.-B. Le, Channel assignment with access contention resolution for cognitive radio networks. IEEE Trans. Veh. Technol. **61**(6), 2808–2823 (2012)
22. C. Ghosh, S. Roy, M.B. Rao, Modeling and validation of channel idleness and spectrum availability for cognitive networks. IEEE J. Sel. Areas Commun. **30**(10), 2029–2039 (2012)
23. S. Kwon, B. Kim, B.h. Roh, Preemptive opportunistic MAC protocol in distributed cognitive radio networks. IEEE Commun. Lett. **18**(7), 1155–1158 (2014)
24. P. Katiyar, K. Rajawat, Channel-aware medium access control in multichannel cognitive radio networks. IEEE Commun. Lett. **19**(10), 1710–1713 (2015)
25. L.-T. Tan, L.-B. Le, Design and optimal configuration of full-duplex MAC protocol for cognitive radio networks considering self-interference. IEEE Access **3**, 2715–2729 (2015)
26. H.-A.-B. Salameh, H. Kasasbeh, B. Harb, A batch-based MAC design with simultaneous assignment decisions for improved throughput in guard-band-constrained cognitive networks. IEEE Trans. Commun. **64**(3), 1143–1152 (2016)
27. C. Passiatore, P. Camarda, A fair MAC protocol for resource sharing in ad-hoc cognitive networks. Telecommun. Syst. **56**(2), 269–283 (2014)
28. S.-K. Sharma, T.-E. Bogale, S. Chatzinotas, B. Ottersten, L.-B. Le, X. Wang, Cognitive radio techniques under practical imperfections: a survey. IEEE Commun. Surv. Tutorials **17**(4), 1858–1884 (2015)
29. S. Lim, T.-J. Lee, A self-scheduling multi-channel cognitive radio MAC protocol based on cooperative communications. IEICE Trans. Commun. **E94-B**(6), 1657–1668 (2011)
30. A. Zappone, C. Zhijiat Chong, E. Jorswieck, S. Buzzi, Green resource allocation in relay-assisted multicarrier IC networks considering circuit dissipated power, in *Proceedings of Third International Conference on Future Energy Systems: Where Energy, Computing and Communication Meet* (e-Energy), Madrid, Spain, 9–11 May 2012, pp. 1–10

Chapter 6
Power Allocation for Optimal Energy Efficiency in MAC Protocol of Cognitive Radio Communication Systems

6.1 Introduction

Energy consumption is an issue of major concern in the present wireless communication scenario. Wireless devices run different services, such as web browsing, gaming, social media and multimedia downloads, which quickly drain the battery, therefore it is important to design an energy efficient user terminal that optimizes battery life. Cognitive radio introduces its own set of energy consumption considerations. The frequency of the return of primary users to the licensed band impacts the energy efficiency of the cognitive radio user, because it may require a restart of the functionalities for spectrum sensing, channel selection and communication over the control and data channels, consuming additional energy [1]. Further, the cognitive users consume considerable energy for exchange of control information and during retransmission when the primary user resumes its transmission. Cognitive radio terminal devices with multiple transceivers give higher sensing accuracy, avoid hidden terminal problems, maximize throughput and are more spectrum efficient than single transceiver devices, but they use more energy. Therefore, there is tradeoff between the number of transceivers and energy efficiency in the cognitive radio MAC protocol design. Qureshi in [1] proposed a reliable and energy-efficient cognitive radio multichannel MAC (RECR-MAC) protocol for ad hoc networks that uses selection criteria to obtain a reliable data channel and also uses one backup channel, resulting in an energy efficient MAC protocol. The cognitive radio user's channel selection criteria in RECR-MAC are based on the amount of interference from the primary users on a channel due to their arrival rate. When the primary user returns to its frequency band during ongoing cognitive user communication, the cognitive user switches to the backup channel instead of restarting the entire process of sensing and then transmission, which results in higher energy efficiency [1]. Moreover, energy saving can be implemented at any layer of the protocol stack, for example at the MAC layer, the network layer and the application layer. However, it is much easier to implement

S. Pandit and G. Singh, *Spectrum Sharing in Cognitive Radio Networks*,
DOI 10.1007/978-3-319-53147-2_6

energy saving mechanisms at lower layers than higher layers because lower layers have relatively more direct access to the medium. Recently, the authors in [2] optimized energy efficiency performance by considering the heterogeneous cloud radio access network. A non-convex fractional optimization problem was formulated, and simulation results showed that the proposed algorithm in the heterogeneous cloud radio access network was more energy efficient than the algorithms proposed in heterogeneous network and cloud radio access network individually. Further, another convex optimization problem was formulated in [3] to maximize energy efficiency in a delayed quality-of-service constrained cognitive radio network. Since it is known that the cognitive user should not interfere with the primary user, and therefore in the spectrum underlay access method where the primary and cognitive user communication is running, simultaneously, the transmit power of the cognitive user should be adjusted. In most of the literature [4–9], location information is assumed to be available, however, the authors in [10] estimated the location or distance to the primary user by considering the received signal strength from the primary at the cognitive user. Further, the authors computed the maximum allowable transmit power of cognitive user based on the estimated distance, shadowing effects and interference temperature constraints at the primary user [10]. The simulation results illustrated that the cooperative method provided improved capacity performance of the cognitive user compared to that of the non-cooperative transmit power method [10]. Another cognitive user underlay model was considered in [11]. The transmit power minimization under the rate constraint was achieved with the help of orthogonal frequency-division multiplexing (OFDM) and a filter bank-based multicarrier scheme. In order to exploit the unused spectrum bands of primary users, the cognitive users employed various spectrum sensing techniques. The spectrum sensing techniques in the cognitive radio network increase the energy consumption while improving the throughput of the networks, and therefore the need for energy savings in cognitive radio networks has attracted a lot of attention from both the government and from network operators. In [12], an energy-efficient transmit power allocation scheme was investigated in cognitive radio networks and the optimization problem was formulated as a ratio of the spectral efficiency to the total energy consumption used in signal transmission and spectrum sensing under the total power constraint. Reference [12] also proposed an optimal energy-efficient power allocation scheme which iteratively improves the energy efficiency and reaches the optimal solution.

This chapter emphasizes the design of an energy efficient MAC protocol for cognitive users. In Chap. 4 and [13], we have only computed the energy efficiency of the proposed cognitive radio MAC protocol. In the present scenario, for purposes of minimizing energy consumption of the terminals, we propose an algorithm to maximize energy efficiency. The energy efficiency issue in cognitive radio communication systems has been discussed in detail in several reported studies [14–20]. Qian et al. [15] maximized the energy efficiency of cognitive radio networks utilizing the frequency of the TV spectrum through the power control for both the centralized and distributed cognitive radio networks. Moreover, in [15] the authors implemented power control in the MAC protocol of a cognitive radio network.

In [16], game theory was used for power allocation to the cognitive users in the MAC protocol. The proposed cost-based algorithm for power allocation minimized the energy consumption of the cognitive radio user's network.

In [17], the authors achieved the optimal sensing and data transmission time in a frame of the cognitive radio user which maximized the energy efficiency, however the throughput was limited due to high detection and low false-alarm probability requirements. This required large sensing time durations, resulting in small data transmission intervals of the fixed frame duration. Therefore, Chatterjee et al. in [18] considered a joint spectrum sensing and data transmission method with the help of cognitive relays which maximized throughput along with reliable sensing performance of the cognitive radio communication system. Further, in [18] cognitive relays amplified and forwarded the cognitive user's source data to the destination in order to deal with energy consumption issues. The optimal strategy for energy efficiency was achieved by considering the interference threshold at the primary receiver, throughput of the cognitive user and high detection and low false alarm probability. However, the proposed method in [18] has delay issues because it has no point-to-point communication among the source and destination cognitive users. However, in OFDM-based cognitive radio networks, the optimal power of the subcarriers has been computed with constraints on the total transmit power and interference. The energy efficiency problem is a fractional programming method, and various methods have been proposed for its solution [14, 19, 20]. In [19], the energy efficiency problem was first converted into a convex programming problem and then an iterative algorithm based on the sequential quadratic problem was used to find the optimal power solution for energy efficiency. The authors in [20] converted the energy efficiency fractional programming problem into a parametric formulation and then dynamic strategy yielded the optimal solution for the problem. For centralized cognitive radio networks, an energy efficient heuristic algorithm was proposed in [14] for optimal energy efficiency. Unfortunately, the methods proposed in [14, 19, 20] for maximizing energy efficiency problems are complex computations, so we propose a very simple method for easy computation of transmit power in order to maximize energy efficiency.

In this chapter, we propose a simple algorithm for computing the optimal transmit power of cognitive radio for different channel gains which maximizes energy efficiency. The exchange of cognitive radio request-to-send (CR-RTS) and cognitive radio clear-to-send (CR-CTS) frames provides the knowledge of channel gain and approximate distances of the cognitive radio transmitter and cognitive radio receiver, which are used to compute optimal transmit power for maximizing energy efficiency. Moreover, the cognitive user energy consumption in different intervals of the proposed MAC protocol, that is, the energy consumption in sensing-sharing, contention, and data transmission interval, are also computed for the proposed algorithm. The simulation results are presented for the energy efficiency variation with the traffic load of licensed channels as well as for different channel gains. In this chapter, the minimization of energy consumption of the cognitive terminal in accessing the licensed channels through the distributed cognitive radio MAC protocol is proposed while simultaneously considering throughput. Further, the algorithm for deciding the optimal transmit power of the cognitive user is based on the channel conditions and the distance metric.

This chapter is organized as follows. Section 6.2 discusses the system model. In Sect. 6.3 the problem is formulated and analysis is presented for the proposed scheme. Section 6.4 explores the results and discusses the proposed system model. Finally, Sect. 6.5 concludes the work.

6.2 System Model

In this chapter, the main aim is to design a self-scheduled-MAC protocol for the cognitive radio network which maximizes the energy efficiency of the cognitive user, and schedules itself for having the highest energy efficiency. The cognitive user's optimal transmit power is computed through the algorithm proposed in Sect. 6.3 which maximizes the energy efficiency of the system. As with the system design proposed in the previous chapters, the MAC protocol has N_{ch} licensed channels, and the idle channels utilized by the cognitive users have different channel characteristics which are defined by the following channel gain set: $H = \{h_1, h_2, \ldots h_{Nch}\}$. Moreover, the N_{CU} number of cognitive users has maximum and minimum limits on the transmit power: P_{max} and P_{min}, respectively. A control channel is available on which the cognitive users share sensing results with each other. We have assumed high detection probability of the licensed channels, such that the probability of detection is almost equal to 1, and false alarm probability is ignored. The RTS frame is transmitted from the cognitive transmitter to the receiver during contention interval in order to reserve the idle licensed channel, and a CTS frame is sent back to the transmitter from the receiver which contains information about the channel gain of the reserved idle licensed channel. We have assumed the flat fading channels and cognitive receiver have information about the channel gain of the licensed channel. The response interval of the CTS frame is used to calculate the distance between the transmitter and receiver of the cognitive user. With the help of this information, the optimal transmit power for cognitive users are computed to maximize the energy efficiency of the cognitive radio communication system. The rest of the system description is similar to the system proposed in Chap. 3.

6.3 Problem Formulation and Performance Analysis

The main aim is to maximize the energy efficiency [20] of a cognitive radio communication system for which we have computed the optimal transit power. The energy efficiency of the proposed cognitive radio communication system is the ratio of the total amount of useful data delivered to the total energy consumed, and is given as:

$$EE = \frac{\text{Total amount of useful data delivered(bits)}}{\text{Total energy consumed(Joule)}} \tag{6.1}$$

where EE is the energy efficiency of the proposed protocol. The total amount of useful data delivered by the ith cognitive user is defined as throughput per cycle time; for a given licensed channel k with probability of detection ≈ 1 ($P_d \approx 1$), this is given as [20]:

$$R_{\rm i} = w_{\rm k} T_{\rm tr} B \, log_2(1 + SNR_{\rm k}/\Gamma) \tag{6.2}$$

where $SNR_{\rm k}$ is the received signal-to-noise ratio at the cognitive receiver on the kth licensed channel and r is the SNR gap to channel capacity and is approximated as $\Gamma \approx -\frac{In(5BER)}{1.5}$ for an uncoded M-QAM with a given bit error rate (BER) [20]. B is the bandwidth of channel k. Further, the $\text{SNR}_{\rm k}$ is given as follows [20]:

$$\text{SNR}_{\rm k} = \frac{\rho_{\rm k} h_{\rm k} Pt_{\rm i}}{L N_0 B N_{\rm f}} \tag{6.3}$$

where $\rho_{\rm k} = \left(\frac{c}{4\Pi d f_{\rm k}}\right)^2$ measures the propagation loss for distance d between the cognitive transmitter and the cognitive receiver at carrier frequency $f_{\rm k}$ of channel k. $Pt_{\rm i}$ is the transmit power calculated for the ith cognitive user over channel k having channel gain $h_{\rm k}$. L is the link margin compensating the hardware process variation and imperfection [20]. N_0 is the noise power spectral density, $N_{\rm f}$ is the receiver noise figure, therefore $N_0 B N_{\rm f}$ is the noise power at the receiver's front end. Moreover, $w_{\rm k}$, which is defined in [20], is the probability of accurately detecting the state of the licensed channel k, and is given as [20]:

$$w_k = \frac{(1-\alpha)(1-P_{f,k})}{(1-\alpha)(1-P_{f,k}) + \alpha(1-P_{d,k})} \tag{6.4}$$

where $P_{f,k}$ and $P_{d,k}$ are the probability of false detection and probability of accurate detection of licensed channel k. Moreover, the energy consumed in the sensing-sharing interval by a cognitive user which senses $Ch_{\max}$ number of channels is:

$$E_{\rm ss_i} = T_{\rm s\_slot} P_{\rm s\_slot} Ch_{\max} + T_{\rm s\_slot} P_{\rm s\_idle} (N_{\rm ch} - Ch_{\max}) \tag{6.5}$$

where $T_{\rm s\_slot}$ is the single sensing-sharing slot duration, $P_{\rm s\_slot}$ and $P_{\rm s\_idle}$ are the sensing and idle mode powers of the cognitive user in a sensing-sharing slot. The first term of Eq. (6.5) computes the amount of energy consumption for sensing and sharing the results by the ith cognitive user. The second term gives the energy consumed by the ith cognitive user for rest of the sensing-sharing interval in which sensing is not performed by the ith cognitive user. The difference in the energy computed in Eq. (4.26) and Eq. (6.5) is that in Eq. (4.26) the whole energy consumed by all cognitive users is computed and in Eq. (6.5) only single cognitive user energy consumption is computed. Further, the energy consumption by the *i*th cognitive user in the contention interval is:

$$E_{\mathrm{ct_i}} = T_{\mathrm{ct\_slot}} P_{\mathrm{ct\_slot}} + T_{\mathrm{ct\_slot}} P_{ct\_c} N[C] + T_{\mathrm{ct\_slot}} P_{\mathrm{ct\_idle}} (CW_{\mathrm{total}} - (N[C] + 1)) \tag{6.6}$$

In (6.6), the first term gives the energy consumed by the *i*th cognitive user during the successful contention slot, the second term represents the energy consumption in collided contention slots, and the third term computes the energy during the idle contention slots. $N[C]$ is the number of collisions of the *i*th cognitive user in the CW_{total} contention slots. Moreover, the *i*th cognitive user energy consumption in the data transmission interval is [20]:

$$E_{\mathrm{tr_i}} = T_{\mathrm{tr}} (\beta P t_{\mathrm{i}} + P_{\mathrm{c}}) \tag{6.7}$$

where $\beta = \frac{\xi}{\varsigma}$ and ξ is the peak-to-average ratio (PAR) of the power amplifier, ς is the drain efficiency of the power amplifier and P_{c} is the amount of power consumed by the transmitter and receiver circuits, with the exception of the power amplifier which is a constant value [20]. Therefore, the total energy consumed by the *i*th cognitive user in the single cycle time is:

$$E_{\mathrm{total}_i} = E_{\mathrm{ss_i}} + E_{\mathrm{ct_i}} + E_{\mathrm{tr_i}} \tag{6.8}$$

Further, in case the data is not transmitted over the transmission interval, then the total energy consumption of the cognitive user is:

$$\mathrm{E}_{\mathrm{total}_i} = E_{\mathrm{ss_i}} + E_{\mathrm{ct_i}} \tag{6.9}$$

Therefore, the energy efficiency defined in Eq. (6.1) is formulated as:

$$EE = \sum\nolimits_{k=1}^{N_{\mathrm{CU}}} \frac{R_{\mathrm{i}}}{E_{\mathrm{total}_i}} \tag{6.10}$$

where R_{i} is the data rate and E_{total_i} is the total energy consumption of the *i*th cognitive user. The assignment of the power to the different cognitive users for maximizing the energy efficiency is performed according to the following proposed algorithm:

Proposed Algorithm

Step 1: *Variable declaration*
N_{CU} = Number of cognitive users in the network
P_{max} = Maximum transmit power allowed by a cognitive user
P_{min} = Minimum transmit power allowed by a cognitive user
h_{k}= Channel gain of the licensed channel *k*
CU_{i} = *ith* cognitive user

Step 2: *Computation of optimal transmit power that maximizes the energy efficiency of CU_i*

$$Pt_{\mathrm{i}} \leftarrow \underbrace{argmax}_{Pt_i} EE$$

The power Pt_{i} is assigned to the CU_i for transmitting data in the data transmission interval. This step is followed for all N_{CU} cognitive users and optimal transmit power is calculated for all the users.

The above algorithm describes a simple linear optimization with constraints on the power, that is, the transmit power of cognitive users, within the defined minimum and maximum transmit power limit. The aim of the proposed linear optimization is to maximize the energy efficiency as defined in Eq. (6.10) and to find the transmit power which results in the maximum energy efficiency within the given constraints. Further, the complexity of the proposed algorithm is $O(N)$, where N is the number of input power levels used for computation of the energy efficiency.

6.4 Simulation Results

The simulation parameters for the method proposed in this chapter are shown in Table 6.1. The energy efficiency variation of a cognitive user transmitting on the idle channel for the different channel gains and having utilization probability $\alpha = 0.5$ is shown in Fig. 6.1. From Fig. 6.1 it is clear that there is an optimal value of transmit power at which the energy efficiency is maximized; this transmit power is computed during the contention interval by the algorithm proposed in Sect. 6.3. The cognitive user transmits at this optimal power in the data transmission interval to maximize energy efficiency. Moreover, Fig. 6.1 illustrates that as the channel condition improves due to increases in the value of channel gain parameter, the energy efficiency is also improving. Further, Fig. 6.2 depicts the energy efficiency with the transmit power for different traffic utilization probabilities. We can see from Fig. 6.2 that with the increase in traffic load probability, the energy efficiency decreases.

In Fig. 6.3 the optimal transmit power computed from the proposed algorithm is simulated for different channel gains and distances between cognitive radio transmitters and receivers. Figure 6.3 shows that with larger distances and for less channel gain, the transmit power requirement is greater than that for small distances and higher channel gain, because the higher channel gains deliver cognitive user information with higher data rates than that for lesser channel gains, hence energy efficiency is high for the same circuit power consumption. Similarly, the distant user needs higher power and vice versa.

Table 6.1 The simulation parameters of the proposed system model

Simulation parameters	Numerical values
Number of licensed channels (N_{ch})	20
Utilization probability of licensed channels (α)	0–1
Number of sensed channels by each cognitive user (Ch_{max})	2
Number of cognitive users (N_{CU})	10
Probability of false alarm ($P_{f,k}$)	0.1
Probability of detection ($P_{d,k}$)	0.9
Cycle time (T_{cycle})	1 s
Channel bandwidth (B)	200 kHz
Carrier frequency (f_{k})	800 MHz
Noise PSD (N_0)	−115 dB
Noise figure	10 dB
Link margin (L)	10 dB
Distance between cognitive transmitter and receiver (d)	100–1000 m
Bit error rate (BER)	10^{-5}
Minimum transmit power limit (P_{min})	100 mW
Maximum transmit power limit (P_{max})	3 W
Circuit power (Pc)	210 mW
Idle interval (T_{idle})	1 ms
Data transmission interval (T_{tr})	862 ms
Sensing-sharing slot interval ($T_{\text{s\_slot}}$)	900 μs
Sensing power ($P_{\text{s\_slot}}$)	110 mW
Idle power ($P_{\text{s\_idle}}$)	50 mW
Contention slot interval ($T_{\text{ct\_slot}}$)	2 ms
Successful contention slot power ($P_{\text{ct\_slot}}$)	110 mW
Idle contention slot power ($P_{\text{ct\_idle}}$)	50 mW
Collided contention slot power (P_{ct_c})	120 mW
Total number of contention slots (CW_{total})	68
PAR (ξ)	6 dB
Drain efficiency (ς)	0.35

Figure 6.4 shows the energy efficiency of a cognitive user with different traffic load values and for different channel gains. In Fig. 6.4, the optimal transmit power, which is computed from the proposed algorithm in Sect. 6.3, is utilized for different channel gains for the computation of energy efficiency. It is clear that the higher channel gain values enhance the energy efficiency of the cognitive user.

In Fig. 6.5 the average value of the energy efficiency is simulated for all 10 cognitive users for traffic load utilizations of 0.1 and 0.5. By utilizing the information about the idle channels' availability, we have computed the energy efficiency of the whole system. The number of idle channels present in the system is

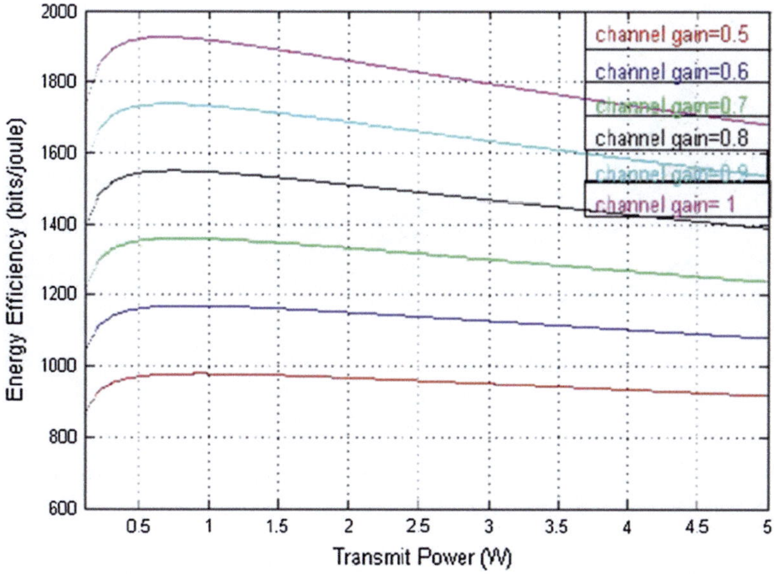

Fig. 6.1 The effect of the variation in transmit power of the cognitive user on the energy efficiency for different channel gains with α = 0.5 and d = 800 m

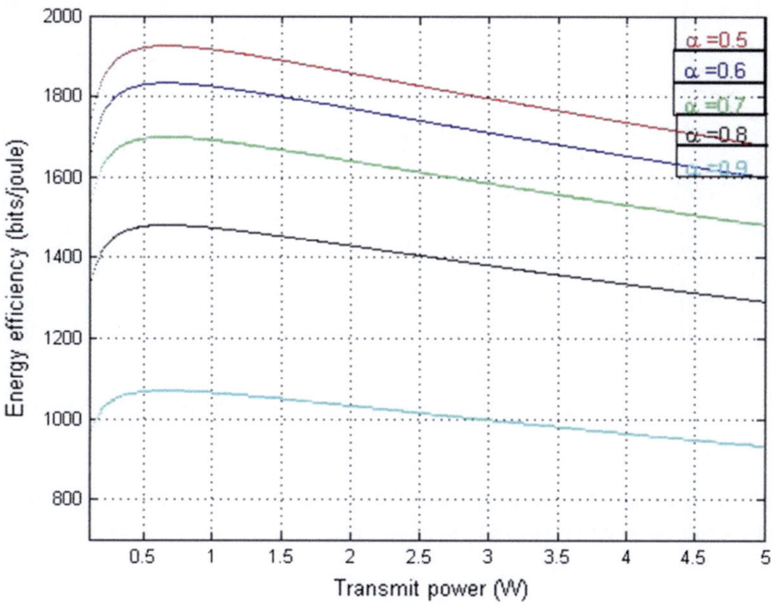

Fig. 6.2 Variation of the transmit power of the cognitive user with energy efficiency for different traffic utilization probabilities and with channel gain 0.8

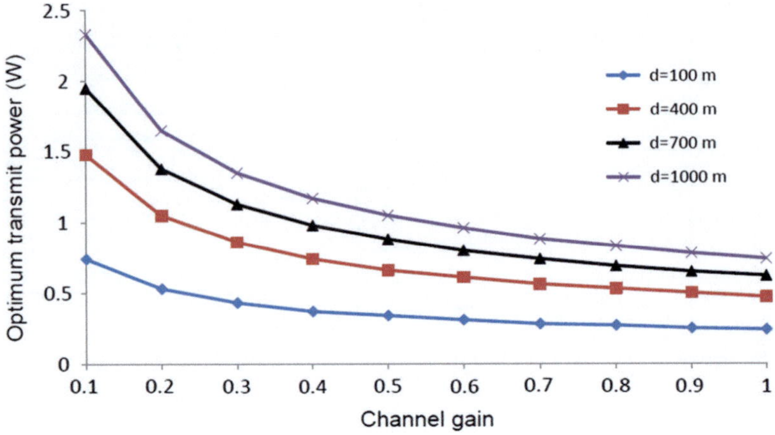

Fig. 6.3 The responses of the channel gain over the optimal transmit power with different cognitive user distances at chosen value of $\alpha = 0.5$

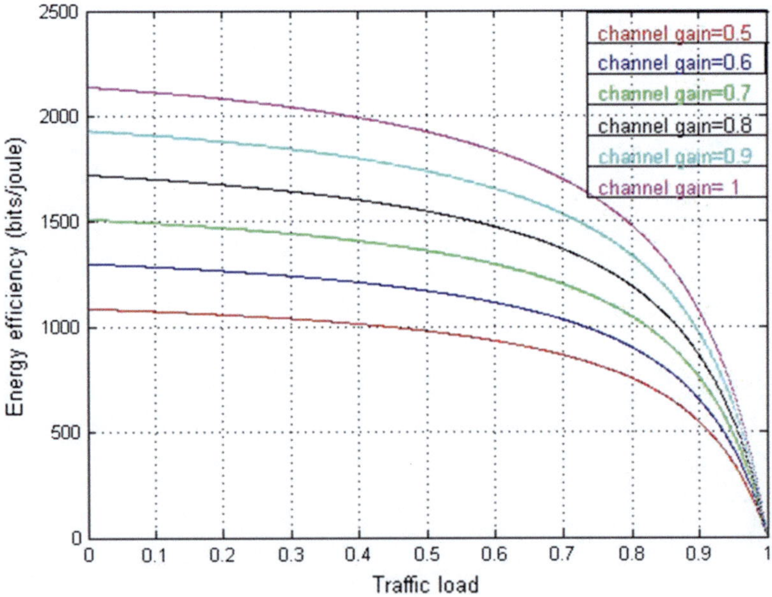

Fig. 6.4 The response of traffic load at optimal transmit power over the energy efficiency for different channel gains for chosen distance of $d = 800$ m

lower for higher traffic load than at lower values, therefore throughput for the latter case is higher than the former, while energy consumption is similar. Hence energy efficiency is higher at low traffic load, as depicted in Fig. 6.5.

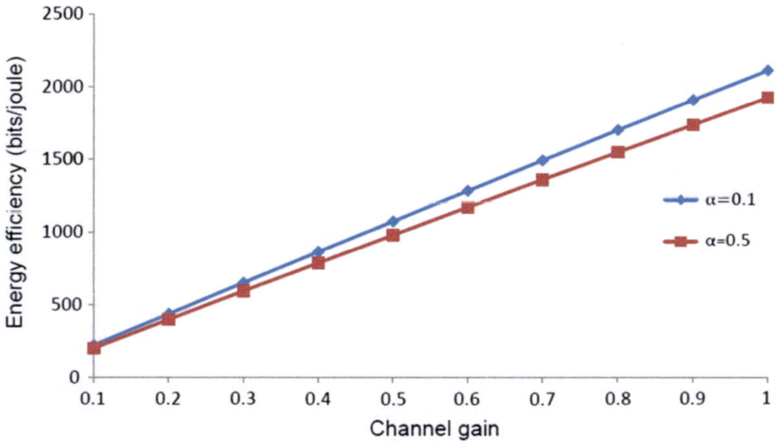

Fig. 6.5 The effect of variation in the channel gain over energy efficiency of 10 cognitive user networks for different traffic loads

6.5 Summary

In this chapter, we addressed the energy efficiency of cognitive radio terminals and obtained the optimal transmit power for cognitive terminals. It was further shown that we could reduce the complexity of proposed algorithm for computing the optimal transmit power. We considered different scenarios of channel conditions at different channel gains and maximized the energy efficiency of the cognitive radio terminal. The significant reduction of the energy utilization becomes more demanding in cognitive radio ad hoc networks where the cognitive users consume a lot of energy during the exchange of control and data frames, and re-transmission if the primary user returns. However, node synchronization is crucial to provide cooperative cognitive communication in decentralized networks. The existing common control channel-based CR-MAC protocols designate cognitive control channels as in-band or out-of-band, which has numerous drawbacks like multi-channel hidden terminals, longer network access delays, and higher control overhead, which results in higher energy consumption and reduced network throughput that severely degrades the performance of the CR-MAC protocol. In a cognitive radio network with energy-harvesting nodes, it is important to improve the energy efficiency as well as spectral efficiency.

References

1. F.-F. Qureshi, A reliable and energy efficient cognitive radio multichannel MAC protocol for ad-hoc networks, Ph.D. thesis, University of Bedfordshire, August 2014
2. M. Peng, K. Zhang, J. Jiang, J. Wang, W. Wang, Energy-efficient resource assignment and power allocation in heterogeneous cloud radio access networks. IEEE Trans. Veh. Technol. **64**(11), 5275–5287 (2015)
3. X.-U. Ding, L.-I. Qun, Energy efficient power allocation for delay-QoS constrained cognitive radio networks. IEICE Trans. Fundam. Electron. Commun. Comput. Sci. **E-99A**(6), 1264–1267 (2016)
4. E. Dall'Anese, S.-J. Kim, G.-B. Giannakis, S. Pupolin, Power control for cognitive radio networks under channel uncertainty. IEEE Trans. Wireless Commun. **10**(10), 3541–3551 (2011)
5. E.-C.-Y Peh, Y.-C. Liang, and Y Zeng, "Sensing and power control in cognitive radio with location information," *Proc. of IEEE International Conference on Communication Systems* (ICCS), Nebraska, USA, 2012, pp. 255–259
6. S. Sorooshyari, C.-W. Tan, M. Chiang, Power control for cognitive radio networks: axioms, algorithms, and analysis. IEEE/ACM Trans. Netw. **20**(3), 878–891 (2012)
7. S.-M. Sanchez, R.-D. Souza, E.-M.-G. Fernandez, V.-A. Reguera, Rate and energy efficient power control in a cognitive radio ad hoc network. IEEE Sig. Process. Lett. **20**(5), 451–454 (2013)
8. S. Gong, P. Wang, Y. Liu, W. Zhuang, Robust power control with distribution uncertainty in cognitive radio networks. IEEE J. Sel. Areas Commun. **31**(11), 2397–2408 (2013)
9. V. Asghari, S.Aïssa, Resource sharing in cognitive radio systems: outage capacity and power allocation under soft sensing, in *Proceedings of IEEE Global Telecommunications Conference* (GLOBECOM), LA, USA, 2008, pp. 1–5
10. K. Umebayashi, J. Lehtomäki, I.-M. Suliman, Analysis of transmit power setting technique for cognitive radio networks. EURASIP J. Wireless Commun. Netw. 6 June 2016
11. J. Denis, M. Pischella, D. Le Ruyet, Efficient power allocation approach for asynchronous cognitive radio networks with FBMC/OFDM. *Cognitive Radio Oriented Wireless Networks*, vol. 172 of the series Lecture Notes of the Institute for Computer Sciences, Social Informatics and Telecommunications Engineering, 29 May 2016, pp. 400–411
12. R. Yadav, K. Singh, A. Kumar, T. Kumar, Energy-efficient power allocation with spectrum sensing in cognitive radio networks, in Proceedings of *IEEE/CIC International Conference on Communications in China (ICCC),* Shenzhen, 2015, pp. 1–6
13. S. Pandit, G. Singh, Medium access control protocol for energy efficiency and optimal power allocation of cognitive radio communication. Wireless Pers. Commun. (under Revision) (2016)
14. S. Bayhan, F. Alagoz, Scheduling in centralized cognitive radio networks for energy efficiency. IEEE Trans. Veh. Technol. **62**(3), 582–595 (2013)
15. L. Qian, X. Li, J. Attia, Z. Gajic, Power control for cognitive radio ad hoc networks, in *Proceedings of 15th IEEE Workshop on Local and Metropolitan Area Networks* (LANMAN), Princeton, NJ, 10–13 June 2007, pp. 7–12
16. F. Wang, M. Krunz, S. Cui, Price-based spectrum management in cognitive radio networks. IEEE J. Sel. Top. Sig. Process. **2**(1), 74–87 (2008)
17. Y. Wu, D.-H.-K Tsang, Energy-efficient spectrum sensing and transmission for cognitive radio system. IEEE Commun. Lett. **15**(5), 545–547 (2011)
18. S. Chatterjee, S.-P. Maity, T. Acharya, Energy efficient cognitive radio system for joint spectrum sensing and data transmission. IEEE J. Emerg. Sel. Top. Circ. Syst. **4**(3), 292–300 (2014)

19. J. Zuo, V.P. Dao, Y. Bao, S. Fang, L. Zhao, C. Zou, Energy-efficiency power allocation for cognitive radio MIMO-OFDM systems. Electron. Telecommun. Res. Inst. (ETRI) J. **36**(4), 686–689 (2014)
20. Y. Pei, Y.-C. Liang, K.-C. The, K.-H. Li, Energy-efficient design of sequential channel sensing in cognitive radio networks: optimal sensing strategy, power allocation, and sensing order. IEEE J. Sel. Areas Commun. **29**(8), 1648–1659 (2011)

Chapter 7
Frame Structure for Throughput Maximization in Cognitive Radio Communication

7.1 Introduction

Spectrum sensing is a key technology of cognitive radio networks, as discussed in Chap. 2. However, spectrum sensing is always not completely perfect and creates a level of uncertainty with regard to spectrum access, The probability of collisions and the probability of interference in secondary user spectrum access impose a significant constraint which definitely affects spectrum decision policy. This constraint along with the throughput and energy or power makes the spectrum access process more complicated. To avoid collision with primary user transmission, an approach known as listen-before-talk (LBT) has been considered [1], which means each time the cognitive user decides to transmit data, it should sense the spectrum in advance. This strategy can waste time and energy and, consequently, reduce the network throughput. This wasting of time is usually referred as sensing time overhead in the literature [2, 3]. Zhao et al. [2] proposed a decentralized cognitive MAC protocol which allows cognitive users to independently search for spectrum opportunities without a central coordinator or a dedicated communication channel. An analytical framework for OSA based on the theory of partially observable Markov decision process (POMDP) was developed. This approach integrated the design of spectrum access protocols at the MAC layer with spectrum sensing at the physical layer and the traffic statistics determined by the application layer of the primary network. It allowed easy integration of spectrum sensing errors and constraints on the probability of collision with the primary users. In [3], the effect of sensing overhead on the system performance for cognitive radio networks with channel bonding was analyzed and analytical expressions for the blocking probability, forced termination probability and throughput were derived. The numerical results revealed that the forced termination probability is unaffected by sensing overhead, while the blocking probability and throughput degrade with the increase in the sensing time.

S. Pandit and G. Singh, *Spectrum Sharing in Cognitive Radio Networks*,
DOI 10.1007/978-3-319-53147-2_7

Nonetheless, spectrum sensing and decisions are two of the most important functions of cognitive radios, which are dependent on each other and are jointly optimized for different constraints as discussed in detail in [1]. These two functions of cognitive radios are considered in a different ways for the single channel and multi-channel access cases in the literature because of some inherent differences between the two scenarios [4, 5]. Hattab and Ibnkahla [4] exhaustively analyzed recent advancements in multiband spectrum sensing techniques, their limitations, and possible future directions for improvement. The cooperative communications for multiband cognitive radio networks to deal with a fundamental limit on diversity and sampling were presented with several limits and tradeoffs of various design parameters. Further, the key multiband cognitive radio networks performance metrics were explored, which are different from the conventional metrics used for single-band-based networks. Masonta et al. [5] provided an up-to-date survey of spectrum decisions in cognitive radio networks and addressed issues of spectrum characterization (including cognitive user's activity modeling), spectrum selection and cognitive radio reconfiguration. The authors also highlighted key open research challenges for each of these issues with practical implementations of spectrum decision in several cognitive radio platforms. In [6–9], to avoid collision with primary user transmission and hence to prevent the retransmission of cognitive user data, the sensing time duration was optimized to have the best sensing precision. In [6], the authors considered additional constraints such as maximum interference level to the PU signal. In [7], the allocation predicament of the sensing period between two primary frequency channels was formulated as a convex optimization problem and an optimal solution was presented. The numerical investigation revealed that the proposed optimal allocation improves the throughput of a cognitive user under a missed detection constraint to protect the primary frequency channels. In [8], the average throughput maximization of a secondary user was formulated by optimizing its spectrum sensing time, assuming a priori knowledge of the presence and absence probabilities of the primary users. In [9], different sensing times were considered for different channel occupancy states. Usually the main problem in these articles is to find the best operating point in the receiver operating characteristic (ROC) curve subject to some constraint. The probability of false alarm and missed detection are the two main criteria in these problems.

Optimizations of sensing period and transmission times are also challenging issues, which were studied in [10, 11]. In [10], an adaptive sensing period optimization scheme based on a multi-objective genetic algorithm formulation was proposed to maximize the spectrum opportunities as well as to minimize the incurred sensing overhead. Kim and Shim [11] developed a sensing-period optimization mechanism and an optimal channel-sequencing algorithm as an environment-adaptive channel-usage pattern estimation method. To maximize the network throughput or transmission efficiency was the main concern of [12]. The authors of [13] explored the potential problem of designing the sensing duration to maximize the achievable throughput for the secondary network under the constraint that the primary users are sufficiently protected. The sensing-throughput trade-off problem was formulated mathematically, and used an energy detection sensing scheme to prove that the formulated problem indeed has one optimal sensing time that yields the highest throughput for the secondary network. Hoang et al. [14] emphasized adaptively scheduled spectrum sensing and data transmission to minimize negative impacts on the performance of cognitive radio networks. Based on the information of channel conditions, the sensing periods were adaptively scheduled to maximize the spectrum efficiency of the cognitive radio operation. In [15], an optimal spectrum sensing framework was developed to solve both the interference avoidance and the spectrum efficiency problem. Initially, a theoretical framework was developed to optimize the sensing parameters in such a way as to maximize the sensing efficiency subject to interference avoidance constraints. In order to exploit multiple spectrum bands, the spectrum selection and scheduling methods were proposed wherein the best spectrum bands for sensing were selected to maximize sensing capacity. An adaptive and cooperative spectrum sensing method was proposed where the sensing parameters were optimized adaptively to the number of cooperating users. Noh et al. [16] analyzed the secondary user throughput of a sensing-based cognitive radio system with Markovian traffic in which an imperfect packet capture occurred upon the random arrival of primary user packets and affected the secondary user throughput. Further, the authors analyzed a joint optimization dealing with both the sensing duration and the sensing period to maximize the secondary user throughput with an interference constraint for the primary user. In [17], a multi-layer spectrum sensing optimization algorithm to maximize sensing efficiency by computing the optimal sensing and transmission durations for a fast changing, dynamic primary user was presented. Joint constraints to correctly reflect interference to the primary user and lost opportunity of the secondary user during the transmission period were formulated. In [18], a novel cognitive radio system that exhibited improved throughput and spectrum sensing capabilities compared to the conventional opportunistic spectrum access cognitive radio systems was investigated. The average achievable throughput of the proposed cognitive radio system under a single high target detection probability constraint as well as its ergodic throughput under average transmits and interference power constraints were analyzed. Moreover, an algorithm that acquired the optimal power allocation strategy and target detection probability was proposed which became an additional optimization variable in the ergodic throughput maximization problem under the imposed average interference power constraint. In [12–18] the joint optimization of sensing and transmission times under different scenarios and constraints were formulated. In most of these articles, the frame structure was divided into two parts, one for spectrum sensing and other for transmission. The problem was to find the best portion of the whole frame for each of the two, because there is an essential trade-off between these two parts; as more time is allocated to sensing, the more accurate is the sensing process and lower the number of collisions that occur, but the less time is left for data transmission. Consequently, there is an obvious trade-off between sensing accuracy and throughput; this trade-off is considered in most of these articles.

It is well known that in the MAC protocol, the data is transmitted in frames. Therefore, in this chapter we consider the frame structure of the cognitive radio user and deal with the sensing-throughput tradeoff problem in cognitive radio systems. Cognitive radio users trying to access the licensed spectrum should consider the impact of their transmissions on the reception quality of the primary licensee. In addition, the secondary access does not affect the primary user (PU) operation as long as the total interference power at the primary receiver remains below a certain threshold. For a wireless receiver, any signal other than the signal originally destined to be received by that receiver is considered as interference [19, 20]. Therefore, one of the main difficulties of allocating resources to cognitive radio communication systems is that the interference power generated by its users at the PU receiver should not exceed the predefined threshold [21] in order to protect the primary users. A potential approach is proposed with the aim of increasing the throughput of the cognitive radio user, in which the cognitive radio user first senses the status (active/idle) of a frequency band and then avoids harmful interference to the PU by adapting transmit power based on the spectrum sensing decision [22, 23]. The significant parameters related to spectrum sensing are: (1) false alarm probability and (2) detection probability. False alarm probability must be low to maximize the opportunity of cognitive user data transmission. On the other hand, higher detection probability provides better PU transmission protection.

A cognitive user that employs conventional frame structure is shown in Fig. 7.1. First, the spectrum sensing and then transmission is performed and the figure depicts that the cognitive user ceases data transmission at the beginning of each frame. The spectrum sensing is performed firstly for τ units of time and then data is transmitted for the remaining frame duration, which is (T-τ). However, there is a potential problem in this scheme because it is well known from classical detection theory [24, 25] that an increase in the sensing time results in higher probability of detection and lower probability of false alarm. However, increased sensing time also results in less data transmission time and hence limits the throughput of the cognitive radio user, causing a sensing-throughput tradeoff problem [26]. Apart from the sensing-throughput trade-off, there is the problem of unpredictable PU transmission during the transmission time of the cognitive user, resulting in data loss.

In order to avoid the sensing-throughput trade-off and to maximize the throughput of spectrum sharing cognitive radio networks, an approach was proposed by Stotas and Nallanathan in [27, 28]. The frame structure of this approach is shown in Fig. 7.2, in which both the spectrum sensing and data transmission are performed at the same time and for the whole frame duration, which increases both the sensing time and data transmission time. This enhancement in the sensing time

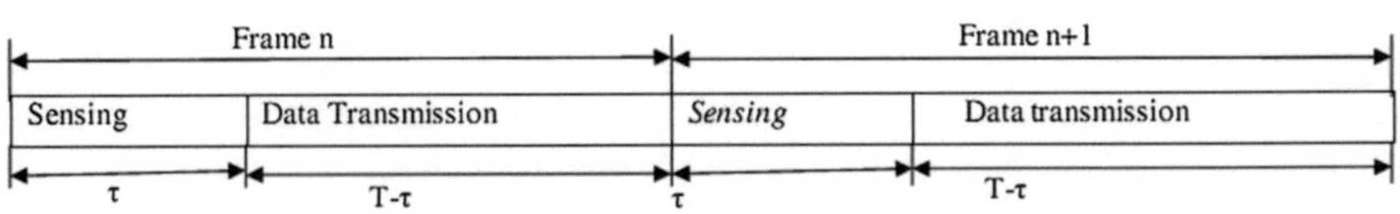

Fig. 7.1 The frame structure of conventional sensing-based spectrum sharing approach for cognitive radio networks

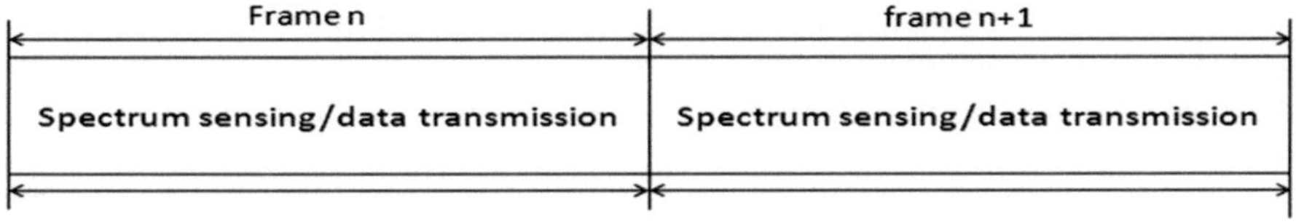

Fig. 7.2 The frame structure of the proposed approach [27, 28]

provides better performance in the form of decreased false alarm probability as well as increased detection probability and, consequently, significant enhancement in the throughput of cognitive radio user [28]. This approach determines the action of cognitive radio user in the next frame which is based on the sensing decision of the previous frame. The cognitive user adapts its transmit power in the next frame to stop transmission, in case the sensing result of the previous frame shows PU transmission, however the cognitive user resumes transmission if the PU is not transmitting. In this way harmful interference to the PUs can be avoided. For example, as shown in Fig. 7.2, the spectrum sensing which has been performed during the frame n is utilized for data transmission in frame $(n + 1)$. The cognitive user during frame $(n + 1)$ transmits data in case sensing in the frame n shows idle PU, and vice versa. However, a potential problem arises if, during the data transmission time of the cognitive user, that is, suppose during frame $(n + 1)$, the PU becomes active from previous frame's (frame n) idle state, but the cognitive user is not aware of this fact since the current frame's (frame $n + 1$) sensing results are not present. Therefore, based on the spectrum sensing decision of frame n, the cognitive user transmits, which results collision of the cognitive user's frame $(n + 1)$ with the PU's frame and all the data carried in the collided frame will be lost. This problem, until recently, was discussed only for case where the spectrum sensing and transmission are performed alternatively [29]. In this chapter, we have focused on this problem.

The remainder of the chapter is organized as follows. Section 7.2 describes the system model of the cognitive user and problem formulation. A novel approach for the cognitive user's data transmission is proposed with the frame structure for data transmission in Sect. 7.2. Further, in Sect. 7.3, throughput and data loss rate for the proposed scheme are discussed and Sect. 7.4 shows the numerically simulated results. Finally, Sect. 7.5 concludes the chapter.

7.2 System Model and Problem Formulation

We have considered a primary user network utilized by the cognitive user. The cognitive user performs an initial spectrum sensing on the allocated spectrum band to determine the current status of the channel. Based on the sensing result, the secondary transmitter communicates when the sensing result detects an absence of primary user data transmission on that spectrum band. If primary user is transmitting, the cognitive user avoids transmission. The secondary receiver decodes the signal sent by the secondary transmitter, strips it away from the received signal and uses the remaining signal to perform spectrum sensing so that the action of the cognitive radio user in the next frame can be determined. Further, at the end of the frame, if the status of primary user has changed after the initial spectrum sensing, the cognitive user adapts its transmit power based on the sensing decision to avoid causing harmful interference to the primary users and to minimize the cognitive user data loss rate.

7.2.1 Cognitive Receiver Structure

The cognitive radio receiver structure for the cognitive radio user in which the spectrum sensing and data transmission are performed simultaneously is shown in Fig. 7.3. The received signal at the cognitive radio user is given by [28]:

$$y = \theta s_{\mathrm{p}} + h_{\mathrm{s}} x_{\mathrm{s}} + w(t) \tag{7.1}$$

where θ denotes the actual status of the frequency band (θ = 1 if the band is active and θ = 0 when it is idle) and s_{p} denotes the signal received from the PU on that frequency band. Further, h_{s} denotes the channel gain between the cognitive transmitter and the cognitive receiver, x_{s} represents the signal from the cognitive transmitter and w(t) denotes the additive white Gaussian noise (AWGN). The received signal is initially passed through the decoder as shown in Fig. 7.3, which decodes the signal from the secondary transmitter. The signal from the cognitive transmitter is cancelled out from the aggregate received signal y, given in Eq. (7.1), therefore the remaining signal is:

$$\tilde{y} = \theta s_{\mathrm{p}} + w(t) \tag{7.2}$$

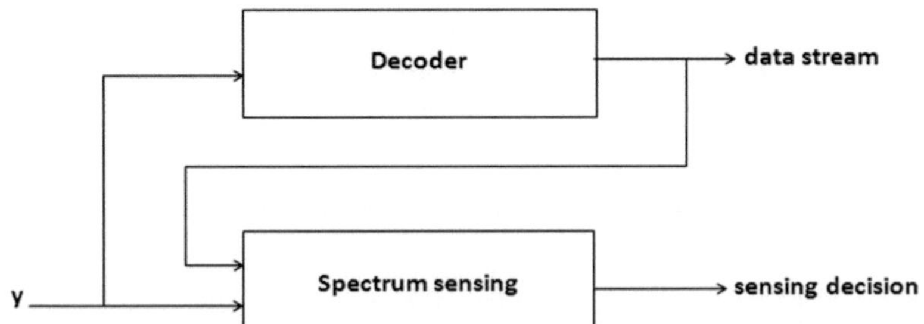

Fig. 7.3 The receiver structure of cognitive user for the frame structure shown in Fig. 7.2 [28]

This signal represented in Eq. (7.2) is used for spectrum sensing. This is the signal that the cognitive receiver would receive if the cognitive transmitter ceases transmission.

7.2.2 Frame Structure

In the frame structure shown in Fig. 7.2, sensing and data transmission are performed simultaneously for whole frame duration T, so that throughput is maximized as compared to the conventional frame structure shown in Fig. 7.1. The frame structure shown in Fig. 7.2 has following advantages:

(1) It enables the detection of very weak PU signals, the detection of which under the frame structure of Fig. 7.1 would significantly reduce the data transmission time due to its large sensing time requirement.
(2) It leads to an improved detection probability, thus better protection of the PUs from harmful interference.
(3) It results in significantly reduced false alarm probability, which enables a better utilization of the available unused spectrum.
(4) The computation of optimal sensing time, as in the conventional frame structure [26], is no longer an issue, since it is maximized and is equal to the frame duration.
(5) Continuous spectrum sensing can be achieved under the proposed cognitive radio system, which ensures better protection of the primary networks.

Apart from the aforementioned advantages of the frame structure shown in Fig. 7.2, there is a technical problem in this frame structure, because the sensing result of the previous frame is used by the next frame for making its data transmission decision on the sensed spectrum. If during the transmission in a frame, the primary user changes state (for example, if θ changes from 0 to 1), the cognitive user's frame collides with the primary user's data and all the data carried in the collided frame will be lost. To reduce data loss due to collision, we propose a novel frame structure (Fig. 7.4), which is a modified form of Fig. 7.2. In this modified frame structure, sensing and data transmission are performed simultaneously; however, instead of sending one long block of data in each frame as shown in Fig. 7.2, we send multiple shorter blocks (sub-frames) of data as shown in Fig. 7.4, and the data transmission is for the whole frame duration T. In addition, the sensing results of the previous frame and current frame that is computed until the start of the sub-frame, are both utilized for transmitting a sub-frame of a frame as shown in Fig. 7.4. The sensing results computed throughout the previous frame until the particular sub-frame in the current frame, are both used to either stop or resume cognitive user's data transmission. The previous frame's whole sensing duration (T ms) results in high detection and low false alarm probabilities and the current

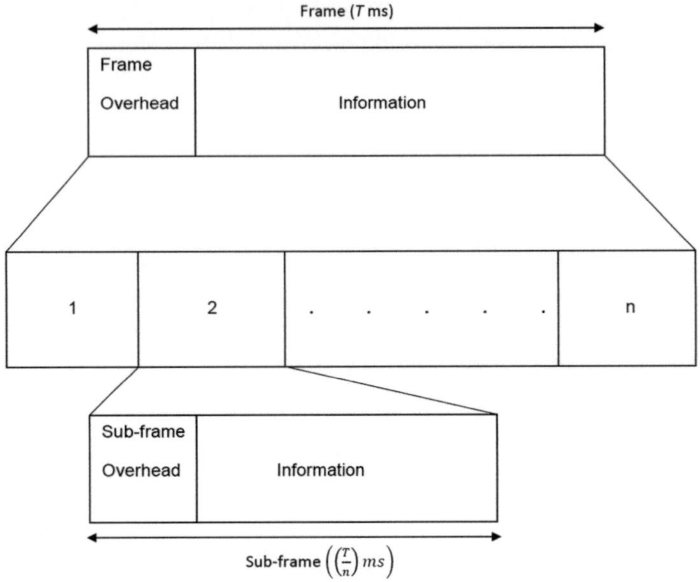

Fig. 7.4 The detailed frame structure of the proposed scheme

frame's sensing duration through the start of the next sub-frame reduces the data loss rate if the PU's presence has been detected in that duration of the current frame.

Now, if during the transmission in a frame, the PU changes from idle to active (θ changes from 0 to 1) and its presence is detected by sensing in the frame, only the data carried in the collided sub-frame of that frame will be lost and all the earlier sub-frames are transmitted successfully, along with avoiding transmission of next sub-frames to prevent collision with primary user. Therefore, it is required that shorter sub-frame durations ((T/n) *ms,* where n is the number of sub-frames in a frame) will reduce the data loss rate and collision with primary users. Some control information is required to be transmitted along with useful data for each frame's successful delivery to its receiver as shown in Fig. 7.4, where frame overhead specifies the control information. In the proposed scheme, we divide each frame into multiple sub-frames, and have to add overhead with each sub-frame of approximately of the same amount as has been added in the single long frame. Therefore, the proposed scheme decreases the data loss rate at the cost of increased overhead. The cognitive user needs to specify how much data loss it can tolerate. Further, the effective throughput, which is the throughput of useful data, that is, information without overhead, and data loss rate both decrease as we increase the number of sub-frames; therefore, there is a tradeoff between the number of sub-frames and effective throughput. Thus, the sensing result of the previous frame and the same frame that is calculated up until the current sub-frame has removed the limitations of Fig. 7.2, in which only previous frame's sensing result is applied to current frame. Further, this method is an efficient method for cognitive user data transmission as compared to that of conventional cognitive user data transmission with alternate sensing and data transmission time.

7.3 Throughput Analysis

There are two probabilities of interest defined under the hypothesis [30, 31] model, which are used for the spectrum sensing:

(1) probability of detection (P_d), which is defined as the probability of the algorithm correctly detecting the presence of primary signal under hypothesis H_1 [30], and
(2) probability of false alarm (P_f), which is defined as the probability of the algorithm falsely declaring the presence of the PU's signal under hypothesis H_0, [30].

As we discussed earlier, from the PU's perspective, if the probability of detection is high, the primary receiver protection is better. However, from the cognitive user's perspective, if the probability of false alarm is low, there are more chances of free spectrum being correctly detected and used by cognitive users. Obviously, for a good detection algorithm, the probability of detection should be as high as possible while the probability of false alarm should be as low as possible. $P(H_0)$ and $P(H_1)$ are the probabilities that a frequency band is idle and active, respectively. Therefore, with the given target probability of detection P_d, for which the PUs are defined as being sufficiently protected, the probability of false alarm is defined as follows [28]:

$$P_f = Q\left(\sqrt{2\gamma+1}Q^{-1}(\bar{P}_d) + \sqrt{\tau f_s \gamma}\right) \tag{7.3}$$

On the other hand, for a target probability of false alarm P_f, the detection probability is given by [28]:

$$P_d = Q\left(\frac{1}{\sqrt{2\gamma+1}}\left(Q^{-1}(\bar{P}_f) - \sqrt{\tau f_s \gamma}\right)\right) \tag{7.4}$$

In Eqs. (7.3) and (7.4), γ is the signal-to-noise ratio (SNR) of the PU's signal at the secondary detector, f_s is the sampling frequency. N is the number of samples used for the spectrum sensing by cognitive user where $N = \tau f_s$. Energy detection is the most popular spectrum sensing technique and its test statistics for received signal y is given as follows:

$$T(y) = \frac{1}{N}\sum_{n=1}^{N}|y(n)|^2$$

where $T(y)$ is a random variable whose value determines the presence and absence of PU by cognitive user's sensing technique. Q is the complementary unit Gaussian distribution function and is defined as [29]:

$$Q(x) = \int_{x}^{\infty} \frac{1}{\sqrt{2\pi}} exp\left(-\frac{u^2}{2}\right) du \tag{7.5}$$

and,

$$Q^{-1}(x) = 1 - Q(x) \tag{7.6}$$

Also,

$$P(H_0) = 1 - P(H_1) \tag{7.7}$$

Therefore, for the conventional scheme the throughput of a cognitive radio user is given by the expression [26]:

$$Th_{\text{conv.}} = \frac{T-\tau}{T} [P(H_1)(1 - P_d) \log_2\left(1 + \frac{SNR_s}{1 + SNR_p}\right) + P(H_0)$$
$$(1 - P_f) \log_2(1 + SNR_s)] \tag{7.8}$$

Equation (7.8) represents the throughput for the frame structure of Fig. 7.1. SNR_s is the SNR of the secondary link, that is, the SNR from cognitive transmitter to cognitive receiver, and SNR_p is the SNR of the primary user signal at the receiver of the cognitive transmission link. The frame structure of Fig. 7.1, whose throughput is given by Eq. (7.8), disrupts the continuous communication in the spectrum sharing cognitive radio network and decreases throughput by the factor of $\left(\frac{T-\tau}{T}\right)$. However, for the proposed approach in which sensing and transmission are performed simultaneously, the expression for the throughput is given by:

$$Th_{\text{prop.}} = P(H_1)(1 - P_d) \log_2\left(1 + \frac{SNR_s}{1 + SNR_p}\right) + P(H_0)(1 - P_f) \log_2(1 + SNR_s) \tag{7.9}$$

From Eq. (7.9), it is clear that throughput is not decreased by the amount $\left(\frac{T-\tau}{T}\right)$ as in the conventional approach, because sensing and transmission are performed simultaneously. Thus, by comparing Eqs. (7.8) and (7.9), it is clear that throughput for the frame structure of Figs. 7.2 and 7.4 is greater than that represented in Fig. 7.1. Further, the effective throughput of a single frame in the proposed scheme, which is defined as the throughput of useful information, is given by:

$$Th_{\text{eff}} = P(H_1)(1 - P_{\text{d}})\log_2\left(1 + \frac{SNR_{\text{s}}}{1 + SNR_{\text{p}}}\right) + P(H_0)(1 - P_{\text{f}})\log_2(1 + SNR_{\text{s}}) - \frac{x \times n}{T} \times \left(P(H_1)(1 - P_{\text{d}})\log_2\left(1 + \frac{SNR_{\text{s}}}{1 + SNR_{\text{p}}}\right) + P(H_0)(1 - P_{\text{f}})\log_2(1 + SNR_{\text{s}})\right) \tag{7.10}$$

where x denotes the overhead duration, n and T denote the number of sub-frames in a frame and frame duration, respectively. Since there is single frame in the frame structure proposed in [28], therefore, in this case we have $n = 1$ that is there is single information block; however in our proposed scheme we have multiple information blocks in each frame. Furthermore, the data loss rate for the proposed scheme in a single frame is given by:

$$Datalossrate\,(\%) = \frac{1}{n} \times 100 \tag{7.11}$$

From Eq. (7.11), it is clear that higher the number of sub-frames in a frame, the lower the data loss rate. Hence, the proposed scheme with multiple sub-frames has less data loss in comparison to that of the earlier scheme proposed by the researchers in [28].

7.4 Simulation Results

In this section, we present the simulation results of the proposed frame structure and compared them with those of the earlier frame structures proposed. For the simulation, the frame duration is set to $T = 100\ ms$ and the probability for the active frequency band is $P(H_1) = 0.2$, therefore from Eq. (7.7), $P(H_0) = 0.8$. The received SNR from the secondary transmitter is $SNR_{\text{s}} = 20\,dB$, and the bandwidth of the channel and the sampling frequency f_{s} are assumed to be 6 MHz. Overhead duration is taken to be *10 ms*, that is, $x = 10\ ms$. In this section, we have numerically simulated the throughput of the cognitive user for the conventional and proposed frame structure by taking different values of SNR from the primary user. With the help of Fig. 4 of [28], we have compared the results for conventional and proposed approaches for low SNR regions. In the proposed scheme, the sensing and data transmission both have been performed simultaneously as also done by Stotas and Nallanathanin [10], therefore the throughput of proposed scheme as shown in Fig. 7.5 for different values of the PU's SNR and the target probability of detection 0.9999 ($P_{\text{d}} = 99.99\%$) is verified with that of Fig. 5 of [28]. Figure 7.5 reveals that the throughput of the cognitive user for higher values of the PU's SNR is much less than that for low values of the received SNR.

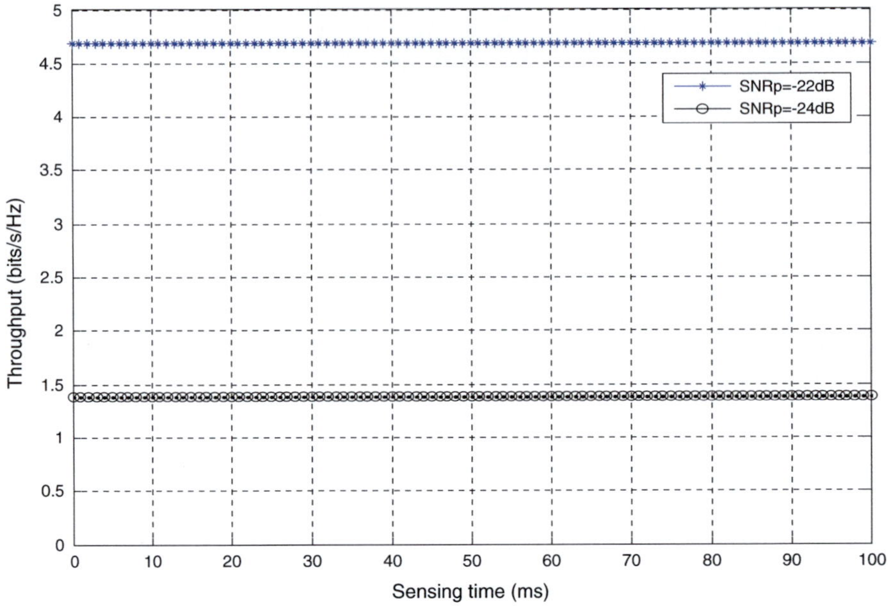

Fig. 7.5 The throughput (bits/second/Hz) of cognitive user versus sensing time (ms) of the current frame for different values of the SNR from the PU

Figure 7.6 compares the effective throughputs of the proposed scheme and the earlier reported scheme [28] for different SNRs from the primary user. It is clear from Fig. 7.6 that the earlier scheme, which has a single sub-frame, allows constant throughput. However, the throughput of the proposed scheme decreases with increases in the number of sub-frames due to the increase in the amount of overhead. However, the higher throughput of the earlier scheme [28] is at the cost of higher data loss if the primary user resumes its transmission in the current frame, which is shown in Fig. 7.7. Figure 7.7 represents the percentage of data loss with respect to the time at which the primary user comes back in a frame, when there are 4 sub-frames in the frame for the proposed scheme. For example, consider in the proposed scheme that the PU resumes transmission during the first sub-frame of *100 ms* frame duration; then only the first sub-frame is lost and the remaining three sub-frames avoid tranmission until the primary user becomes inactive. In this example, only 25% of the data is lost in the proposed scheme. However in the earlier reported scheme [28], the whole frame of duarion *100 ms* is lost when the primary user resumes transmission during this frame. Furthermore, only a single sub-frame is lost in the proposed scheme, irrespective of the time at which the primary user comes back into transmission, which reduces the data loss rate of the proposed scheme in comparison to that of earlier scheme, as shown in Fig. 7.7.

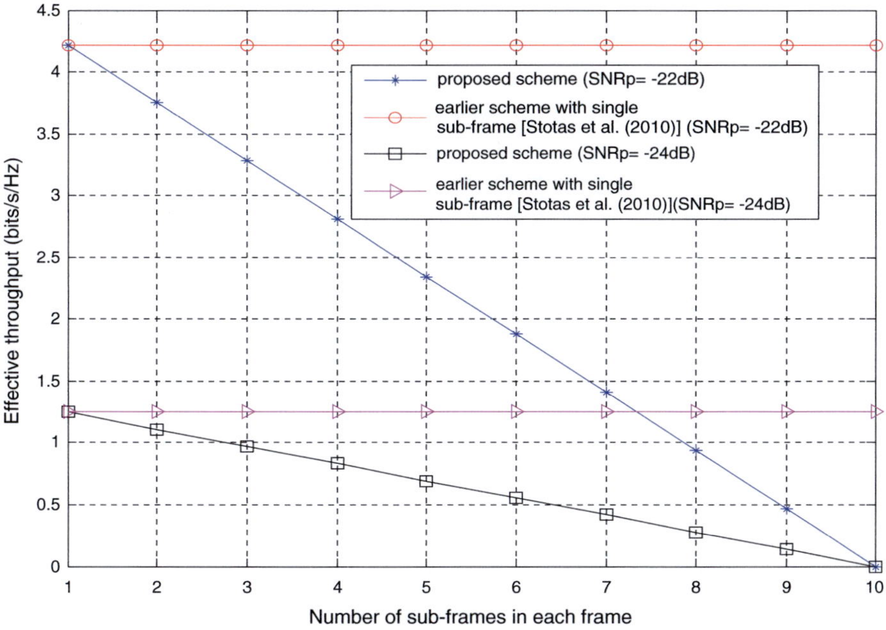

Fig. 7.6 The response of the effective throughput with number of sub-frames in the proposed scheme and the earlier scheme [28] having a single sub-frame, in 100 ms frame duration, 10 ms overhead and different SNRs from primary users

Figure 7.8 represents the throughput versus $P(H_0)$, that is, the probability of frequency band being idle for the chosen target probability of detection 99.99%. It is clear that as the probability of a frequency band being idle increases, the throughput of the cognitive users also increases and is greater for the proposed approach as compared to that of the conventional approach, where the sensing and data transmission are performed alternatively in a frame [26]. Figure 7.9 shows the variation of throughput of the cognitive users with the target probability of detection for the proposed scheme with different SNRs from the primary user. It is further depicted in Fig. 7.9 that as the target probability of detection increases, the throughput of the cognitive user decreases slightly, however in the conventional frame structure the throughput degradation rate is high with a slight change in the target probability of detection, as is clear from Fig. 6 of [28]. Thus, in the proposed approach, we have obtained high protection of data in a frame against the interference for the PU, and we have significantly enhanced the throughput of cognitive users, simultaneously.

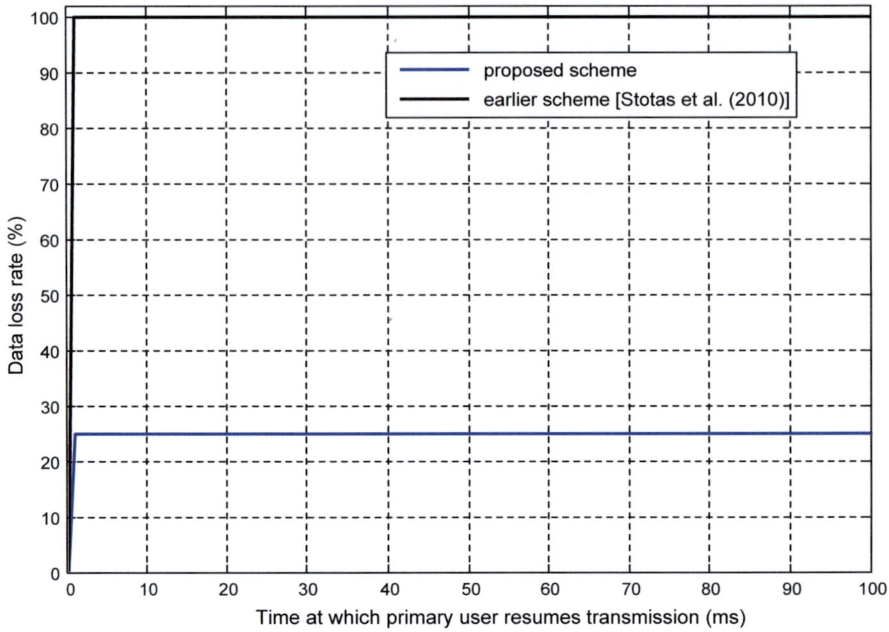

Fig. 7.7 The response of the data loss rate in the proposed and earlier scheme [28] with the time at which primary user resumes transmission in a 100 ms frame duration with 4 sub-frames in the proposed scheme

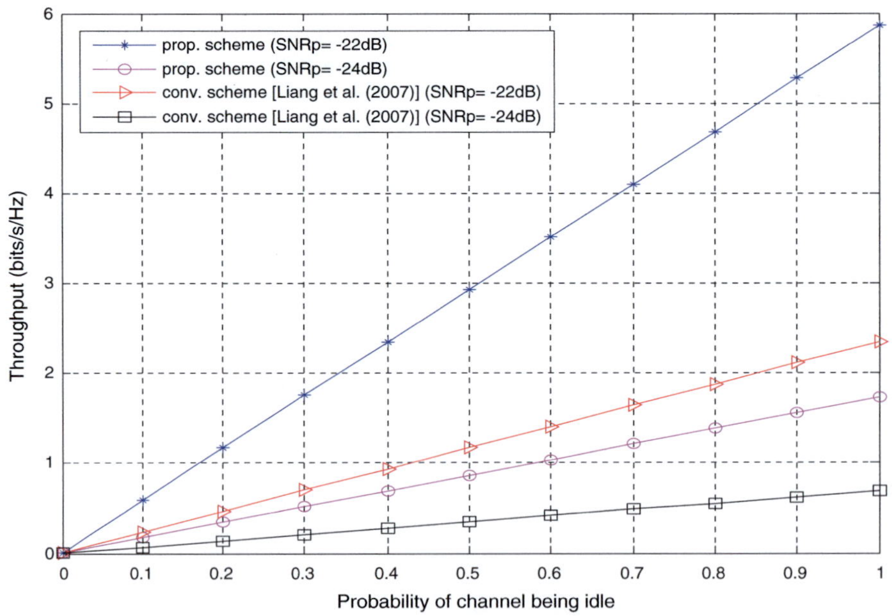

Fig. 7.8 The throughput (bits/second/Hz) of the cognitive users versus probability of the primary user being idle $P(H_0)$ for different values the SNR from the primary user

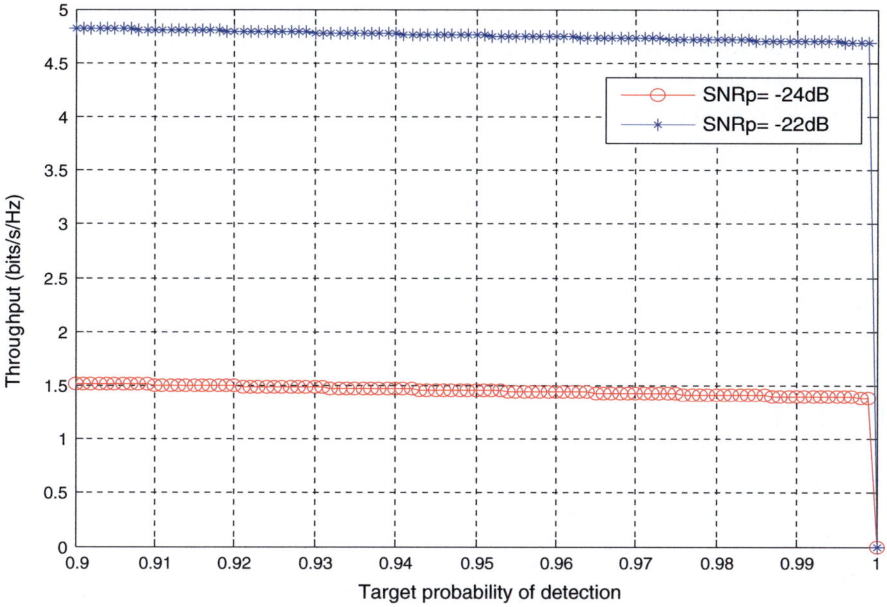

Fig. 7.9 The throughput (bits/second/Hz) of the cognitive user versus target probability of detection for $SNR_p = -22\,dB$

7.5 Summary

This chapter dealt with the throughput maximization of the cognitive radio user with reduced data loss rate. We have compared numerically simulated results of throughput of the cognitive user for the proposed approach with that of the earlier approaches. The simulation results reveal a significant improvement in the throughput of the cognitive user for the proposed approach. The method of simultaneous sensing and data transmission presented in [28] has a drawback that is solved by an enhancement in the frame structure discussed in this chapter. The frame structure enhancement decreases the data loss rate in comparison to that of the earlier scheme. Thus, the data loss rate has been minimized by dividing the transmission time into small segments consisting of multiple sub-frames in a frame. Moreover, the effect of dividing a frame into multiple sub-frames on the effective throughput was also shown and the number of sub-frames versus effective throughput tradeoff problem was discussed. Therefore, in the proposed frame structure, primary users are also adequately protected against the harmful interference by the cognitive user's transmission in the same frequency band. To maximize some important metrics for instance transmission/energy efficiency or throughput with respect to the network parameters, spectrum sensing and transmission time are key studies in cognitive radio networks. However, there is always a tradeoff between the sensing time and transmission time. For example, longer sensing times result in higher detection accuracy, but the cognitive user loses transmission opportunities and hence experiences decreases in transmission efficiency. Various reported studies are available to find the best transmission and sensing time to maximize the efficiencies, such as transmission and energy, however the interference and hence the probability of collision was not defined and formulated correctly. To reduce the overhead and increase the network throughput which usually addressed the joint optimization of spectrum sensing and spectrum access is a very challenging task. Application of a broad area of mathematical modeling including traffic parameter estimation, spectrum prediction and optimization algorithms is still in progress [31, 32].

References

1. M.R. Amini, F. Hemati, A. Mirzavandi, Trilateral tradeoff of sensing, transmission, and contention times in a multiuser split-phase CR networks. IEEE Sens. J. **15**(10), 6044–6055 (2015)
2. Q. Zhao, L. Tong, A. Swami, Y. Chen, Decentralized cognitive MAC for opportunistic spectrum access in ad hoc networks: A POMDP framework. IEEE J. Sel. Areas Commun. **25** (3), 589–600 (2007)
3. H. Katayama, H. Masuyama, S. Kasahara, Y. Takahashi, Effect of spectrum sensing overhead on performance for cognitive radio networks with channel bonding. J. Ind. Manage. Optim. **10**(1), 21–40 (2014)
4. G. Hattab, M. Ibnkahla, Multiband spectrum access: Great promises for future cognitive radio networks. Proc. IEEE **102**(3), 282–306 (2014)
5. M.T. Masonta, M. Mzyece, N. Ntlatlapa, Spectrum decision in cognitive radio networks: A survey. IEEE Commun. Surveys Tuts. **15**(3), 1088–1107 (2013)
6. P. Wang, L. Xiao, S. Zhou, J. Wang, Optimization of detection time for channel efficiency in cognitive radio systems, in *Proceedings IEEE Communications and Networking Conference (WCNC)*, pp. 111–115
7. I. Kim, D. Kim, Optimal allocation of sensing time between two primary channels in cognitive radio networks. IEEE Commun. Lett. **14**(4), 297–299 (2010)
8. H. Shokri-Ghadikolaei, Y. Abdi, M. Nasiri-Kenari, Analytical and learning-based spectrum sensing time optimisation in cognitive radio systems. Commun. IET **7**(5), 480–489 (2013)
9. Y. Liu, S. Liang, X. Zhang, A sensing duration optimization algorithm in cognitive radio, Proc. SPIE, **8**, 878417-1–878417-9 (Mar. 2013)
10. A. Balieiro, P. Yoshioka, K. Dias, C. Cordeiro, D. Cavalcanti, Adaptive spectrum sensing for cognitive radio based on multi-objective genetic optimisation. Electron. Lett. **49**(17), 1099–1101 (2013)
11. H. Kim, K.G. Shin, Efficient discovery of spectrum opportunities with MAC-layer sensing in cognitive radio networks. IEEE Trans. Mobile Comput. **7**(5), 533–545 (2008)
12. C. Cormio, K.R. Chowdhury, A survey on MAC protocols for cognitive radio networks. Ad Hoc Netw. **7**(7), 1315–1329 (2009)
13. Y.-C. Liang, Y. Zeng, E.C.Y. Peh, A.T. Hoang, Sensing-throughput tradeoff for cognitive radio networks. IEEE Trans. Wireless Commun. **7**(4), 1326–1337 (2008)
14. A.T. Hoang, Y.-C. Liang, Y. Zeng, Adaptive joint scheduling of spectrum sensing and data transmission in cognitive radio networks. IEEE Trans. Commun. **58**(1), 235–246 (2010)
15. W.-Y. Lee, I.F. Akyildiz, Optimal spectrum sensing framework for cognitive radio networks. IEEE Trans. Wire. Commun. **7**(10), 3845–3857 (2008)

16. G. Noh, J. Lee, H. Wang, S. Kim, S. Choi, D. Hong, Throughput analysis and optimization of sensing-based cognitive radio systems with Markovian traffic. IEEE Trans. Veh. Technol. **59** (8), 4163–4169 (2010)
17. K. Chang, B. Senadji, Spectrum sensing optimisation for dynamic primary user signal. IEEE Trans. Commun. **60**(12), 3632–3640 (2012)
18. S. Stotas, A. Nallanathan, On the throughput and spectrum sensing enhancement of opportunistic spectrum access cognitive radio networks. IEEE Trans. Wire. Commun. **11**(1), 97–107 (2012)
19. K. Navaie, On the interference management in wireless multi-user network. Telecommun. Syst. **46**(2), 135–148 (2011)
20. L. Tang, Q. Chen, G. Wang, X. Zeng, H. Wang, Opportunistic power allocation strategies and fair subcarrier allocation in OFDM-based cognitive radio networks. Telecommun. Syst. **52**(4), 2071–2202 (2013)
21. Y. Zhang, C. Leung, Cross-layer resource allocation for real-time services in OFDM-based cognitive radio systems. Telecommun. Syst. **42**(1–2), 97–108 (2009)
22. S. Kapoor, S.V.R.K. Rao, G. Singh, Opportunistic spectrum sensing by employing matched filter in cognitive radio network, in *Proceedings International Conference on Communication System and Networking Technology (CSNT-2011),* India, 3–5 June 2011, pp. 580–583
23. S.-V.-R.-K. Rao, G. Singh, Wavelet based spectrum sensing techniques in cognitive radio. Procedia Eng. **38**, 880–888 (2012)
24. S.-M. Kay, *Fundamentals of Statistical Signal Processing, Volume II: Detection Theory.* (Prentice Hall, Englewood Cliffs, 1998)
25. H.-V. Poor, *An Introduction to Signal Detection and Estimation,* 2nd edn. (Springer, New York 1998)
26. Y.-C. Liang, Y. Zeng, E. Peh, and A.-T. Hoang, Sensing throughput trade-off for cognitive radio networks, in *Proceedings of IEEE International Conference on Communications (ICC 2007),* Glasgow, June 2007, pp. 5330–5335
27. S. Stotas, A. Nallanathan, On the throughput maximization of spectrum sharing cognitive radio networks, in *Proceedings of IEEE. Global Telecommunications Conference (GLOBECOM 2010),* Miami, FL, 6–10 Dec 2010, pp. 1–5
28. S. Stotas, A. Nallanathan, Overcoming the sensing throughput tradeoff in cognitive radio networks, in *Proceedings of IEEE International Conference on Communication (ICC)*, Cape Town, 23–27 May 2010, pp. 1–5
29. S.-S. Tzeng and C.-W. Huang, Effective throughput maximization for in-band sensing and transmission in cognitive radio networks. Wirel Netw. **17**(4), 1025–1029 (2011)
30. A. Ghasemi, E. S. Sousa, Collaborative spectrum sensing for opportunistic access in fading environment, in *Proceedings of IEEE Int. Symposium on New Frontiers in Dynamic Spectrum Access Networks*, Baltimore, MD, USA, November 2005, pp. 131–136
31. P. Thakur, A. Kumar, S. Pandit, G. Singh, S.N. Satashia, Advanced frame structures for hybrid spectrum access strategy in cognitive radio communication systems. IEEE Commun. Lett. **21**(2), 410–413 (2017)
32. P. Thakur, A. Kumar, S. Pandit, G. Singh, S.N. Satashia, Performance analysis of high-traffic cognitive radio communication system using hybrid spectrum access, prediction and monitoring techniques. Wireless Networks, Published Online, (2017). DOI: 10.1007/s11276-016-1440-7

Chapter 8
Capacity Limits Over Fading Environment with Imperfect Channel State Information for Cognitive Radio Networks

8.1 Introduction

An important goal in the design of next generation wireless networks is the seamless connectivity of all types of devices/networks and communication protocols that have been established by the regulatory authorities to fulfill consumer requirements. To yield high data rates and low latencies, the next generation communication network must incorporate revolutionary approaches to using the radio spectrum, which is becoming very scarce as almost all the frequencies are already occupied by the licensed users [1]. A proposed solution to this apparent spectrum scarcity is cognitive radio with spectrum sharing, an approach that can enhance spectrum utilization. Spectrum sharing requires that the primary (licensed spectrum) users allow secondary (unlicensed) users to access the licensed spectrum, which is feasible provided that the primary users (PUs) are protected from the interference of cognitive users (CUs) [2]. Cognitive radio communication technology reduces the crowd of unlicensed users by using the large portion of unused licensed spectrum, within contextual constraints regarding time and location [2, 3]. Various standardization bodies have developed different standards for cognitive radio networks and dynamic spectrum access in order to integrate with existing wireless technologies such as IEEE 802.22 standard wireless regional area networks, IEEE 802.11af standard for TV White Space operations and IEEE 1900 standard for spectrum access networks [4]. Further, design of wireless systems requires the collaborative efforts of various research communities such as communication theory, network engineering, signal processing, game theory, reconfigurable antennas and radio frequency design [5–7]. This chapter elaborates on the following points.

- A simple optimal power allocation scheme for efficient spectrum sharing with imperfect channel state information (CSI) between the CU and PU in the Rayleigh fading environment

S. Pandit and G. Singh, *Spectrum Sharing in Cognitive Radio Networks*,
DOI 10.1007/978-3-319-53147-2_8

- An understanding of the average power consumption of the cognitive transmitter under the joint peak transmit power and peak interference power constraints to achieve the lower limits of ergodic and outage capacities
- The outage probability and consumption of power under the individual peak interference power constraint.

This chapter is structured as follows. We review the related work in Sect. 8.2. Section 8.3 describes the proposed system model and in Sect. 8.4, the power constraint under which the transmission power is allocated to the CU is discussed. In Sect. 8.5 the ergodic and outage capacity under the Rayleigh fading channel of the proposed communication system is evaluated and numerical simulation results are discussed. Finally Sect. 8.6 concludes the work and recommends the future scope.

8.2 Related Work

The spectrum access strategy to provide efficient spectrum allocation to the CU is an important issue in cognitive radio communication network research. Various spectrum sharing approaches are discussed in references [8–10]. In [8], the CSI between CU transmitter and the PU receiver was employed to compute the maximum allowable CU transmit power within interference limits. The authors derived a closed-form mathematical expression for the CU capacity under the peak received interference power constraint and CU transmit power constraint. In [9] the authors explored cooperative communications for spectrum sharing in a cognitive wireless relay network, and a cognitive space-time-frequency coding technique which can opportunistically adjust its coding structure by adapting itself to the dynamic spectrum environment, is also exploited to maximize the spectrum opportunities. In addition, Wang and Zhang [10] investigated an opportunistic spectrum access technique in cognitive radio networks when a decode and forward relay is employed. Two cognitive spectrum access approaches were proposed based on white space modeling, referred to as successive sensing based spectrum access and simultaneous sensing based spectrum access [10].

In addition to efficient spectrum allocation, the energy efficiency of the network is also an emerging concern, therefore several novel power allocation strategies have been proposed by various researchers under different spectrum sharing approaches [11–13]. In [11], the authors studied optimal power allocation strategies to achieve the ergodic as well as outage capacity of the CU fading channel under different types of power constraints. In addition to the interference power constraint of the PU, the transmit power constraint of the CU is also considered because the transmit power and the interference power can be limited either by a peak or an average power constraint. Kim et al. [12] examined the dynamic spectrum sharing problem among the PUs and secondary users (SUs) in a cognitive radio network by considering the scenario where the PUs exhibit on-off behavior and the SUs are able to dynamically measure/estimate sum interference from the PUs at their receiving ends, and solve the problem of fair spectrum sharing among the SUs subject to their quality-of-service (QoS) constraints as well as interference power constraints for the PUs. In [13], a resource allocation framework considering both the interference power constraints for the PUs and QoS constraints for the SUs was presented for a spectrum underlay approach in cognitive wireless networks. The interference from the SUs to PUs was controlled below a tolerable limit. Furthermore, the admission control algorithms were used during high network load conditions which performed jointly with the power control so that QoS requirements of all the admitted SUs were satisfied while keeping the interference to PUs below the tolerable limit. Li [14] proposed an efficient power allocation algorithm for centralized as well as distributed cognitive radio networks, when a pair of the PUs and multiple pairs of CUs are in the network. In [15], effective capacity of the CU link is evaluated under signal-to-interference noise ratio (SINR) and QoS constraints. In [16], the geometric programming approach was used for the optimal power allocation to the CU under different channel conditions in order to compute the secondary link capacity. Parsaeefard and Sharafat [17] proposed an algorithm for distributed uplink power allocation in an underlay cognitive radio network with imperfect CSI and illustrated that the robustness is introduced into the network at the cost of social utility.

Channel capacity is the best performance metric for analyzing any cognitive radio network model, and several capacity notions are expressed for different fading channels such as ergodic capacity for the fast-fading channel and outage capacity for the slow-fading channel [18]. Various researchers have analyzed the capacity limits of the CU link over different fading channels with perfect and imperfect CSI [19, 20]. Rezki and Alouini [19] presented a cognitive radio communication system in which the CU is aware of the instantaneous CSI of the secondary link but knows only the statistics and an estimated version of the secondary transmitter-primary receiver link. The mathematical expression for optimum power profile and ergodic capacity of the secondary link were derived for general fading channels with a continuous probability density function (pdf) under the average and peak transmit power constraints. In [20], the authors analyzed the capacity gains of an opportunistic spectrum sharing channels in the fading environments with perfect and imperfect CSI and derived the ergodic and outage capacities along with their optimum power allocation policies for the Rayleigh flat-fading channels, and provided closed-form expressions for these capacity metrics considering the average received-power constraint. Recently, Farraj and Ekin [21] illustrated that capacity and bit error rates (BERs) are independent of the transmitted power of the CU, however these are affected by the environmental considerations such as shaping parameters. In [22], the authors reported the ergodic sum capacity limits of CU under transmit power and interference power constraints, when a multiple PU network and SU network is present. Son et al. [23] presented power allocation policies in OFDM-based cognitive radio networks under the availability of inter-system (between CU-Tx and PU-Rx) CSI at different capabilities of licensed users, particularly the peak interference power tolerable to the PU and the average interference power tolerable to the PU. For this PU model, two optimization problems were discussed regarding how to maximize the capacity of the CU while maintaining the QoS of the PU, under the assumption that both intra (between CU-Tx and CU-Rx)-and inter-system CSI are fully available. Due to loose cooperation between SU and PU, it may be difficult or even infeasible for the SU to obtain the full inter-system CSI; thus, under partial CSI, the authors also formulated another optimization problem by introducing interference power outage constraints. The extensive numerical results illustrated that the spectral efficiency achieved by the SU with partial inter-system CSI fell within a reasonable range of outage probability. Also, the spectral efficiency achieved by the CU with partial-CSI was less than half of that achieved with full-CSI with a reasonable change of the outage probabilities [23]. If the CU shares the bandwidth of a channel with the PU using dynamic spectrum access techniques, then the outage capacity with N number of multiple carriers has a variance which is N times smaller than that of the single carrier [24]. The CUs support the PU in improving its QoS by using inactive unlicensed users as cooperative relay nodes for the PU [25, 26]. The CU link with high channel gain achieves better channel capacity when multiple CUs share the spectrum with a single PU [27]. The joint congestion control and power control problem via effective network utility maximization with the link outage constraints was explored in [28]. In [29], it was reported that the average transmit power consumption of the CU is significantly greater under the interference temperature constraint. The variations in outage probability under the peak transmit power and peak/average interference power with noise error variance is presented in [30]. Recently, Pandit and Singh [31] achieved significantly more capacity by an adaptive power transmission technique in comparison to that of the adaptive rate and power transmission policy at the cost of BER. In [32], the authors derived the closed-form expressions for the ergodic capacities of the SU with imperfect CSI under the average interference power constraint and peak interference power constraint. It was illustrated that the ergodic capacity of the CU was robust to the channel estimation errors and feedback delay. Further, it was also shown that decreasing the distance between CU-Tx and CU-Rx or increasing the distance between CU-Tx and PU-Rx can mitigate the impact of the imperfect CSI and significantly increase the ergodic capacity of the CU.

Li and Goldsmith [33] studied three types of capacity regions: the ergodic capacity region, the zero-outage capacity region, and the outage capacity region with nonzero outage for the fading broadcast channels and obtained their corresponding optimal resource allocation strategies. In [34], the authors derived an expression for the outage capacity for fading broadcast channels, considering both the transmitter and receivers had perfect CSI, and specified a strategy which bounds the outage probability of different spectrum-sharing techniques for specified required rates of each user. The corresponding optimal power allocation scheme was a multiuser generalization of the threshold-decision rule for a single-user fading channel. The numerical results for different outage capacity regions were obtained for the Nakagami-*m* fading model. Gastpar [35] investigated the behavior of capacity when constraints were placed on the channel output signal. This investigation was motivated by questions arising in spectrum sharing and dynamic spectrum allocation—multiple independent networks share the same frequency band, but are spatially mostly disjoint. Sboui et al. [36] considered a spectrum sharing communication scenario in which the primary and secondary users are communicating simultaneously with their respective destinations using the same carrier frequency. The mathematical expression for both the optimal power profile and ergodic capacity are derived for fading channels, under the average transmit power and instantaneous interference outage constraints. After deriving the expression for capacity considering a noisy version of the cross-link and secondary-link CSI, the authors provided an ergodic capacity generalization, through a unified expression, that encompassed several previously studied spectrum sharing settings. Musavian and Aissa [37] presented the fundamental capacity limits of opportunistic spectrum-sharing channels in the fading environment and derived the fading channel capacity of a CU subject to both the average and peak received-power constraints at the PU receiver. In addition, the mathematical expressions were derived for the capacity and optimum power allocation schemes for three different capacity notions, namely, ergodic, outage, and minimum-rate, considering the flat Rayleigh fading.

8.3 System Model

In thIn the proposed system model, we have consider multiple PUs and a single CU as shown in Fig. 8.1 which transmit data at the same time. The CU shares the spectrum with one of the PUs without affecting its QoS. For interference-free spectrum sharing, the optimal power is allocated to the CU under the joint transmit power and received interference power constraints. We also consider the discrete time flat fading channel where the received signal of CU depends on the transmitted signal, which is mathematically expressed as [18]:

$$y_{ss}(n) = x_{ss}(n)h_{ss}(n) + \sum\nolimits_{i=1}^{n} x_p(n)h_{ps_i}(n) + w_{ss}(n) \tag{8.1}$$

where, n, $h_{ss}(n)$, $h_{sp}(n)$ and $h_{ps_i}(n)$ are the time index, channel gain of the CU link, channel gain between CU-Tx and PU-Rx and i^{th} PU-Tx and CU-Rx, respectively. $h_{ss}(n), h_{sp}(n)$ and $h_{ps_i}(n)$ are the independent and identically distributed (iid) channel gain with exponential distribution. $w_{ss}(n)$ is the zero-mean complex symmetric additive white Gaussian noise (AWGN) (Figure 8.1).

In the proposed system model, the capacity of the CU has been maximized while maintaining the QoS of the PU under the assumption that both the intra-and inter-system channel state information are partially or imperfectly available due to loose cooperation between the CU and PU. The imperfect CSI is provided to CU-Tx by i^{th} PU, which is represented as $\breve{h}_{sp_i}$ (n). Thus, the ergodic capacity and outage capacity have been computed under the imperfect CSI. The CU estimates the channel gain by the minimum mean square error (MMSE) channel estimation

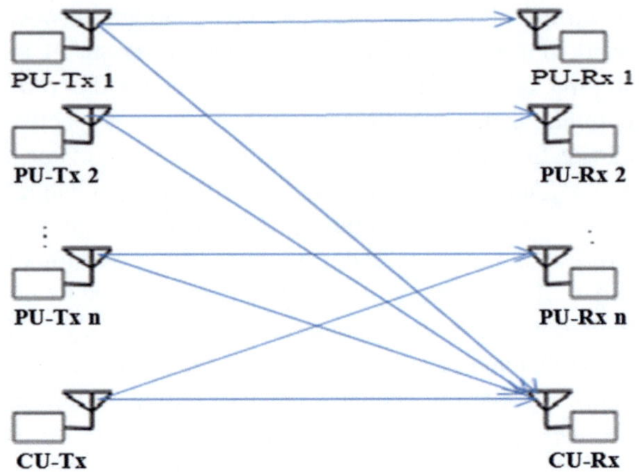

Fig. 8.1 The spectrum sharing system model of a cognitive radio network

technique. The imperfect CSI in the proposed cognitive radio communication system ca be described as follows. The CU-Tx has knowledge only about the average channel gain over all the subchannels, instead of individual channel gain for each subchannel. In order to keep the interference at the PU-Rx below a desired level, in reported literature [38] it is assumed that CU-Tx is fully aware of the channel from the CU-Tx to PU-Rx. However, as compared to the intra-system CSI between the CU-Tx and the CU-Rx, which is relatively easy to obtain, it would be difficult for the CU-Tx to obtain full inter-system CSI because the PU and CU systems are usually loosely coupled (no explicit communication between them). Even if they are tightly coupled, to yield inter-system CSI is difficult for the CU due to the large amount of feedback overhead. Therefore considering only imperfect CSI between the CU and PU seems to be a reasonable approach. Zhang et al. [39] presented a vigorous cognitive beam-forming difficulty with imperfect CSI in multi-input-single-output and multi-input-multi-output environments. There are several studies on the capacity analysis of cognitive radio network with imperfect channel knowledge in flat-fading environment, considering that the CSI obtained by the CU experiences channel estimation error [40]. The channel estimation error is represented as:

$$\tilde{h}_{spi}(n) = h_{spi}(n) - \hat{h}_{spi}(n) \tag{8.2}$$

where, $\tilde{h}_{spi}(n)$ and $\hat{h}_{spi}(n)$ are the zero-mean circularly symmetric complex Gaussian distributed random variable with variance $(\sigma^2/2)$ and $(1-\sigma^2)/2$, respectively. For simplicity, we have ignored the time index. Due to the MMSE estimation characteristics, $\tilde{\mathrm{h}}_{spi}$ and $\hat{\mathrm{h}}_{spi}$ are the uncorrelated channel gain. The channel power gain is given by $|h_{sp}|^2$. The channel power gain of the CU link, between CU-Tx and PU-Rx link and i^{th} PU-Tx and CU-Rx link are represented by $g_{ss}, g_{spi}\, and\, g_{ps}$, respectively.

8.4 Ergodic and Outage Capacity

The ergodic capacity is an effective metric for fast fading channels or delay insensitive applications, where the block of information can experience different fading states of the channel during transmission. However, for the slower fading channels or delay sensitive applications like voice and video transmission, the outage capacity comprises a more suitable metric for the capacity of the system due to the fact that only a cross-section of the channel characteristics is experienced during the transmission period of a block of information.

8.4.1 Power Constraints

We have considered P_{pk} and Q_{pk} as the peak transmit power of CU and peak interference power of PU-Rx, respectively. The instantaneous transmitted power of CU-Tx depends on the channel power gain g_{ss} and the estimated value g_{sp}, which is denoted by $\hat{g}_{sp}$. The instantaneous power at the CU-Tx is expressed as [11]:

$$P(\hat{g}_{sp1}, \hat{g}_{sp2}..\hat{g}_{spn}, g_{ss}) > 0, \forall(\hat{g}_{sp1}..\hat{g}_{spn}, g_{ss}) \quad (8.3)$$

The peak transmit power constraint is represented as [11]:

$$P(\hat{g}_{sp1}, \hat{g}_{sp2}..\hat{g}_{spn}, g_{ss}) \le P_{pk}, \forall(\hat{g}_{sp1}..\hat{g}_{spn}, g_{ss}) \quad (8.4)$$

and the peak interference power constraint is provided as [11]:

$$g_{spi}P(\hat{g}_{sp1}, \hat{g}_{sp2}..\hat{g}_{spn}, g_{ss}) \le Q_{pki}, \forall(\hat{g}_{sp1}..\hat{g}_{spn}, g_{ss}), \quad i = 1..n \quad (8.5)$$

However, the instantaneous peak interference power constraint is valid only for a short time. For this reason, the interference outage concept was introduced by Musavian and Aissa [20]. Thus, the outage interference power constraint is represented as [20]:

$$P_r\{g_{spi}(P(\hat{g}_{sp1}, \hat{g}_{sp2}..\hat{g}_{spn}, g_{ss})) \ge Q_{pki}\} \le P_0 \quad (8.6)$$

where, $P_r\{.\}$ and P_0 are the probability of function and outage interference level, respectively. Therefore, Eq. (8.6) can be simplified as:

$$P(\hat{g}_{sp1}, \hat{g}_{sp2}..\hat{g}_{spn}, g_{ss}) \le \min\left(\frac{Q_{pki}}{\hat{g}_{spi} - \sigma^2 lnP_0}\right), i = 1..n \quad (8.7)$$

In addition to this, the average interference power constraint is expressed as:

$$E\left[g_{spi}P\left(\hat{g}_{sp1},\hat{g}_{sp2}..\hat{g}_{spn},g_{ss}\right)\right]\leq Q_{avgi}, i=1..n \tag{8.8}$$

Due to the imperfect channel state information, the g_{spi} is unknown. Therefore, the estimated value of g_{spi} is expressed as:

$$\hat{g}_{spi}=g_{spi}-\hat{g}_{spi} \tag{8.9}$$

where, $\hat{g}_{spi}$, g_{spi} and $\hat{g}_{spi}$ are the estimated, ideal (true) and estimated error values of the g_{sp}, respectively. Therefore, the average interference power constraint is expressed as [20]:

$$E\left[\hat{g}_{spi}P\left(\hat{g}_{sp1},\hat{g}_{sp2}..\hat{g}_{spn},g_{ss}\right)\right]\leq Q_{avgi}-\sigma^2E\left[P\left(\hat{g}_{sp1},\hat{g}_{sp2}..\hat{g}_{spn},g_{ss}\right)\right] \tag{8.10}$$

For the optimal transmit power computation, the combination of instantaneous CU-Tx power, peak transmit power of CU and outage constraints is represented by R_1 and the combination of instantaneous CU-Tx power, peak transmit power of CU and average interference power constraints is represented by R_2.

8.4.2 Ergodic Capacity

The ergodic capacity is the maximum achievable rate averaged over all the fading states [11]. Therefore, the ergodic capacity of a cognitive link is computed by solving the optimization problem [11]:

$$C_{ergodic}=\max_{P(\hat{g}_{sp1},\hat{g}_{sp2},\ldots\hat{g}_{spn},g_{ss})\in R}E\left[\log_2\left(1+\frac{g_{ss}.P\left(\hat{g}_{sp},g_{ss}\right)}{N_o.B+\sum_{i=1}^{n}g_{ps}*P_i)}\right)\right] \tag{8.11}$$

where E{.} is the expected value and g_{ss},g_{ps} and $\hat{g}_{sp}$ follow the Rayleigh distribution whose probability density function (pdf) is specified as: $e^{-g_{ss}}$, $e^{-g_{ps}}$ and $e^{-\hat{g}_{sp}/(1-\sigma^2)}/(1-\sigma^2)$, respectively [19]. When the multiple PUs are considered, then pdf of the channel power gain between the cognitive transmitter and primary receivers is evaluated as follows.

Let $\hat{g}_{spi}$ (i = 1…n) be iid random variables. It is assumed that the channel gain of a cognitive link is independent from the channel gain between the cognitive transmitter and primary receivers. Therefore, $\hat{g}_{sp}$ is expressed as:

$$\hat{g}_{sp}=\max\left(\hat{g}_{spi}\right)i=1..n$$

Then the cumulative distribution function of $\hat{g}_{sp}$ is expressed as:

$$F_{\hat{g}_{sp}}\left(\hat{g}_{sp}\right) = \prod\nolimits_{i=1}^{n} F_{\hat{g}_{spi}}\left(\hat{g}_{sp}\right) = \left(1 - e^{-\frac{\hat{g}_{sp}}{1-\sigma^2}}\right)^n \tag{8.12}$$

On differentiating Eq. (8.12), pdf of $\hat{g}_{sp}$ is written as:

$$f_{\hat{g}_{sp}}\left(\hat{g}_{sp}\right) = n\frac{e^{-\frac{\hat{g}_{sp}}{1-\sigma^2}}}{1-\sigma^2}\left(1 - e^{-\frac{\hat{g}_{sp}}{1-\sigma^2}}\right)^{n-1} \tag{8.13}$$

In a similar way, the pdf for multiple primary transmitter and cognitive receiver is expressed as:

$$f_{g_{ps}}\left(g_{ps}\right) = ne^{-g_{ps}}\left(1 - e^{-g_{ps}}\right)^{n-1} \tag{8.14}$$

However, both channels are considered as Rayleigh fading channels and the probability density functions of $\widehat{g_{sp}}$ and g_{ss} are represented as $e^{-\hat{g}_{sp}/(1-\sigma^2)}/(1-\sigma^2)$ and $e^{-g_{ss}}$, respectively, as discussed in [19]. N_0 and B are the noise power spectral density at the primary receiver and total available bandwidth, respectively. Therefore, the ergodic capacity of a cognitive link can be maximized by allocating the optimal power to SU-Tx.

8.4.2.1 Optimal Power Allocation Under Peak Transmit Power and Peak Interference Power Constraints

The joint peak transmit power and peak interference power constraints are combined as follows:

$$P(\hat{g}_{sp1}, \hat{g}_{sp2}, \ldots \hat{g}_{spn}, g_{ss}) \leq P\left(min\left(P_{pk}, \frac{Q_{pki}}{\widehat{g_{spi}} - \sigma^2 . \ln P_0}\right)\right)$$

Therefore, to maximize the ergodic capacity, the optimal power allocation to the CU is expressed as follows:

$$P(\hat{g}_{sp1}, \hat{g}_{sp2}, \ldots \hat{g}_{spn}, g_{ss}) = \begin{cases} P_{pk}, & \widehat{g_{sp_i}} \leq \frac{Q_{pki}}{P_{pk}} + \sigma^2 \ln P_0 \\ \frac{Q_{pki}}{\widehat{g_{spi}} - \sigma^2 \ln P_0}, & otherwise \end{cases} \tag{8.15}$$

From Eq. (8.15), we observe that when the outage interference constraint is satisfied, then the CU transmits with peak power, otherwise power has to be reduced according to the channel power gain, error variance and outage constraint.

8.4.2.2 Optimal Power Allocation Under Peak Transmit Power and Average Interference Power Constraints

The optimal power under the peak transmit power and the average interference power constraints are computed by the Lagrangian method [41] as follows:

$$\begin{aligned} L\big(P(\hat{g}_{sp1}, \hat{g}_{sp2}, \cdots \hat{g}_{spn}, g_{ss}), \lambda\big) = E\bigg(&\log_2(1 + P(\hat{g}_{sp1,sp2}, \cdots \hat{g}_{spn}, g_{ss})\frac{g_{ss}}{N_0 B}\bigg) \\ &- \lambda\Big(E\Big(\hat{g}_{\rm sp} P(\hat{g}_{sp1}, \hat{g}_{sp2}, \cdots \hat{g}_{spn}, g_{ss})\Big) \\ &- Q_{av} + \sigma^2 E(P(\hat{g}_{sp1}, \hat{g}_{sp2}, \cdots \hat{g}_{spn}, g_{ss})) \end{aligned} \tag{8.16}$$

For a particular fading state, Eq. (8.16) can be represented as:

$$\begin{aligned} &\max_{P(\hat{g}_{\rm sp}, g_{ss})} \log_2\left(1 + \frac{P(\hat{g}_{sp1}, \hat{g}_{sp2}, \cdots \hat{g}_{spn}, g_{ss}) g_{ss}}{\sum_{i=1}^{n} g_{psi} * P_i + \mathrm{N_0 B}}\right) - \lambda(\hat{g}_{\rm sp} P(\hat{g}_{sp1}, \hat{g}_{sp2}, \cdots \hat{g}_{spn}, g_{ss}) \\ &- Q_{av} + P(\hat{g}_{sp1}, \hat{g}_{sp2}, \cdots \hat{g}_{spn}, g_{ss})) - \mu\big(P(\hat{g}_{sp1}, \hat{g}_{sp2}, \cdots \hat{g}_{spn}, g_{ss}) - P_{pk}\big) \\ &+ v P(\hat{g}_{sp1}, \hat{g}_{sp2}, \cdots \hat{g}_{spn}, g_{ss}) \end{aligned} \tag{8.17}$$

s.t $P(\hat{g}_{sp1}, \hat{g}_{sp2}, \cdots \hat{g}_{spn}, g_{ss}) \geq 0$, $P(\hat{g}_{sp1}, \hat{g}_{sp2}, \cdots \hat{g}_{spn}, g_{ss}) \leq P_{pk}$.

The dual function of Eq. (8.17) is represented as:

$$\begin{aligned} L\big(P(\hat{g}_{sp1}, \hat{g}_{sp2}, \cdots \hat{g}_{spn}, g_{ss}), \lambda, \mu, v\big) = \log\bigg(&1 + \frac{P(\hat{g}_{sp1}, \hat{g}_{sp2}, \cdots \hat{g}_{spn}, g_{ss})}{N_0 B}\bigg) \\ &- \lambda\big(\hat{g}_{\rm sp} P(\hat{g}_{sp1}, \hat{g}_{sp2}, \cdots \hat{g}_{spn}, g_{ss}) - Q_{av} + \sigma^2 P(\hat{g}_{sp1}, \hat{g}_{sp2}, \cdots \hat{g}_{spn}, g_{ss})\big) \\ &- \mu\big(P(\hat{g}_{sp1}, \hat{g}_{sp2}, \cdots \hat{g}_{spn}, g_{ss}) - P_{pk}\big) + v P(\hat{g}_{sp1}, \hat{g}_{sp2}, \cdots \hat{g}_{spn}, g_{ss}) \end{aligned} \tag{8.18}$$

By using Karush-Kuhn-Tucker (KKT) conditions, the optimal power is computed as:

$$\begin{aligned} \frac{\partial L\big(P(\hat{g}_{sp1}, \hat{g}_{sp2}, \cdots \hat{g}_{spn}, g_{ss}), \lambda, \mu, v\big)}{\partial P(\hat{g}_{sp1}, \hat{g}_{sp2}, \cdots \hat{g}_{spn}, g_{ss})} = &\frac{g_{ss}}{\sum_{i=1}^{n} g_{psi} * P_i + \mathrm{N_0 B} + P(\hat{g}_{sp1}, \hat{g}_{sp2}, \cdots \hat{g}_{spn}, g_{ss}) g_{ss}} \\ &- \lambda\Big(\hat{g}_{\rm spi} + \sigma_p^2\Big) - \mu + v = 0 \end{aligned} \tag{8.19}$$

$$\mu\big(P(\hat{g}_{sp1}, \hat{g}_{sp2}, \cdots \hat{g}_{spn}, g_{ss}) - P_{pk}\big) = 0 \tag{8.20}$$

$$v P\big(\hat{g}_{sp1}, \hat{g}_{sp2}, \cdots \hat{g}_{spn}, g_{ss}\big) = 0 \tag{8.21}$$

From Eq. (8.19), we get:

$$P\left(\hat{g}_{sp1},\hat{g}_{sp2},\ldots\hat{g}_{spn},g_{ss}\right)=\frac{K}{\lambda\left(\hat{g}_{\mathrm{spi}}+\sigma^2\right)+\mu-v}-\frac{\sum_{i=1}^{n}g_{psi}*P_i+\mathrm{N_0B}}{g_{ss}} \tag{8.22}$$

If we consider, $P(\hat{g}_{sp1},\hat{g}_{sp2},\ldots\hat{g}_{spn},g_{ss})<P_{pk}$, it is possible only if $\hat{g}_{\mathrm{sp}}\geq\frac{K}{\lambda\left(P_{pk}+\frac{\sum_{i=1}^{n}g_{psi}*P_i+\mathrm{N_0B}}{g_{ss}}\right)}-\sigma^2$, so it contradicts the assumption. Therefore,

$$P(\hat{g}_{sp1},\hat{g}_{sp2},\ldots\hat{g}_{spn},g_{ss})=P_{pk}$$

if

$$\hat{g}_{\mathrm{sp}}\leq\frac{K}{\lambda\left(P_{pk}+\frac{\sum_{i=1}^{n}g_{psi}*P_i+\mathrm{N_0B}}{g_{ss}}\right)}-\sigma^2.$$

Suppose, if $P(\hat{g}_{sp1},\hat{g}_{sp2},\ldots\hat{g}_{spn},g_{ss})>0$ when $\hat{g}_{\mathrm{sp}}\geq\frac{K}{\lambda\left(P_{pk}+\frac{\sum_{i=1}^{n}g_{psi}*P_i+\mathrm{N_0B}}{g_{ss}}\right)}$ $-\sigma^2$ from Eq. (8.21), $v=0$, then Eq. (8.22) becomes:

$P(\hat{g}_{sp1},\hat{g}_{sp2},\ldots\hat{g}_{spn},g_{ss})=\frac{K}{\lambda(\hat{g}_{\mathrm{spi}}+\sigma^2)+\mu}-\frac{\sum_{i=1}^{n}g_{psi}*P_i+\mathrm{N_0B}}{g_{ss}}$ then $P(\hat{g}_{sp1},\hat{g}_{sp2},\ldots$ $\hat{g}_{spn},g_{ss})>0$ which results in $\frac{K}{\lambda(\hat{g}_{\mathrm{spi}}+\sigma^2)+\mu}-\frac{\sum_{i=1}^{n}g_{psi}*P_i+\mathrm{N_0B}}{g_{ss}}>0$ since $\mu\geq0$,

$$\frac{K}{\lambda\left(\hat{g}_{\mathrm{spi}}+\sigma^2\right)}-\frac{\sum_{i=1}^{n}g_{psi}*P_i+\mathrm{N_0B}}{g_{ss}}>\frac{K}{\lambda\left(\hat{g}_{\mathrm{spi}}+\sigma^2\right)+\mu}-\frac{\sum_{i=1}^{n}g_{psi}*P_i+\mathrm{N_0B}}{g_{ss}}>0,$$

Therefore, $P\left(\hat{g}_{sp1},\hat{g}_{sp2},\ldots\hat{g}_{spn},g_{ss}\right)=0$ if $\hat{g}_{\mathrm{spi}}\geq\frac{Kg_{ss}}{\lambda(\sum_{i=1}^{n}g_{psi}*P_i+\mathrm{N_0B})}-\sigma^2$. and

$$P\left(\hat{g}_{sp1},\hat{g}_{sp2},\ldots\hat{g}_{spn},g_{ss}\right)=\frac{K}{\lambda(\hat{g}_{\mathrm{spi}}+\sigma^2)}-\frac{N_0}{g_{ss}}\,if\,\frac{K}{\lambda\left(P_{pk}+\frac{\sum_{i=1}^{n}g_{psi}*P_i+\mathrm{N_0B}}{g_{ss}}\right)}-\sigma^2\leq\hat{g}_{\mathrm{spi}}\leq\frac{Kg_{ss}}{\lambda(\sum_{i=1}^{n}g_{psi}*P_i+\mathrm{N_0B})}-\sigma^2.$$

Therefore, the optimal power allocations under the peak transmit power and average interference power constraints are expressed as:

$$P(\hat{g}_{sp1}, \hat{g}_{sp2}, \cdots \hat{g}_{spn}, g_{ss}) = \begin{cases} P_{pk} & if \frac{K}{\lambda\left(P_{pk} + \frac{N_0 B}{g_{ss}}\right)} - \sigma^2 \geq \hat{g}_{\mathrm{spi}} \\ \frac{K}{\lambda(\hat{g}_{\mathrm{sp}} + \sigma^2)} - \frac{N_0}{g_{ss}} & if \frac{K}{\lambda\left(P_{pk} + \frac{N_0 B}{g_{ss}}\right)} - \sigma^2 \leq \hat{g}_{\mathrm{sp}} \leq \frac{K g_{ss}}{\lambda N_0} - \sigma^2 \\ 0, & otherwise \end{cases} \tag{8.23}$$

8.4.2.3 Power Consumption of Cognitive Transmitter Without Primary User's Interference

The average power consumption of CU-Tx under the peak transmit power and peak interference power constraints are expressed as:

$$\begin{aligned} E\left[P\left(\hat{g}_{spi} g_{ss}\right)\right] &= P_{pk} - \frac{P_{pk}}{1-\sigma^2}\exp\left(\frac{Q_{pki}}{P_{pk}} + \sigma^2 \log(P_0)\right) \\ &\quad - \frac{Q_{pki}}{1-\sigma^2}\exp\left(-\sigma^2 \log(P_0)\right) Ei\left(-\frac{Q_{pki}}{P_{pk}(1-\sigma^2)}\right) \end{aligned} \tag{8.24}$$

8.4.3 Outage Capacity

The outage capacity is the maximum transmission rate that can be maintained over the fading blocks with a given outage probability [11]. The objective function of outage capacity is expressed as:

$$\min_{\mathrm{P}\left(\hat{g}_{sp1}, \hat{g}_{sp2}, \ldots \hat{g}_{spn}, g_{ss}\right) \in \mathrm{R}} \Pr\left\{\log_2\left(1 + \frac{g_{\mathrm{ss}} \mathrm{P}\left(\hat{g}_{sp1}, \hat{g}_{sp2}, \cdots \hat{g}_{spn}, g_{ss}\right)}{\sum_{i=1}^{n} g_{psi} * P_i + \mathrm{N_0 B}}\right)\right\} \leq r_0 \tag{8.25}$$

where, $R \in \{R_1, R_2\}$. N_0, B, g_{psi} and P_i are the noise power spectral density at PU-Rx, total available bandwidth, channel power gain between i^{th} PU-Tx and CU-Rx and power transmitted by PU-Tx, respectively.

8.4.3.1 Optimal Power Allocation Under Peak Transmit Power and Peak Interference Power Constraints

For the optimal power allocation $R \in R_1$ and two dimensional truncated channel inversions (2D-TCI) strategy is used over $\hat{g}_{sp}$ and g_{ss}. Therefore, the optimal transmit power of CU is expressed as:

$$\mathrm{P}\left(\hat{g}_{sp1}, \hat{g}_{sp2}, \hat{g}_{sp3} \cdots \hat{g}_{spn}, g_{ss}\right) = \begin{cases} \frac{(N_oB + \sum_{i=1}^{n} g_{ps} * P_i)(2^{r_o} - 1)}{g_{ss}}, g_{ss} \geq \frac{(2^{r_0} - 1)(N_oB + \sum_{i=1}^{n} g_{ps} * P_i)}{P_{pk}} and \\ \hat{g}_{spi} \leq \frac{g_{ss} Q_{pk}}{(N_oB + \sum_{i=1}^{n} g_{ps} * P_i)(2^{r_0} - 1)} + \sigma^2 \ln(P_0) \\ 0, otherwise \end{cases} \quad (8.26)$$

Let $\frac{g_{ss} Q_{pk}}{(N_oB + \sum_{i=1}^{n} g_{ps} * P_i)(2^{r_0} - 1)} + \sigma^2 ln(P_0)$ and $\frac{(2^{r_0} - 1)(N_oB + \sum_{i=1}^{n} g_{ps} * P_i)}{P_{pk}}$ be denoted by the auxiliary variables u and z, respectively. By substituting Eqs. (8.26) in (8.25), we yield the outage probability as:

$$P_{out} = 1 - \iiint f_{\hat{g}_{sp}}(\hat{g}_{sp}) f_{g_{ss}}(g_{ss}) f_{g_{ps}}(g_{ps}) d\hat{g}_{sp} dg_{ss} dg_{ps} \quad (8.27)$$

Where, $f_{\hat{g}_{sp}}(\hat{g}_{sp}), f_{g_{ps}}(g_{ps})$ and $f_{g_{ss}}(g_{ss})$ are the probability density functions of $\hat{g}_{sp}$, g_{ps} and g_{ss}, respectively. The outage capacity for Rayleigh fading channel is computed as:

$$C_{outage} = \log_2(1 + F^{-1}(1 - P_{out})\gamma) \quad (8.28)$$

where $F(x) = \Pr(g_{ss} > x)$ is the complementary cumulative distribution function of g_{ss} and γ is the signal-to-noise ratio (SNR) [18].

8.4.3.2 Optimal Power Allocation Under Peak Interference Power Constraints

$$P\left(\hat{g}_{sp1}, \hat{g}_{sp2} \cdots \hat{g}_{spn}, g_{ss}\right) = \min\left(\frac{Q_{pki}}{\hat{g}_{spi} - \sigma^2 \ln P_0}\right), i = 1..n \quad (8.29)$$

The outage probability is computed as stated earlier. This equation illustrates that the power allocation to a CU with respect to different PUs has been evaluated and then the minimum power among the calculated set has been allocated to the CU.

8.4.3.3 Power Consumption of Cognitive Transmitter Without Primary User's Interference

The average consumption of power of a cognitive transmitter when the optimal power is allocated under peak transmit power and peak interference power constraints without PU interference is expressed as:

$$E\left[P\left(\hat{g}_{spi}g_{ss}\right)\right] = Ei\left(1, \frac{1}{P_{pk}}\right) - Ei\left(1, \frac{Q_{pki}+1}{P_{pk}(1-\sigma^2)}\right)\exp\left(\frac{-\sigma^2\log(P_0)}{1-\sigma^2}\right) \quad (8.30)$$

The average expenditure of CU power when the optimal power is allocated by considering only the peak interference power constraint:

$$E\left[P\left(\hat{g}_{spi}g_{ss}\right)\right] = Q_{pki}\exp\left(\frac{-\sigma^2\log(P_0)}{1-\sigma^2}\right)\left(-\log\left(\frac{1-\sigma^2}{Q_{pk}}+1\right) - Ei\left(\frac{\sigma^2\log(P_0)}{1-\sigma^2}\right)\right) \quad (8.31)$$

8.5 Simulation and Analysis

In this section, we present the numerically simulated results of the ergodic capacity and outage capacity with and without the interference of PU-Tx to the CU link of the proposed cognitive radio network model. The performance is analyzed under the average as well as peak interference power constraints. The simulation result of the proposed model depicts a significant improvement in the data transmission rate. In Fig. 8.2, the ergodic capacity without PU-Tx interference to the CU link under

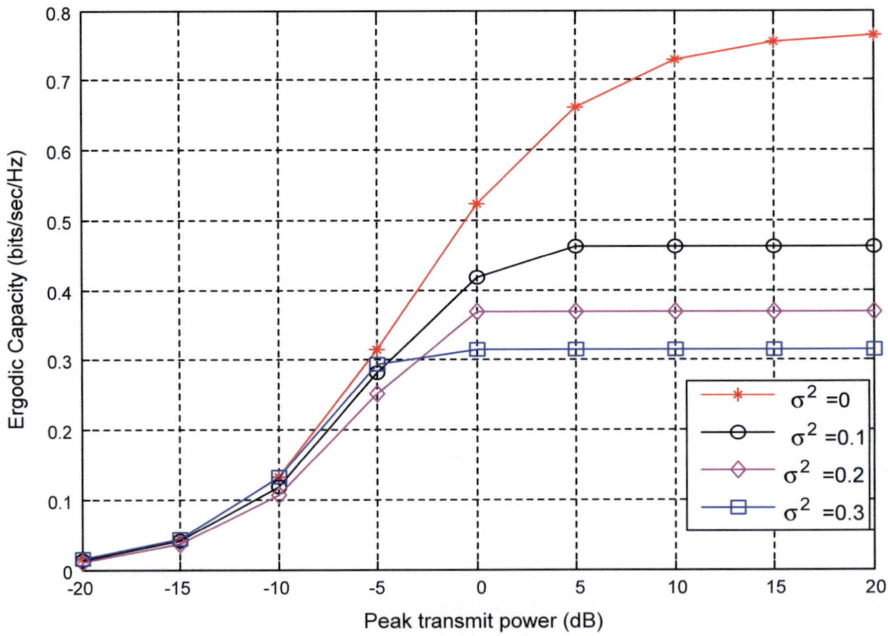

Fig. 8.2 The response of peak transmit power (*dB*) on the ergodic capacity (*bits/s/Hz*) for different values of variance at arbitrary chosen values of the peak interference power (−5 dB)

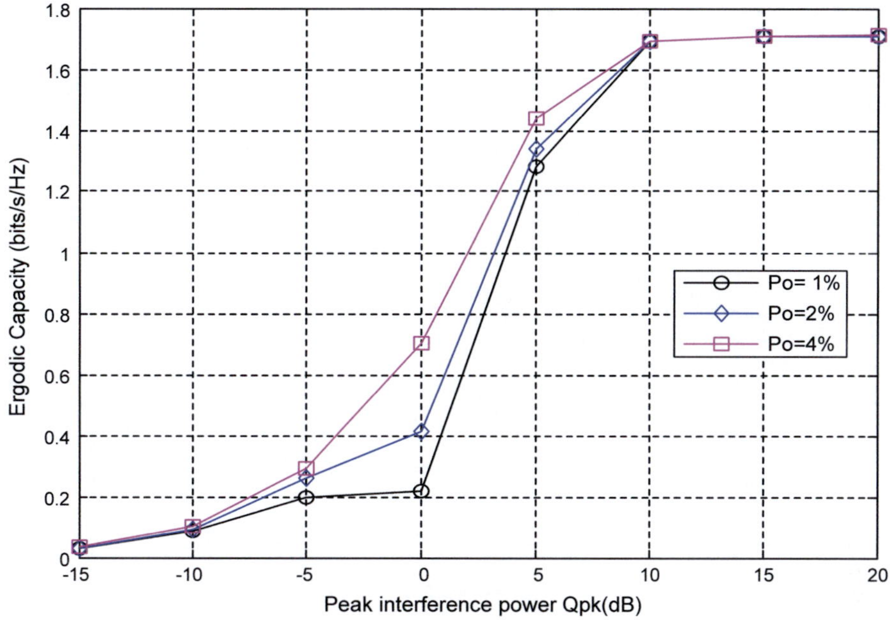

Fig. 8.3 The response of peak interference power on the ergodic capacity (bits*/s/Hz*) for different values of the interference outage level at arbitrary chosen values of the peak transmit power (10 dB) and error variance 0.2

peak transmit power for different values of the error variance at arbitrary chosen peak interference power (−5 dB) is computed. The numerically simulated result for ergodic capacity of cognitive link with perfect CSI between CU-Tx and PU-Rx is validated with reported literature [11]. However, if the peak transmit power is below the peak interference power, then the ergodic capacity for different channel conditions (various values of error variance) increases monotonically. Above the peak interference power, the ergodic capacity gradually becomes constant as shown in Fig. 8.2. In addition, the figure shows that with the increase of noise variance, the ergodic capacity declines in comparison to the perfect channel state information. In Fig. 8.3, the ergodic capacity is analyzed for different interference outage levels with a fixed noise variance of 0.2. The figure also shows that with the increase of interference outage level, the ergodic capacity level rises, but when the peak transmit power is greater than that of peak interference power, there is no significant effect of increasing the interference outage level.

The average power expenditure of CU-Tx is investigated in Fig. 8.4 with the peak interference power for different combinations of interference outage level and noise error variance. Here, there is significantly more power consumption if the interference outage level rises for a fixed error variance. On the other hand, if the

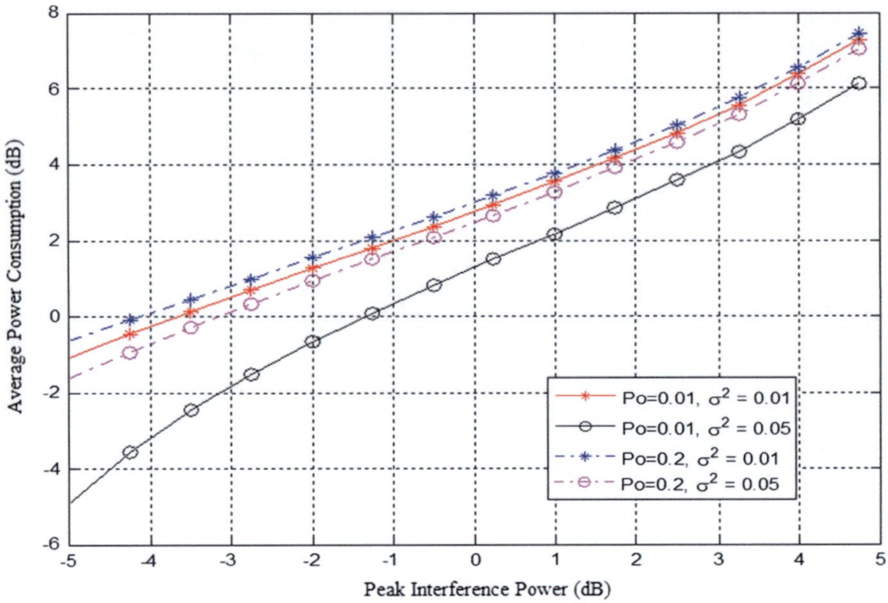

Fig. 8.4 The average power consumption to achieve ergodic capacity limits of a cognitive transmitter with different combinations of interference outage level and error variance

interference level is fixed, then with the increase of noise variance, the average power consumption of the CU decreases. Moreover, it reveals that with the increase of the interference power constraint, the power consumption of CU-Tx is monotonically increased. The variation in ergodic capacity under the joint peak transmit power and average interference power constraints without consideration of interference of the PU is illustrated in Fig. 8.5. It illustrates that when the peak transmit power is more than the average interference power, there is variation in ergodic capacity with error variance. However, when the peak transmit power becomes less than that of the average interference power, the ergodic capacity for different error variance values remain the same. In addition, the numerically simulated result of the ergodic capacity with perfect channel state information is validated with literature reported in [11].

A comparison of Figs. 8.2 and 8.5 reveals that the average interference power constraint is better than that of the peak interference power constraint. The effects of the interference of PUs on the ergodic capacity of the CU link under joint peak transmit power and peak interference power for different error variances is illustrated in Fig. 8.6. It is shown that as the interference of the PU to CU link increases, the ergodic capacity of the CU link decreases. With the comparison of Figs. 8.2 and 8.6, it is revealed that there are significant reductions in the ergodic capacity at the

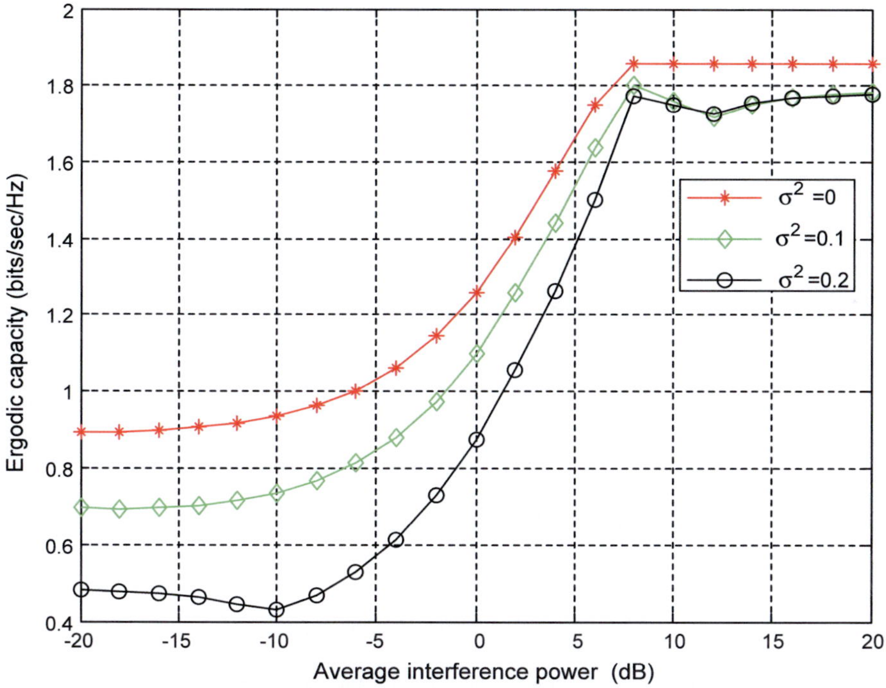

Fig. 8.5 The response of average interference power (*dB*) on the ergodic capacity (*bits/s/Hz*) for different values of variance at arbitrary chosen value of the peak transmit power (5 dB)

peak transmit power, the peak interference power and noise error variance, of 5,−5 dB and 0.1, respectively. In addition, it is analyzed that as the number of PUs increases above two then reliable communication cannot be achieved. The variations in outage probability without and with PU interference under the peak transmit power and peak interference power constraints are presented in Figs. 8.7 and 8.8a, respectively.

If the PUs interfere with the CU link, the outage probability level rises. In cases without PU-Tx interference, it has been illustrated that as the peak transmit power is less than that of the peak interference power, the outage probability level remains the same for different noise error variances. But with PU interference, the outage probability is constant with different outage probability levels. Figure 8.8a, shows that the outage probability increases with the increase of PU's interference. The figure also shows that if the number of PUs increases above four, then the data rate of the CU link drops considerably; therefore communication cannot be established efficiently. Further, Fig. 8.8b demonstrates the effect of interference of the PU to the CU link when the transmit power to the CU is allocated under the peak interference power constraint only. A comparison between Figs. 8.8a and b reveals that the

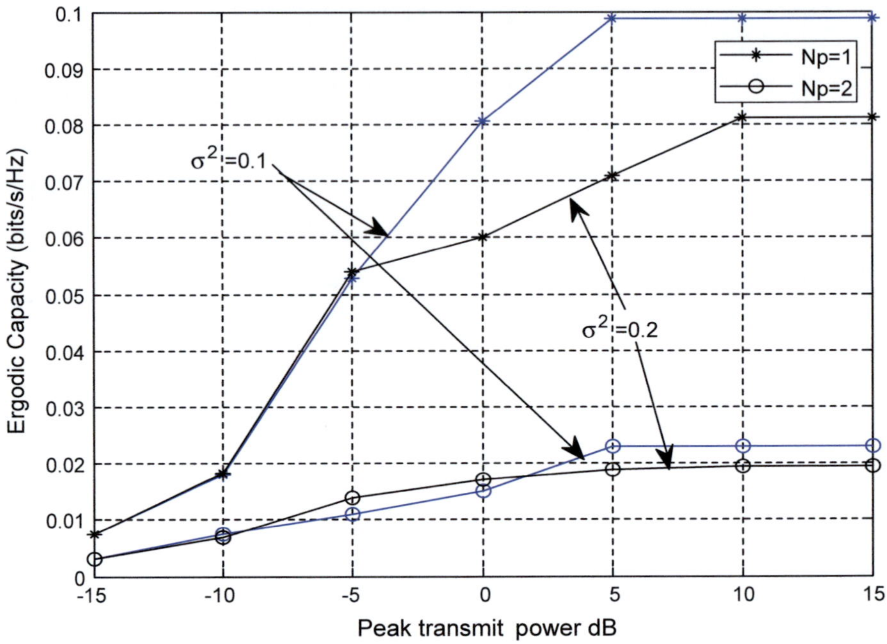

Fig. 8.6 The response of peak transmit power (*dB*) on the ergodic capacity (*bits/s/Hz*) of cognitive user links with multiple primary users interference of arbitrary chosen values of the fixed peak interference power (−5 dB), with fixed noise variances $\sigma^2 = 0.1$ and $\sigma^2 = 0.2$

outage probabilities under the joint peak transmit power and peak interference power are greater in comparison to the individual peak interference power constraint. If the multiple numbers of PUs interfere with the CU link, then the peak interference constraint provides better result.

The consumption of power of CU-Tx under the joint peak transmit power and peak interference power, as well as under the peak interference power only, is portrayed in Fig. 8.9. It is shown that the consumption of power under the joint constraints (the peak transmit power and peak interference power) is much lower in comparison to that of the peak interference power constraint only. The average power consumption of CU under the peak interference power constraint only is validated with the reported literature [19]. From Fig. 8.9, it is clear that it is significantly much better to allocate power to the CU under the joint constraints (the peak transmit power and peak interference power) as compared to the peak interference power constraint only.

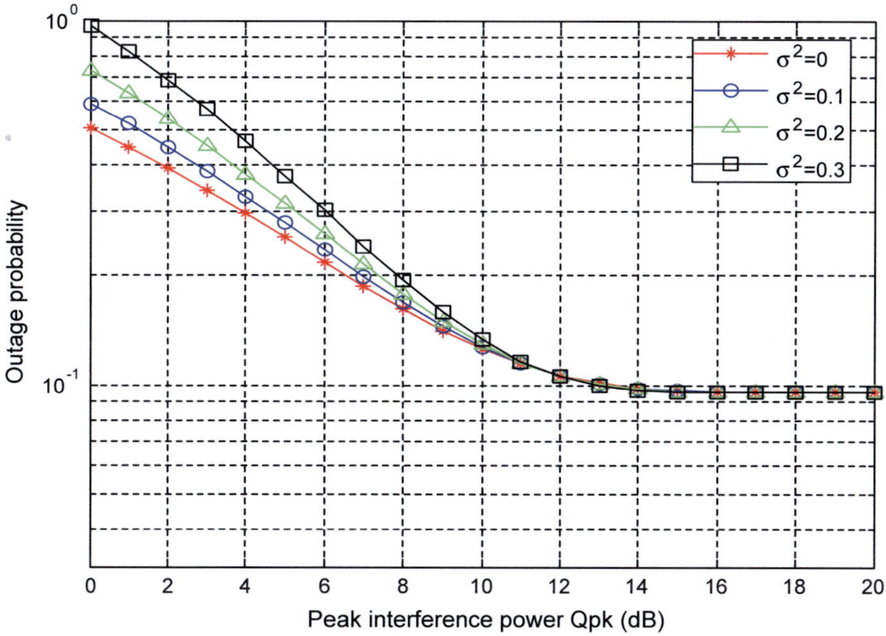

Fig. 8.7 The response of the peak interference power (9 dB) on the outage probability at arbitrary chosen values of the peak transmit power (10 dB) and fixed data transmission rate (r_0 = 1 bits/sec/Hz) under the peak interference power and the peak transmit power constraint

8.6 Summary

In this chapter, we presented the analysis of ergodic capacity and outage probability of the CU link, with and without PU interference. It was shown that with the numerically increasing value of the noise error variance and interference from the PU to CU link, the ergodic capacity of the CU link decreases and the outage probability increases significantly. The power consumption of CUs under the joint constraint (peak transmit power and peak interference power) is very much lower as compared to that of the peak interference power constraint. The capacity limits under the joint peak/average transmit power and average interference power constraint is also a very important issue to analyze the proposed system model, which will be reported in a future communication. In practice, the available system parameters (CSI and interference power) to enable power control and beam-forming could be uncertain due to various factors such as estimation error and/or measurement error, thus the robustness of the designed algorithms should be considered in order to overcome the effects of parametric uncertainty. The stochastic Gaussian model includes mean and variance side information of the fading coefficients at the transmitter. The channel estimation acquired independently by the transmitter may

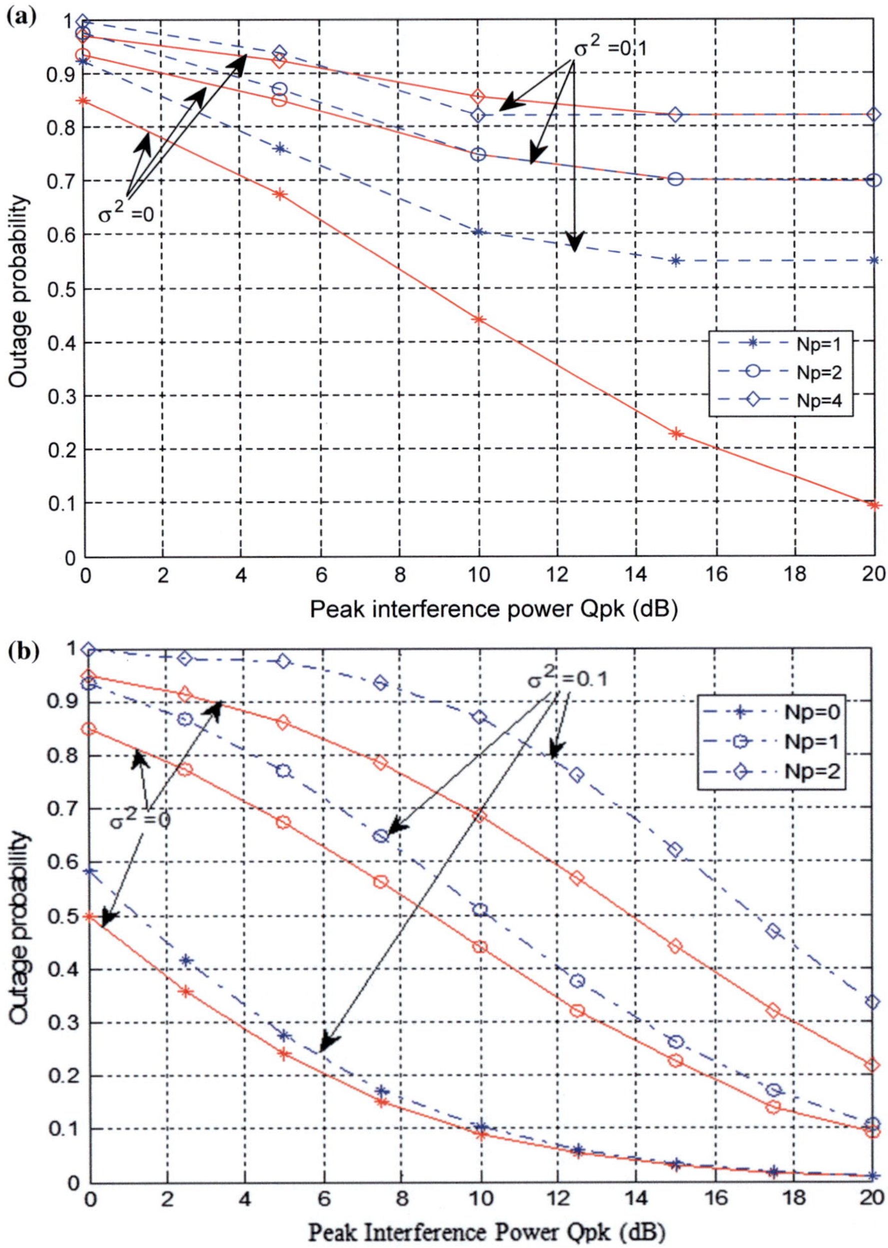

Fig. 8.8 The outage probability of a cognitive user link with multiple primary users interference with variation in the peak interference power for chosen values of peak transmit power (10 dB) with fixed $\sigma^2 = 0$ and $\sigma^2 = 0.1$ under (**a**) peak transmit power and peak interference constraints and (**b**) peak interference only

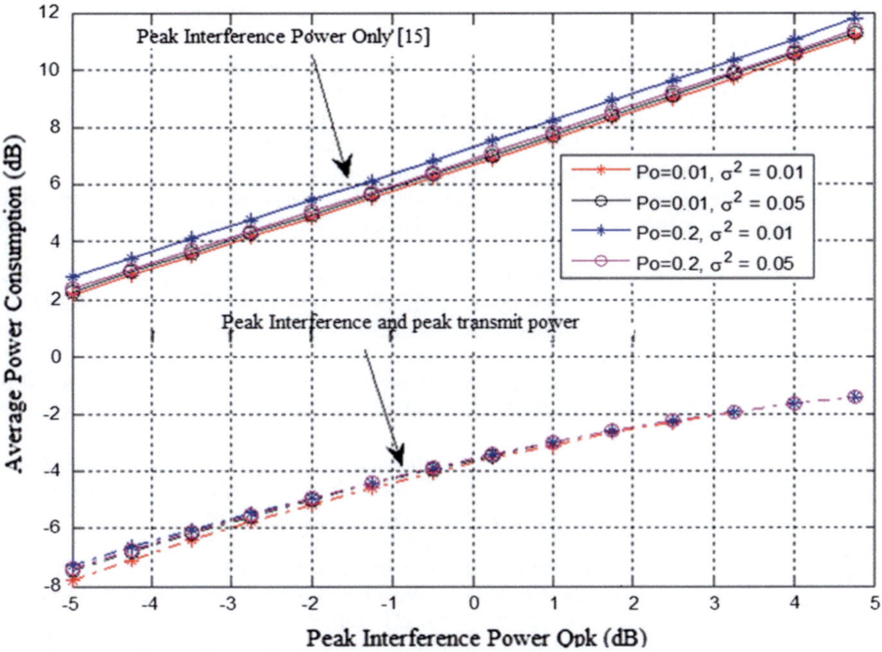

Fig. 8.9 The average power consumption to achieve outage capacity limit of a cognitive user transmitter with different combinations of interference outage level and error variance under the peak transmit power and peak interference constraints, as well as under the peak interference power constraint only

suffer from the channel estimation accuracy due to RF chain impairment, which limits channel estimation reciprocity. In addition, the causality requires acquiring CSI prior to transmission, while the channel may change when actual transmission takes place. These give rise to the practical stochastic Gaussian model with mean and variance. The practical transmit channel state information model is a stochastic Gaussian model with mean and variance information which is commonly used for modeling channel estimation error. The extensive numerical results illustrate that spectral efficiency is achieved by the SU with partial inter-system CSI within a reasonable range of outage probability.

References

1. P. Kolodzy, Spectrum policy task force report federal Commun. Comm. Washington, DC, Rep. ET Docket 02-135, Nov 2002
2. J. Marinho, E. Monterio, Cognitive radio: Survey on communication protocols, spectrum decision issues and future research directions. Wireless Netw. **18**, 147–164 (2011)
3. J. Mitolla, Cognitive radio architecture evolution: Annals of telecommunication. Ann. Telecommun. **64**, 419–441 (2009)

4. A. Khattab, M.A. Bayoumi, Standardization of cognitive radio networking: A comprehensive survey. Ann. Telecommun. **70**(11), 465–477 (2015)
5. Y.-C. Liang, K.-C. Chen, G.Y. Li, P. Mahonen, Cognitive radio networking and communication: An overview. IEEE Trans. Veh. Tech. **60**(7), 3386–3407 (2011)
6. R.C. Santiago, Michał Szydełko, Adrian Kliks, Fotis Foukalas, Yoram Haddad, Keith E. Nolan, Mark Y. Kelly, Moshe T. Masonta, Ilangko Balasingham, 5G: The convergence of wireless communications. Wireless Pers. Commun. **83**(3), 1617–1642 (2015)
7. Shweta Pandit, G. Singh, An overview of spectrum sharing techniques in cognitive radio communication system. Wireless Network (Online), Dec 2015. doi:10.1007/s11276-015-1171-1
8. H.A. Suraweera, P.J. Smith, M. Shafi, Capacity limits and performance analysis of cognitive radio with imperfect channel knowledge. IEEE Trans. Veh. Tech. **59**(4), 1811–1822 (2010)
9. K.B. Letaief, W. Zhang, Cooperative communications for cognitive radio networks. Proc. IEEE **97**(5), 878–893 (2009)
10. Z. Wang, W. Zhang, Opportunistic spectrum access in cognitive relay networks based on white space modeling. in *Proc. IEEE Global Telecommunications Conference (GLOBECOM 2010)*, Miami, FL, Dec 2010, p. 1–5
11. X. Kang, Ying-Chang Liang, Arumugam Nallanathan, Hari Krishna Garg, and Rui Zhang, Optimal power allocation for fading channels in cognitive radio networks: Ergodic capacity and outage capacity. IEEE Trans. Wireless Commun. **8**(2), 940–950 (2009)
12. D.I. Kim, L.B. Le, E. Hossain, Joint rate and power allocation for cognitive radios in dynamic spectrum access environment. IEEE Trans. Wireless Commun. **7**(12), 5517–5527 (2008)
13. L.B. Le, E. Hossain, Resource allocation for spectrum underlay in cognitive radio networks. IEEE Trans. Wireless Commun. **7**(12), 5306–5315 (2008)
14. D. Li, Efficient power allocation for multiuser cognitive radio networks. Wireless Pers. Commun. **59**, 589–597 (2010)
15. S. Vassaki, M.I. Poulakis, A.D. Panagopoulous, P. Constantinou, QoS driven power allocation under peak and average interference power constraints in cognitive radio networks. Wireless Pers. Commun. **78**, 449–474 (2014)
16. S. Singh, D. Teal Paul, A. Dmochowski Pawel, and Alan J. Coulson, Interference management in cognitive radio systems with feasibility detection. IEEE Trans. Veh. Tech. **62** (8), 3711–3720 (2013)
17. S. Parsaeefard, A.R. Sharafat, Robust distributed power control in cognitive radio networks. IEEE Trans. Mob. Comput. **12**(4), 609–620 (2013)
18. D. Tse, P. Viswanath, *Fundamentals of Wireless Communication* (Cambridge University Press, U.K., 2005)
19. Z. Rezki, M.-S. Alouini, Ergodic capacity of cognitive radio under imperfect channel-state information. IEEE Trans. Veh. Tech. **61**(5), 2108–2119 (2012)
20. L. Musavian, S. Aissa, Fundamental capacity limits of cognitive radio in fading environments with imperfect channel state information. IEEE Trans. Commun. **57**(11), 3472–3480 (2009)
21. A.K. Farraj, S. Ekin, Performance of cognitive radios in dynamic fading channels under primary outage constraint. Wireless Pers. Commun. **73**(8), 637–649 (2013)
22. R. Zhang, S. Cui, Y.-C. Liang, On ergodic sum capacity of fading cognitive multiple access and broadcast channels. in *Proc. IEEE Annual Allerton Conf. on Communication, Control and Computing* (Urbana-Champaign, IL, 2008), p. 879–886
23. K. Son, B.C. Jung, S. Chong, D.K. Sung, Power allocation policies with full and partial inter-system channel state information for cognitive radio networks. Wireless Netw. **19**, 99–113 (2012)
24. F.R. Mireles, Outage capacity for secondary users due to bandwidth fluctuations: multicarrier versus single carrier. Wireless Pers. Commun. **78**, 915–925 (2014)
25. Md. F. Kader, Asaduzzaman, M.M. Hoque, Outage Capacity analysis of a cooperative relaying scheme in interference limited cognitive radio networks. Wireless Pers. Commun. **79**, 2127–2140 (2014)

26. V. Bhaskar, B. Dutta, Ergodic and outage capacity maximization of cognitive radio networks in cooperative relay environment using optimal power allocation. Ann. Telecommun. **69**, 621–632 (2014)
27. A.K. Farraj, S.L. Miller, Scheduling in a spectrum sharing cognitive environment under outage probability constraint. Wireless Pers. Commun. **70**, 785–805 (2012)
28. F. Wang, X. Liao, S. Guo, H. Huang, Jointly optimal congestion and power control for Rayleigh faded channels with outage constraints. Wireless Pers. Commun. **77**, 101–125 (2013)
29. A.K. Farraj, E.M. Hamma, Impact of quality of service constraints on the performance of spectrum sharing cognitive users. Wireless Pers. Commun. **69**, 673–688 (2012)
30. B. Bharti, G. Singh, Analysis of outage capacity of cognitive radio network with imperfect channel state information. in *Proc. IEEE Int. Conf. on Signal Processing and Communication* India, March 16–18, 2015, pp. 1–6
31. S. Pandit, G. Singh, Channel capacity in fading environment with CSI and interference power constraints for cognitive radio communication system. Wireless Netw. **21**(4), 1275–1288 (2015)
32. D. Xu, Z. Feng, P. Zhang, On the impacts of channel estimation error and feedback delay on the ergodic capacity for the spectrum sharing cognitive radio. Wireless Per. Commun. **72**(4), 1875–1887 (2013)
33. L. Li, A. Goldsmith, Capacity and optimal resource allocation for fading broadcast channels-Part I: Ergodic capacity. IEEE Trans. Inf. Theory **47**(3), 1083–1102 (2001)
34. L. Li, A. Goldsmith, Capacity and optimal resource allocation for fading broadcast channels-Part II: Outage capacity. IEEE Trans. Inf. Theory **47**(3), 1103–1127 (2001)
35. M. Gastpar, On capacity under receive and spatial spectrum-sharing constraints. IEEE Trans. Inf. Theory **53**(2), 471–487 (2007)
36. L. Sboui, Z. Rezki, M.-S. Alouini, A unified framework for the ergodic capacity of spectrum sharing cognitive radio system. IEEE Trans. Wireless Commun. **12**(2), 877–887 (2013)
37. L. Musavian, S. Aissa, Capacity and power allocation for spectrum-sharing communications in fading channels. IEEE Trans. Wireless Commun. **8**(1), 148–156 (2009)
38. A. Ghasemi, E.S. Sousa, Fundamental limits of spectrum-sharing in fading environment. IEEE Trans. Wireless Commu. **6**(3), 649–658 (2007)
39. L. Zhang, Y.C. Liang, Y. Xin, Robust cognitive beam forming with partial channel state information. IEEE Trans. Wireless Commun. **8**(8), 4143–4153 (2009)
40. L. Musavian, S. Aissa, Fundamental capacity limits of cognitive radio in fading environment with imperfect channel information. IEEE Trans. Wireless Communication **57**(11), 3472–3480 (2009)
41. S. Boydand, L. Vandenberghe, *Convex Optimization*. (Cambridge University Press, Cambridge, UK, 2004)

Chapter 9
Channel Capacity of Cognitive Radio in a Fading Environment with CSI and Interference Power Constraints

9.1 Introduction

In general, channel capacity is used as a basic performance measurement tool for the analysis and design of new and more efficient techniques to improve the spectral efficiency of wireless communication systems. An adaptive power transmission scheme that achieves the Shannon capacity under the fading environment was discussed in [1], and average transmit power constraints along with the availability of channel state information (CSI) at the cognitive transmitter were initially considered in [2]. The power optimization problem with peak and average transmit power constraints was investigated [3]. In spectrum sharing systems, CSI can be used at the cognitive/secondary transmitter to adaptively adjust the transmission resources as discussed in [4, 5]. In [5], knowledge of the secondary link CSI and information at the secondary transmitter (ST) (CR transmitter) about the channel between the secondary transmitter and the primary receiver (PR) was used to obtain the optimal power transmission policy of the secondary user (SU) under constraints on the peak and average received-power at the primary receiver. Ghasem and Sousa [6] demonstrated that the secondary user may take advantage in the fading environment between the primary and secondary user by opportunistically transmitting with high power when the signal received by the licensed receiver is deeply faded.

One of the most efficient ways to determine the spectrum occupancy is to sense the activity of primary users operating in the secondary user's range of communication [7]. Practically, it is difficult for a secondary user to have direct access to the CSI pertaining to the primary user link. Recent work on spectrum sharing systems has concentrated on sensing the primary transmitter's activity, and is based on local processing at the secondary user side [8]. In this context, the sensing ability is provided by a sensing detector mounted on the secondary user's equipment, which scans the spectrum for specific times [9]. The activity statistics of the primary user's signal in the shared spectrum is computed and, based on the sensing information [10], the cognitive user is capable of determining the local presence of the

S. Pandit and G. Singh, *Spectrum Sharing in Cognitive Radio Networks*,
DOI 10.1007/978-3-319-53147-2_9

primary transmitter in a specific spectrum band. For instance, the received signals at the energy-based detector [11, 12] were used to detect the presence of unknown primary transmitters. However, by using this sensing information obtained from the spectrum sensor and considering that the secondary transmitter does not have information about the state of its corresponding channel, the power adaptation strategy that maximizes the channel capacity of the secondary user's link is investigated in [13]. Rezki and Alouini in [14] considered the limited/imperfect CSI at the secondary transmitter and computed the Ergodic channel capacity. Further, in [15] the power allocation for erroneous estimated channel gain between the secondary user and primary base station was performed through a geometric programming problem which was solved by Lagrange dual decomposition. However, only the underlay spectrum sharing model was considered in [15]. Parsaeefard and Sharafat in [16] considered the cognitive nodes as relay nodes and illustrated the power and channel allocation strategy to the cognitive users in the Rayleigh fading environment. In [17], the rate loss constraint (RLC) is considered instead of conventional interference power constraints in order to protect the primary user, and the channel capacity of a cognitive user that utilizes primary users' OFDM (orthogonal frequency division multiplexing) subcarriers, is maximized by RLC and cognitive user transmit power constraints. In [14–18] the authors computed the channel capacity of the cognitive user without considering the channel sensing information available at the secondary transmitter.

In this chapter, we focus on a cognitive radio wireless communication system with maximum achievable Ergodic channel capacity, considering a single cognitive user. In a collaborative communication framework, either extra relay terminals assist the communication between some dedicated sources and their corresponding destinations, and/or they allow the users in a network to help each other to achieve higher communication system capacity than the single point-to-point communication between source and destination [19, 20]. In this chapter we have considered point-to-point communication between the cognitive users without any kind of cooperation/collaboration among them. Therefore, if more than one cognitive users are competing to access the primary user's spectrum hole, then due to probable inter-cognitive users' interference, the maximum achievable channel capacity is upper bounded by only the single cognitive user's case. The proposed spectrum sharing system has a pair of primary transmitter (PT) and PR as well as a pair of ST and secondary receiver (SR), as shown in Fig. 9.1. Further, the small-scale fading effects over the transmit power of the secondary transmitter in the proposed system has been explored. However, in [21] this type of system model is considered without fading in the link channel between the ST and the PR. Therefore, the Ergodic channel capacity for the Nakagami-*m* fading channel in the secondary and primary links is the basic motivation of this chapter. The power of the secondary transmitter is controlled based on the:

(i) Sensing information about the primary user's activity, and
(ii) CSI of the secondary and primary link.

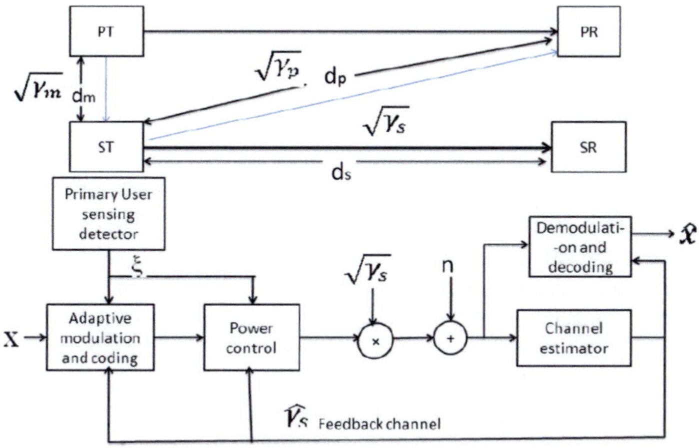

Fig. 9.1 The proposed spectrum sharing system model

Moreover, the constraint on average interference at the primary radio receiver is considered for the channel capacity. Since the cognitive user is able to adapt any modulation strategy, it can change its modulation strategy according to the fading environment, and hence both policies in the rate and power are established [22], which is referred as the variable rate and power transmission scheme. In this context, we have also considered the variable rate and power *M*-QAM transmission strategy in the cognitive radio communication system where the rate and power of the ST is adaptively controlled based on the availability of the secondary user's link CSI and sensing information about the primary user's activity. Therefore, in this chapter we have numerically computed the channel capacity in the fading environment under the average interference power constraint with two adaptation policies for spectrum sharing. The channel capacity is maximized for these two policies by considering the Lagrange optimization problem for average interference power constraint. The small-scale fading effect over the transmit power of the secondary transmitter is also presented.

The remainder of this chapter is organized as follows. Section 9.2 concerns the spectrum sharing system model. Section 9.3 discusses the power and rate adaptation policy, and in Sect. 9.4 Ergodic channel capacity of the adaptation policies under Nakagami-*m* fading is computed. In Sect. 9.5, the numerical simulation results of the proposed spectrum sharing model are presented, and finally, Sect. 9.6 summarizes the work.

9.2 Spectrum Sharing System

9.2.1 System Model

This proposed spectrum sharing system consists of a PT and PR pair as well as an ST and SR pair, as shown in Fig. 9.1. In this scenario, the secondary user is allowed to use the spectrum band assigned to the primary user as long as the interference power imposed by secondary transmitter on the primary receiver is less than a predefined threshold value, which is the interference temperature limit. We consider the primary user link that is the channel between the PT and PR to be a stationary block-fading channel. According to the definition of block-fading, the channel gain remains constant over some block length T and after that time, the channel gain changes to a new independent value based on its distribution [21].

The average transmit power of the PT is assumed to be P_t, its average ON/active time is α, and its average OFF/inactive time is $\bar{\alpha} = 1 - \alpha$ [13]. In addition, we have assumed a discrete-time flat-fading channel with perfect CSI at the receiver and transmitter of the secondary user. As shown in Fig. 9.1, the secondary/cognitive receiver generates and estimates the channel power gain $(\widehat{\gamma_s})$ between the secondary transmitter and secondary receiver (SR). We have assumed that the channel power gain is fed back to the secondary transmitter error-free and without delay. Further, the channel gain between the transmitter and receiver of the secondary user, ST and PR as well as between the PT and ST, are given by $\sqrt{\gamma_s}$, $\sqrt{\gamma_p}$, and $\sqrt{\gamma_m}$, respectively. The channel power gains γ_s, γ_p, and γ_m are independent of each other. We have obtained the cognitive radio communication system's Ergodic channel capacity by considering the distribution of γ_s and γ_p as the Nakagami-m distribution. d_m, d_s and d_p are the distances between ST to PR, ST to SR, and ST to PR, respectively. Moreover, the channel between the PT and SR is considered an additive white Gaussian noise (AWGN) channel, denoted as n, and can be modeled as a zero-mean Gaussian random variable with variance N_0B, where N_0 and B denote the noise power spectral density and the signal bandwidth, respectively. x is the data transmitted from ST and $\hat{x}$ is the estimated transmitted data at SR as shown in Fig. 9.1.

9.2.2 Spectrum Sensing Module

As is clear from Fig. 9.1, the secondary transmitter is equipped with a spectrum sensing detector whose function is to sense the frequency band of the primary user for the secondary user's transmission. Based on the received signals, the detector computes a single sensing metric denoted by ξ, [12]. The sensing metric is the total primary signal power in the number of independent signal samples [13]. We consider that the statistics of ξ conditioned on the primary user being active or idle are known prior to the ST. Using the energy detection method for sensing information on the primary user being active or idle, the sensing parameter ξ is modeled according to Chi-square probability distribution functions (pdfs) with ν degrees of freedom as discussed in [11], where ν is related to the number of samples used in the sensing period, *N* We define the pdf of ξ, given that the PT is active or idle, by $f_1(\xi)$ and $f_0(\xi)$, respectively, that is, $f_1(\xi)$ and $f_0(\xi)$ are conditional probabilities. According to [23, pp. 941], for a large number of ν (for example ≥ 30), one can approximate the Chi-square distribution with a Gaussian pdf. Since the number of observation samples can be large enough for the approximation to be valid, we choose $f_1(\xi) \sim \mathcal{N}(\mu_{1,}\delta_1^2)$ and $f_0(\xi) \sim \mathcal{N}(\mu_{0,}\delta_0^2)$, where $(\mu_{1,}\delta_1^2)$ and $(\mu_{0,}\delta_0^2)$ are given by [8]. The probability distribution of ξ depends on [13]:

$$\left.\begin{array}{ll} \left.\begin{array}{l} \mu_1 = N\left(\frac{P_t}{d_m^2} + 1\right) \\ \delta_1^2 = 2N\left(\frac{P_t}{d_m^2} + 1\right)^2, \text{ and} \end{array}\right\} & \text{when PT is active} \\ \left.\begin{array}{l} \mu_0 = N \\ \delta_0^2 = 2N \end{array}\right\} & \text{when PT is idle} \end{array}\right\} \tag{9.1a}$$

and the probability distributions of ξ are given as [8]:

$$\left.\begin{array}{l} f_0(\xi) = \frac{1}{\sqrt{2\pi\delta_0^2}} exp\left(\frac{-(\xi-\mu_0)^2}{2\delta_0^2}\right) \\ f_1(\xi) = \frac{1}{\sqrt{2\pi\delta_1^2}} exp\left(\frac{-(\xi-\mu_1)^2}{2\delta_1^2}\right) \end{array}\right\} \tag{9.1b}$$

In this chapter, we have used the energy detector for spectrum sensing due to its easy implementation and low computational complexity, as discussed in [11]. The other sensing detectors can also be used for spectrum sensing since the authors' main motive is to compute the sensing metric ξ, which represents the total signal power observed or the correlation between the observed signal and a known signal pattern [13]. However, the main difference lies in the number of samples required for the same performance in different detectors, and that depends on the required signal-to-noise ratio [11]. In addition, the cognitive radio user transmission should be limited so that it does not cause harmful interference to the primary user. Therefore, a limit or constraint is set at PR called the average interference power constraint or simply interference constraint. When PU is active, ST cannot transmit at a power which exceeds the average interference power constraint at the primary receiver, which is given as [21]:

$$E_{\gamma_{s},\xi,\gamma_{p}}\left[P(\gamma_{s},\gamma_{p},\xi)\gamma_{p}|PUisON\right] \le Q_{\mathrm{int}};\forall\gamma_{s},\gamma_{p},\xi \tag{9.2}$$

where the transmit power of SU is $P(\gamma_{s},\gamma_{p},\xi)$ and expectation over the joint pdf of random variables γ_s, γ_p and ξ is denoted by $E_{\gamma_{s},\xi,\gamma_{p}}[.]$. Q_{int} is the interference limit set at PR, that is, the maximum interference power that it can tolerate without degrading its own performance. The constraint defined in Eq. (9.2) is used to compute the Ergodic channel capacity. However, the average interference power constraint is considered only because we have assumed that the licensed user performance is measured by the average signal-to-noise ratio (SNR) and not by instantaneous SNR. Moreover, the Ergodic channel capacity under the average received power constraint is, in general, higher than that of the peak received power constraint due to the more restrictive nature of the peak power, as opposed to the average interference power constraint.

9.3 Rate and Power Adaptation Policy for *M*-QAM

The data rate and power adaptation is a potential transmission strategy which adjusts the transmit power and data rate of a cognitive radio system to improve the spectrum efficiency for utilizing the shared spectrum [21, 24–26]. Data rate adaptation is a spectrally efficient technique, and its adaptation can be achieved either through variation of the symbol time duration [27] or by varying the constellation size [28]. However, the former method is spectrally inefficient and requires variable-bandwidth system design as discussed in [29]. The variable data rate adaptation policy using varying constellation size is fixed bandwidth with a spectrally efficient method [29]. The Ergodic channel capacity under adaptation policy of the variable data rate and power transmission strategy in *M*-QAM signal constellation is considered with the knowledge of CSI and spectrum sensing information at the secondary transmitter side, which satisfies the predefined bit-error-rate (BER) requirements and adheres to the constraints on the average interference power at the primary user. In this case, the cognitive radio adapts the transmit power according to:

(i) the primary and secondary channel power gain γ_p and γ_s, respectively,
(ii) the primary user's activity states ξ, subjected to the average interference, and
(iii) the instantaneous bit-error-rate constraint $P_b(\gamma_s,\xi) = P_{\mathrm{b}}$.

The P_b bound for each value of γ_s and ξ is given as [21]:

$$P_b(\gamma_s,\xi) \le 0.2exp\left(\frac{-1.5}{M-1} \times \frac{P(\gamma_{s},\gamma_{p},\xi)\gamma_s}{N_0 B}\right) \tag{9.3}$$

where M is the constellation size or the number of symbols in the particular modulation format. $P(\gamma_{s},\gamma_{p},\xi)$ is the transmit power of ST. To satisfy the conditions as discussed in Eq. (9.3), we can adjust the values of M and $P(\gamma_{s,}\gamma_{p,}\xi)$. However, the instantaneous bit error rate constraint given by Eq. (9.3) holds for $M \geq 4$ [21]. We can also express Eq. (9.3) by the following mathematical expression:

$$P_b(\gamma_s, \xi) \leq 0.2exp\left(\frac{-1.5}{M-1} SNR_{ss}\right) \tag{9.3a}$$

where SNR_{ss} is the signal-to-noise power ratio of the ST to SR. For both the adaptive data rate and adaptive power transmission policy, Eq. (9.3) should be satisfied for the following constraint on average interference power:

$$\frac{P(\gamma_{s,}\gamma_{p,}\xi)\gamma_p}{N_0 B} \leq Q_{int} \tag{9.3b}$$

or

$$SNR_{sp} \leq Q_{int}$$

where SNR_{sp} is the signal-to-noise power ratio of secondary transmitter to primary receiver. After some mathematical manipulation of Eq. (9.3), we obtain the following maximum constellation size for a given $P_b(\gamma_s, \xi)$:

$$M(\gamma_s, \xi) = 1 + K\left(\frac{P(\gamma_{s,}\gamma_{p,}\xi)\gamma_s}{N_0 B}\right) \tag{9.3c}$$

Moreover, we can achieve the constellation size that is the value of M in M-QAM modulation format for an arbitrary chosen bit-error-rate, the average interference power and the ratio of $\frac{\gamma_s}{\gamma_p}$, and is given by the following expression:

$$M = 1 + K\left(\frac{\gamma_s}{\gamma_p}\right) Q_{int}$$

and,

$$M = 2^n = 2^{log_2\left(1 + K\left(\frac{\gamma_s}{\gamma_p}\right) Q_{int}\right)} \tag{9.4}$$

where

$$K = \frac{-1.5}{In(5P_b)} < 1 \tag{9.5}$$

and n is the number of bits per symbol. However, for $M < 4$ which is assumed for BPSK, the error rate is given in [29]. Therefore, the Ergodic channel capacity under average interference power constraint and given P_b is:

$$\frac{C_{er}}{B} = \max_{P(\gamma_s,\gamma_p,\xi)} \iint log_2\left(1 + \frac{K\gamma_s P(\gamma_{s,}\gamma_{p,}\xi)}{N_0 B}\right) f_s(\gamma_s) f_p(\gamma_p)(\alpha f_1(\xi) + \bar{\alpha} f_0(\xi)) d\gamma_s d\gamma_p \tag{9.6}$$

With the constraint:

$$\iint \gamma_p P(\gamma_{s,}\gamma_{p,}\xi) f_s(\gamma_s) f_p(\gamma_p) f_1(\xi) d\gamma_s d\gamma_p \leq Q_{int} \tag{9.7}$$

The transmitter power $P(\gamma_{s,}\gamma_{p,}\xi)$ of the cognitive transmitter is the joint function of secondary channel gain, primary channel gain and sensing metric. Asghari and Aissa [21] provided a mathematical expression for the channel capacity of the secondary user's link for power adaptation policies under the interference and peak power constraint with the sensing pdf's. However, the primary user's link channel power gain γ_p, which is presented in Eq. (9.6), was not considered in [30]. Now, we have to maximize the Ergodic capacity of the system as given by Eq. (9.6) by simultaneously satisfying the constraint given in Eq. (9.7). Therefore, to yield the optimal power allocation $P(\gamma_{s,}\gamma_{p,}\xi)$, we form the Lagrangian multiplier, λ [31] and construct the following Lagrangian function:

$$\begin{aligned} & L(P(\gamma_{s,}\gamma_{p,}\xi), \lambda) \\ & = \iint log_2\left(1 + \frac{K\gamma_s P(\gamma_{s,}\gamma_{p,}\xi)}{N_0 B}\right) f_s(\gamma_s) f_p(\gamma_p)(\alpha f_1(\xi) + \bar{\alpha} f_0(\xi)) d\gamma_s d\gamma_p \\ & - \lambda\left(\iint \gamma_p P(\gamma_{s,}\gamma_{p,}\xi) f_s(\gamma_s) f_p(\gamma_p) f_1(\xi) d\gamma_s d\gamma_p - Q_{int}\right) \end{aligned} \tag{9.8}$$

$L(P(\gamma_{s,}\gamma_{p,}\xi), \lambda)$ is the concave function of $P(\gamma_{s,}\gamma_{p,}\xi)$, and the interference constraint defined in Eq. (9.7) is convex, therefore the first order condition that is the derivative of $L(P(\gamma_{s,}\gamma_{p,}\xi), \lambda)$ with respect to $P(\gamma_{s,}\gamma_{p,}\xi)$ is a sufficient KKT condition for the optimality [32] and the sufficient condition allows us to obtain a solution. Now, the optimization problem being convex (i.e. this problem is a maximization problem with a concave cost function and a convex set of constraints), there is a unique solution. Hence, the solution given by the sufficient condition is the only solution and is given by:

$$\begin{aligned} \frac{\partial L(P,\lambda)}{\partial P} = & \frac{1}{1 + \frac{K\gamma_s P(\gamma_{s,}\gamma_{p,}\xi)}{N_0 B}} \frac{K\gamma_s}{N_0 B} (\alpha f_1(\xi) + \bar{\alpha} f_0(\xi)) f_s(\gamma_s) f_p(\gamma_p) \\ & - \lambda \gamma_p f_1(\xi) f_s(\gamma_s) f_p(\gamma_p) = 0 \end{aligned}$$

or

$$\frac{\partial L(P,\lambda)}{\partial P}=\frac{K\gamma_{\mathrm{s}}}{N_0B+K\gamma_sP(\gamma_{\mathrm{s}},\gamma_{\mathrm{p}},\xi)}(\alpha f_1(\xi)+\bar{\alpha}f_0(\xi)) \\ -\lambda\gamma_{\mathrm{p}}f_1(\xi)=0 \tag{9.9}$$

and

$$P(\gamma_{\mathrm{s}},\gamma_{\mathrm{p}},\xi)=\frac{\gamma_{\mu}(\xi)}{\lambda\gamma_{\mathrm{p}}}-\frac{N_0B}{\gamma_{\mathrm{s}}K} \tag{9.10a}$$

If we assume $P(\gamma_{\mathrm{s}},\gamma_{\mathrm{p}},\xi)=0$ for some values of $\gamma_{\mathrm{s}},\gamma_{\mathrm{p}}$, and ξ, which take place in the condition defined below and after putting $P(\gamma_{\mathrm{s}},\gamma_{\mathrm{p}},\xi)=0$ in Eq. (9.10a), we get:

$$\frac{\gamma_{\mathrm{p}}}{\gamma_{\mathrm{s}}}>\frac{\gamma_{\mu}(\xi)K}{\lambda N_0B} \tag{9.10b}$$

Therefore, from Eqs. (9.10a) and (9.10b), the power $P(\gamma_{\mathrm{s}},\gamma_{\mathrm{p}},\xi)$ is adapted to maximize the Ergodic channel capacity as defined in Eq. (9.6), which is given as:

$$P(\gamma_{\mathrm{s}},\gamma_{\mathrm{p}},\xi)=\begin{cases}\frac{\gamma_{\mu}(\xi)}{\lambda\gamma_{\mathrm{p}}}-\frac{N_0B}{\gamma_sK}, & \frac{\gamma_{\mathrm{p}}}{\gamma_{\mathrm{s}}}\leq\frac{\gamma_{\mu}(\xi)K}{\lambda N_0B}\\ 0, & \frac{\gamma_{\mathrm{p}}}{\gamma_{\mathrm{s}}}>\frac{\gamma_{\mu}(\xi)K}{\lambda N_0B}\end{cases} \tag{9.10c}$$

where

$$\gamma_{\mu}(\xi)=\alpha+\bar{\alpha}\frac{f_0(\xi)}{f_1(\xi)}. \tag{9.11}$$

The optimal power allocation obtained by Eq. (9.10a) represents the greater transmission power, which can be used when γ_{s} increases and γ_p decreases and the average interference constraint at the primary receiver is satisfied. This is due to the primary user's fading channel advantage which enhances the cognitive user's capacity. The sensing decision is considered in Eq. (9.11), where we observe that when the conditional probability that the PU is idle $(f_0(\xi))$ gets higher than that of being active $(f_1(\xi))$, then the value of $\gamma_{\mu}(\xi)$ has an ascending behavior and $\gamma_{\mu}(\xi)>1$, otherwise, $\gamma_{\mu}(\xi)<1$. Therefore, as the conditional probability distribution of the primary user being idle gets higher than being active, $\gamma_{\mu}(\xi)$ increases and, consequently, we can increase the secondary user's transmission power without causing harmful interference to the PR. Note that when $\gamma_{\mu}(\xi)=1$, the ST has no information about the primary user's activity. Accordingly, it considers that the primary user is always active $\left(\frac{f_0(\xi)}{f_1(\xi)} = 1\right)$ and continuously transmits with the same power level with which it is already transmitting. For $\gamma_\mu(\xi)$, the values of $f_0(\xi)$ and $f_1(\xi)$ should be taken at that value of ξ which is computed by the sensing detector for a given detection and false alarm probabilities. A higher value of ξ as compared to threshold that is the energy computed in a particular time interval over a spectrum, indicates the presence of PU signal, and vice versa [13]. However, if we modify the probability of false alarm, the value of ξ is also modified. By substituting Eq. (9.10a) in Eq. (9.7), we get:

$$\iint_0^{\frac{K\gamma_\mu(\xi)}{\lambda_0 N_0 B}} \left(\frac{\gamma_\mu(\xi)}{\lambda_0} - \frac{N_0 B \gamma_p}{\gamma_s K}\right) f_s(\gamma_s) f_p(\gamma_p) f_1(\xi) d\gamma_s d\gamma_p = Q_{int}$$

where λ_0 is determined in such a way that the average interference power constraint in Eq. (9.7) is equal to Q_{int}

$$\iint_0^{\frac{K\gamma_\mu(\xi)}{\lambda_0 N_0 B}} \left(\frac{\gamma_\mu(\xi)}{\lambda_0 N_0 B} - \frac{\gamma_p}{\gamma_s K}\right) f_s(\gamma_s) f_p(\gamma_p) f_1(\xi) d\gamma_s d\gamma_p = \frac{Q_{int}}{N_0 B} = \Phi$$

or

$$\iint_0^{K\gamma_\mu(\xi)\gamma_0} \left(\gamma_\mu(\xi)\gamma_0 - \frac{\gamma_p}{\gamma_s K}\right) f_s(\gamma_s) f_p(\gamma_p) f_1(\xi) d\gamma_s d\gamma_p = \Phi \tag{9.12}$$

where $\gamma_0 = \frac{1}{\lambda_0 N_0 B}$, and $\Phi = \frac{Q_{int}}{N_0 B}$ is the average SNR [4]. By substituting Eq. (9.10a) in Eq. (9.6), gives the following Ergodic channel capacity expression:

$$\frac{C_{er}}{B} = \int_{\frac{\gamma_s}{\gamma_p} \geq \frac{N_0 B \lambda_0}{K\gamma_\mu(\xi)} = \frac{1}{K\gamma_0\gamma_\mu(\xi)}}^{\infty} log_2\left(1 + \frac{K\gamma_s}{N_0 B}\left[\frac{\gamma_\mu(\xi)}{\lambda_0 \gamma_p} - \frac{N_0 B}{\gamma_s K}\right]\right) f_s(\gamma_s) f_p(\gamma_p)(\alpha f_1(\xi) + \bar{\alpha} f_0(\xi)) d\gamma_s d\gamma_p$$

or

$$\frac{C_{er}}{B} = \int_{\frac{\gamma_s}{\gamma_p} \geq \frac{N_0 B \lambda_0}{K\gamma_\mu(\xi)} = \frac{1}{K\gamma_0\gamma_\mu(\xi)}}^{\infty} log_2\left(\frac{K\gamma_s\gamma_\mu(\xi)}{N_0 B \lambda_0 \gamma_p}\right) f_s(\gamma_s) f_p(\gamma_p)(\alpha f_1(\xi) + \bar{\alpha} f_0(\xi)) d\gamma_s d\gamma_p$$

or

$$\frac{C_{er}}{B} = \int_{\frac{\gamma_s}{\gamma_p} \geq \frac{N_0 B \lambda_0}{K\gamma_\mu(\xi)} = \frac{1}{K\gamma_0\gamma_\mu(\xi)}}^{\infty} log_2\left(\frac{K\gamma_s\gamma_\mu(\xi)\gamma_0}{\gamma_p}\right) f_s(\gamma_s) f_p(\gamma_p)(\alpha f_1(\xi) + \bar{\alpha} f_0(\xi)) d\gamma_s d\gamma_p \tag{9.13}$$

or

$$\frac{C_{er}}{B} = \underset{\frac{\gamma_s}{\gamma_p} \geq \frac{N_0 B \lambda_0}{K\gamma_\mu(\xi)}}{E_{\gamma_s,\gamma_p,\xi}} \left[log_2 \left(\frac{K\gamma_u(\xi)\gamma_s}{\lambda_0 N_0 B \gamma_p} \right) \right] \tag{9.14}$$

where C_{er} denotes the Ergodic capacity and $E[.]$ denotes the expectation operator. Equation (9.14) is similar to that presented in [21, Eq. (30)] except the term γ_p, which is due to the consideration of the primary channel gain in the cognitive user's system capacity. However, when only the power adaptation policy is considered instead of power and rate adaptation policy, then the additional constraint of Eq. (9.5) is not needed, and the Ergodic channel capacity of adaptive power transmission policy is given by the following mathematical expression, substituting $K = 1$ in Eq. (9.14):

$$\frac{C_{er}}{B} = \underset{\frac{\gamma_s}{\gamma_p} \geq \frac{N_0 B \lambda_0}{\gamma_\mu(\xi)}}{E_{\gamma_s,\gamma_p,\xi}} \left[log_2 \left(\frac{\gamma_u(\xi)\gamma_s}{\lambda_0 N_0 B \gamma_p} \right) \right] \tag{9.15}$$

Comparing the Ergodic capacity of power adaptation policy as given by Eq. (9.15) and rate and power adaptation policy for *M*-QAM modulation format in Eq. (9.14), Eq. (9.14) reveals that there is an effective power loss of *K* for adaptive *M*-QAM compared to that of Eq. (9.15). However, for the adaptive power transmission policy, the probability of error is significantly greater and is fixed, at 0.0446, in comparison to that of the adaptive rate and power transmission policy, where the probability of bit error can vary according to the quality-of-service requirement.

9.4 Effect of Channel Conditions

In this section, we explore the fading channel effect on the cognitive radio communication system performance and numerically compute the Ergodic channel capacity in different fading environments.

- Nakagami-*m* fading

The Nakagami-*m* distribution often provides the best fit to the urban [33] and indoor [34] multipath propagation and gives AWGN, Rayleigh and Rician fading channel models by adjusting the fading parameter *m*, which is the ratio of line-of-sight (LOS) signal power to the multipath signal power. The channel fading model based on Nakagami distribution, both γ_s and γ_p, would be distributed according to the following Gamma distribution [6]:

$$f(\gamma) = \frac{m^m \gamma^{m-1}}{\Gamma(m)} e^{-m\gamma}$$

where m and γ are shape parameter and channel power gain, respectively. Therefore, the pdf $f_s(\gamma_s)f_p(\gamma_p)$ is given as:

$$f_s(\gamma_s)f_p(\gamma_p) = \left(\frac{m_0}{m_1}\right)^{m_0} \frac{z^{m_1-1}}{\beta(m_0, m_1)\left(x + \frac{m_0}{m_1}\right)^{m_0+m_1}} \tag{9.16}$$

where m_0 and m_1 are m parameters [6] for γ_p and γ_s, respectively. $\frac{\gamma_p}{\gamma_s} = z$, and z is a random variable. $\beta(.)$ is the beta function. When $m_0 = m_1 = m$, the Eq. (9.16) becomes:

$$f_s(\gamma_s)f_p(\gamma_p) = \frac{z^{m-1}}{\beta(m, m)(z+1)^{2m}} \tag{9.17}$$

By substituting Eq. (9.17) in (9.12), we yield the following value of secondary transmit power, which satisfies the average interference constraint for the Nakagami-m fading channel:

$$\iint_0^{K\gamma_\mu(\xi)\gamma_0} \left(\gamma_\mu(\xi)\gamma_0 - \frac{\gamma_p}{\gamma_s K}\right) \frac{z^{m-1}}{\beta(m, m)(z+1)^{2m}} f_1(\xi)d\gamma_s d\gamma_p = \frac{Q_{int}}{N_0 B} \tag{9.18}$$

and the Ergodic channel capacity from Eq. (9.13), for the Nakagami-m fading environment is given by:

$$\frac{C_{er}}{B} = \int_{\frac{\gamma_s}{\gamma_p} \geq \frac{N_0 B \lambda_0}{K\gamma_\mu(\xi)} = \frac{1}{K\gamma_0\gamma_\mu(\xi)}}^{\infty} log_2\left(\frac{K\gamma_s\gamma_\mu(\xi)\gamma_0}{\gamma_p}\right) \frac{z^{m-1}}{\beta(m, m)(z+1)^{2m}} (\alpha f_1(\xi) + \bar{\alpha} f_0(\xi))d\gamma_s d\gamma_p \tag{9.19}$$

9.4.1 Rayleigh Fading

The Nakagami-m distribution with fading parameter equal to 1 represents the Rayleigh fading channel, and the pdf $f_s(\gamma_s)f_p(\gamma_p)$ will have log-logistic distribution [6]. By substituting $m = 1$ in Eq. (9.18), we get:

$$\int_0^{K\gamma_\mu(\xi)\gamma_0} \left(\gamma_\mu(\xi)\gamma_0 - \frac{z}{K}\right) \frac{1}{(1+z)^2} f_1(\xi) dz = \frac{Q_{\text{int}}}{N_0 B}$$

or

$$f_1(\xi)\left(-\frac{1}{K}\log_2(1 + K\gamma_\mu(\xi)\gamma_0) + \gamma_\mu(\xi)\gamma_0\right) = \frac{Q_{\text{int}}}{N_0 B} = \Phi \tag{9.20}$$

Therefore the capacity of the cognitive radio communication system in the Rayleigh fading environment is achieved by putting $m = 1$ in Eq. (9.19):

$$\frac{C_{\text{er}}}{B} = \int_{\frac{1}{\gamma_0\gamma_\mu(\xi)}}^{\infty} log_2(K\gamma_0\gamma_\mu(\xi)z) \frac{1}{(1+z)^2} (\alpha f_1(\xi) + \bar{\alpha} f_0(\xi)) dz$$
$$\text{or } \frac{C_{\text{er}}}{B} = (\alpha f_1(\xi) + \bar{\alpha} f_0(\xi)) \log_2(1 + K\gamma_\mu(\xi)\gamma_0(\Phi)) \tag{9.21}$$

where $\gamma_0(\Phi)$ is from the Eq. (9.20) for a given Φ. Equation (9.21) gives the Ergodic channel capacity of adaptive rate and power transmission policy under the Rayleigh fading environment. Further, the capacity of adaptive power transmission policy under the Rayleigh fading environment is as given below:

$$\frac{C_{\text{er}}}{B} = (\alpha f_1(\xi) + \bar{\alpha} f_0(\xi)) log_2(1 + \gamma_\mu(\xi)\gamma_0(\alpha)) \tag{9.22}$$

9.4.2 Rician Fading

The Nakagami-*m* distribution with the fading parameter greater than or equal to 2 represents the Rician fading channel. Now, by substituting $m = 2$ in Eq. (9.18), we get the following expression for the Rician fading channel:

$$\int_0^{K\gamma_\mu(\xi)\gamma_0} \left(\gamma_\mu(\xi)\gamma_0 - \frac{z}{K}\right) \frac{6z}{(1+z)^4} f_1(\xi) dz = \frac{Q_{\text{int}}}{N_0 B}$$

or

$$f_1(\xi)\left(\frac{3K\gamma_0\gamma_\mu(\xi) + 2}{6K(1 + K\gamma_0\gamma_\mu(\xi))^2} + \frac{\gamma_0\gamma_\mu(\xi)}{6} - \frac{2}{6K}\right) = \frac{Q_{\text{int}}}{N_0 B} = \Phi \tag{9.23}$$

Therefore, for the spectrum sharing system operating under the predefined power constraints and a target BER value P_b, the Rician fading channel capacity expression of the secondary user's link, based on the adaptive rate and power *M*-QAM transmission policy, is obtained by putting $m = 2$ in Eq. (9.19):

$$\frac{C_{\text{er}}}{B} = \int_{\frac{1}{K\gamma_0\gamma_\mu(\xi)}}^{\infty} log_2\left(K\gamma_0(\Phi)\gamma_\mu(\xi)z\right)\frac{6z}{(1+z)^4}(\alpha f_1(\xi) + \bar{\alpha} f_0(\xi))dz \tag{9.24}$$

where $\gamma_0(\Phi)$ is from Eq. (9.23) for a given Φ. Furthermore, the Ergodic channel capacity of adaptive power transmission policy in the Rician fading environment is given by the following expression:

$$\frac{C_{\text{er}}}{B} = \int_{\frac{1}{\gamma_0\gamma_\mu(\xi)}}^{\infty} log_2\left(\gamma_0\gamma_\mu(\xi)z\right)\frac{6z}{(1+z)^4}(\alpha f_1(\xi) + \bar{\alpha} f_0(\xi))dz \tag{9.25}$$

Similarly, we can compute the channel capacity for different fading parameter values, however it leads to cumbersome mathematical expressions.

9.5 Simulation Results

In this section, we numerically simulate the proposed spectrum sharing system model that operates under the constraints on the average received-interference power in the Nakagami-*m* fading environment for adaptation strategies such as variable power and variable rate and power, as presented in the preceding Sects. 9.3 and 9.4.

The position of terminals as shown in Fig. 9.1 is assumed in such a way that $d_s = d_p = 1$ (unit) and $d_{\text{m}} = 3$ (unit). The channel gains $(\gamma_{\text{s}})^{1/2}$ and $\left(\gamma_{\text{p}}\right)^{1/2}$ are distributed according to the Nakagami-*m* fading pdf. Furthermore, we assumed $N_0B = 1$ and the sensing detector computes the sensing-information metric for $N = 30$ observation samples. We suppose that the primary user remains active at 50% of the time ($\alpha = 0.5$) and have set the PU's transmit power $P_t = 1$. Figure 9.2a illustrates the distribution of conditional probabilities $f_0(\xi)$ and $f_1(\xi)$ corresponding to the different values of energy detected by sensing detector in the particular number of samples. Moreover, these distributions are used for the computation of $\gamma_\mu(\xi)$ for different detected energy values in a particular interval as shown in Fig. 9.2b. Three regions have been recognized for the parameter $\gamma_\mu(\xi)$, namely, $\gamma_\mu(\xi) > 1$, $\gamma_\mu(\xi) = 1$ and $\gamma_\mu(\xi) < 1$. In Fig. 9.2b, when $\gamma_\mu(\xi) > 1$ represent that the probability of the PU to be idle is higher than that of being active otherwise, $\gamma_\mu(\xi) < 1$. The power and rate are adapted according to the channel gains and the sensing information. Moreover, the higher power levels are used by secondary users when the probability of the primary user being inactive is significantly more (higher values of $\gamma_\mu(\xi)$) in comparison to the case for which $\gamma_\mu(\xi)$ is less. We have considered the bit-error-probability 10^{-2}, 10^{-4} and 10^{-6} for the adaptive rate and power transmission policy for these two cases: $\left(\gamma_\mu(\xi) > 1 \text{ and } \gamma_\mu(\xi) < 1\right)$.

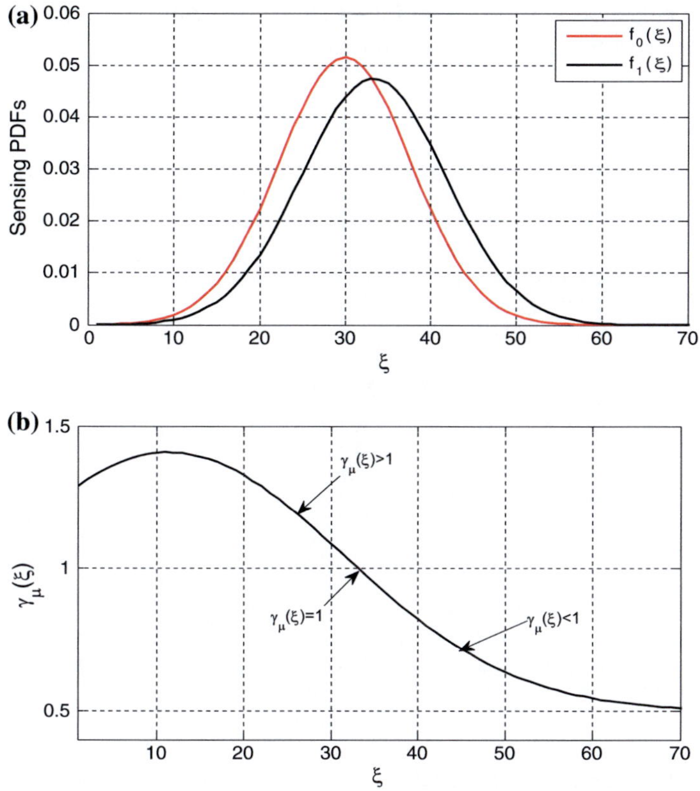

Fig. 9.2 The soft sensing information **a** Spectrum sensing probability density functions given that the primary user is idle $f_0(\xi)$ and active $f_1(\xi)$ [21], and **b** $\gamma_\mu(\xi)$ variation for $N = 30$, $P_t = 1$, $\alpha = 0.5$ and $d_m = 3$ [21]

For the Rayleigh fading environment or Nakagami-*m* distribution with $m = 1$, Fig. 9.3a, b shows the variation of the Lagrangian parameter λ and Ergodic channel capacity with Q_{int} for the adaptive power and adaptive rate and power transmission policy, while considering the sensing information metric available at the cognitive user. The simulation results in Fig. 9.3 are presented for the value of parameter $\gamma_\mu(\xi) < 1$. Moreover, Fig. 9.3a shows the optimum value of the Lagrangian parameter for the given Q_{int} and $\gamma_\mu(\xi)$, which satisfy (9.20) and provide the adaptation in transmit power needed for the Rayleigh fading channel. It is clear from Fig. 9.3b that as the interference tolerance (Q_{int}) at the primary receiver increases, the capacity of the secondary user increases due to the increase in transmit power of the secondary user. The Ergodic capacity of adaptive rate and power transmission policy is less in comparison to that of the adaptive power transmission policy, since there is an additional constraint on target BER in the former policy. In addition, as the required BER decreases, the Ergodic capacity of the system is less, as depicted from Fig. 9.3b. For example, the capacity for P_b of 10^{-6} is less than that for $P_b = 10^{-2}$ due to the stricter constraint on the required error rate.

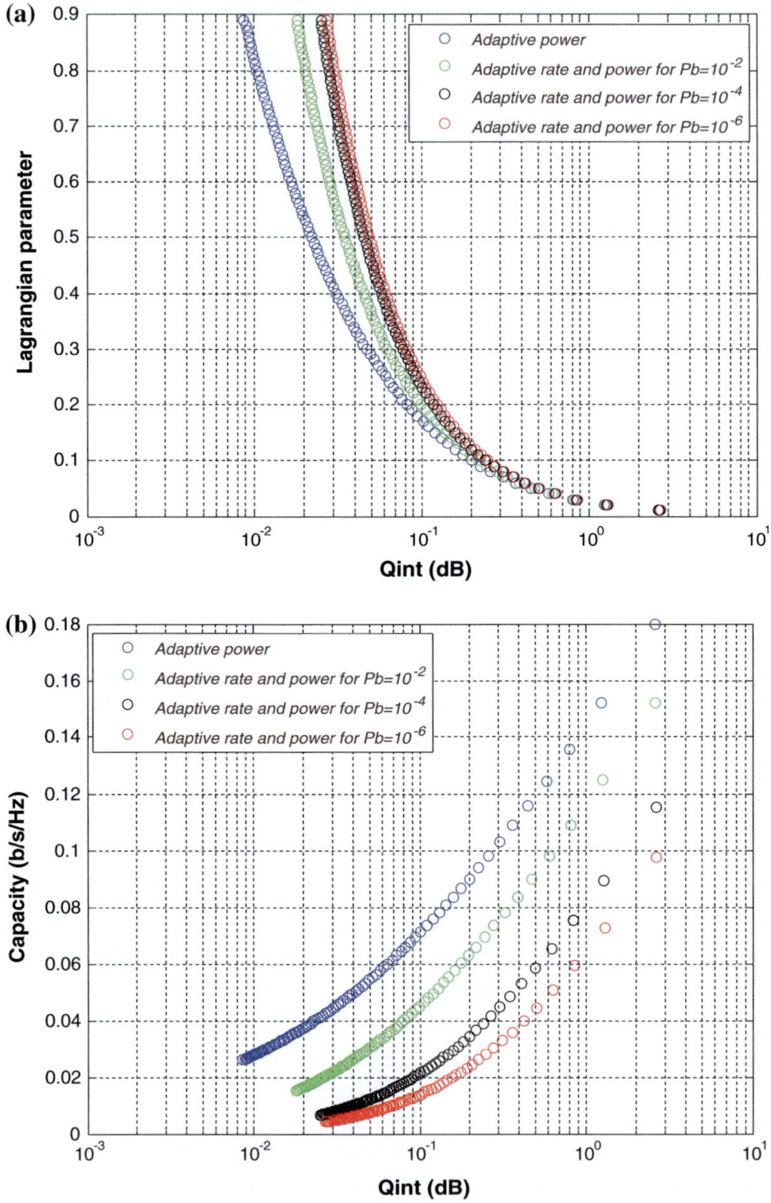

Fig. 9.3 The response of primary receiver interference power constraint for the adaptive power and adaptive rate and power transmission policies in the Rayleigh fading channel for *M*-QAM modulation and $\gamma_{\mu}(\xi) = 0.8$ over **a** the Lagrangian parameter, and **b** Ergodic channel capacity

In Fig. 9.4a, b, we have considered the value of the parameter $\gamma_{\mu}(\xi) > 1$, which shows that the probability of the primary user being active is greater than that of it being inactive so it leads to an increase in the transmit power; consequently the

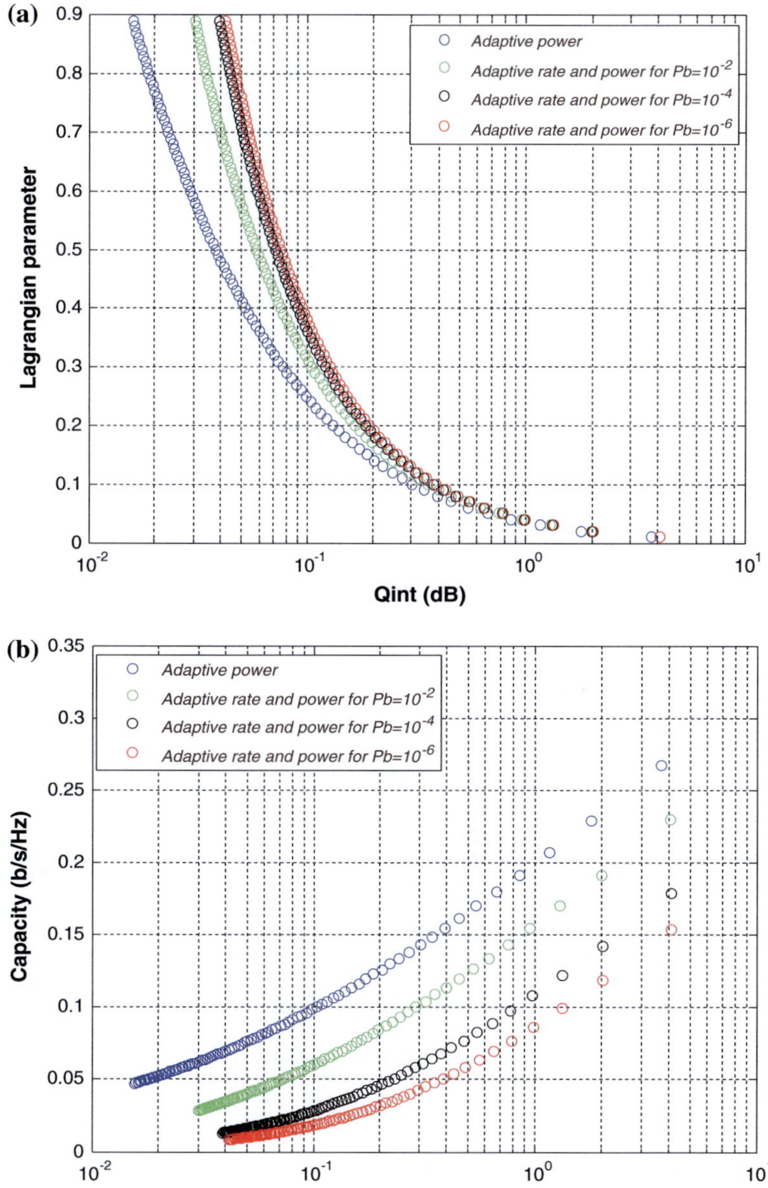

Fig. 9.4 The response of primary receiver interference power constraint for the adaptive power and adaptive rate and power transmission policies in the Rayleigh fading channel for *M*-QAM modulation and $\gamma_{\mu}(\xi) = 1.2$ over **a** the Lagrangian parameter, and **b** Ergodic channel capacity

result is an increase in capacity of the secondary user in comparison to the capacity that is shown in Fig. 9.3b, where $\gamma_{\mu}(\xi) < 1$. Further, without considering the sensing information available at the secondary user, the capacity variations with Q_{int} presented in Fig. 9.5 have been validated with Fig. 3 of [6], which is the case when only the average interference power constraint is considered. The effect of average interference power constraint Q_{int} on the capacity and Lagrangian parameter λ in the Nakagami-m fading environment with $m = 2$, that is, for the Rician fading channel for the adaptive power and adaptive rate and power transmission, is shown in Fig. 9.6a, b for the case when $\gamma_{\mu}(\xi) < 1$. Moreover, for the adaptive power and adaptive rate and power transmission policy, the comparison of the capacity for three cases of BER that is 10^{-2}, 10^{-4} and 10^{-6} is presented in Fig. 9.6b.

Figure 9.7a, b present the Lagrangian parameter and capacity in the Rician fading environment (Nakagami-m distribution with $m = 2$) for $\gamma_{\mu}(\xi) > 1$. The comparison of Fig. 9.6b with 9.7b reveals that the significant enhancement in the capacity is due to the higher power adaptation of the secondary transmitter. Moreover, the capacity comparison between Rayleigh and Rician fading environments demonstrates that the capacity of the cognitive radio network for the latter case is less than that of the former for a given Q_{int}. The reason lies in the fact that

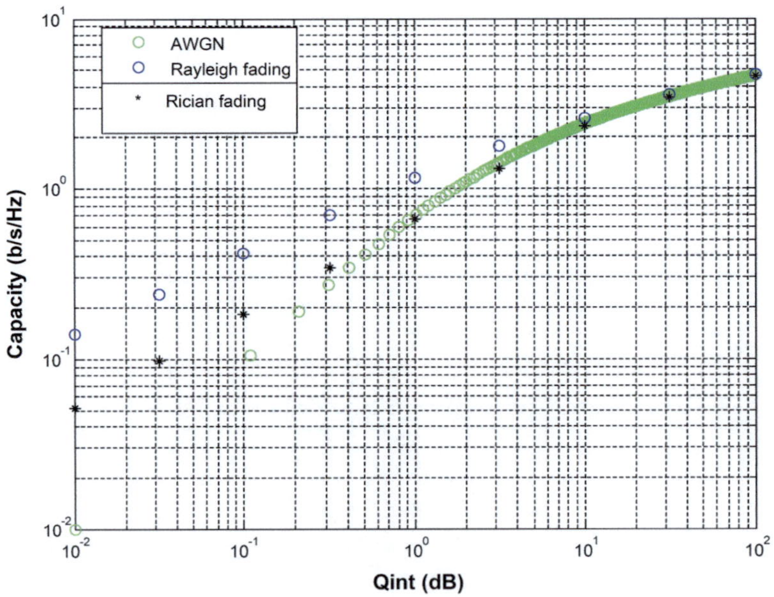

Fig. 9.5 The capacity under the average interference-power constraint as reported in [6]

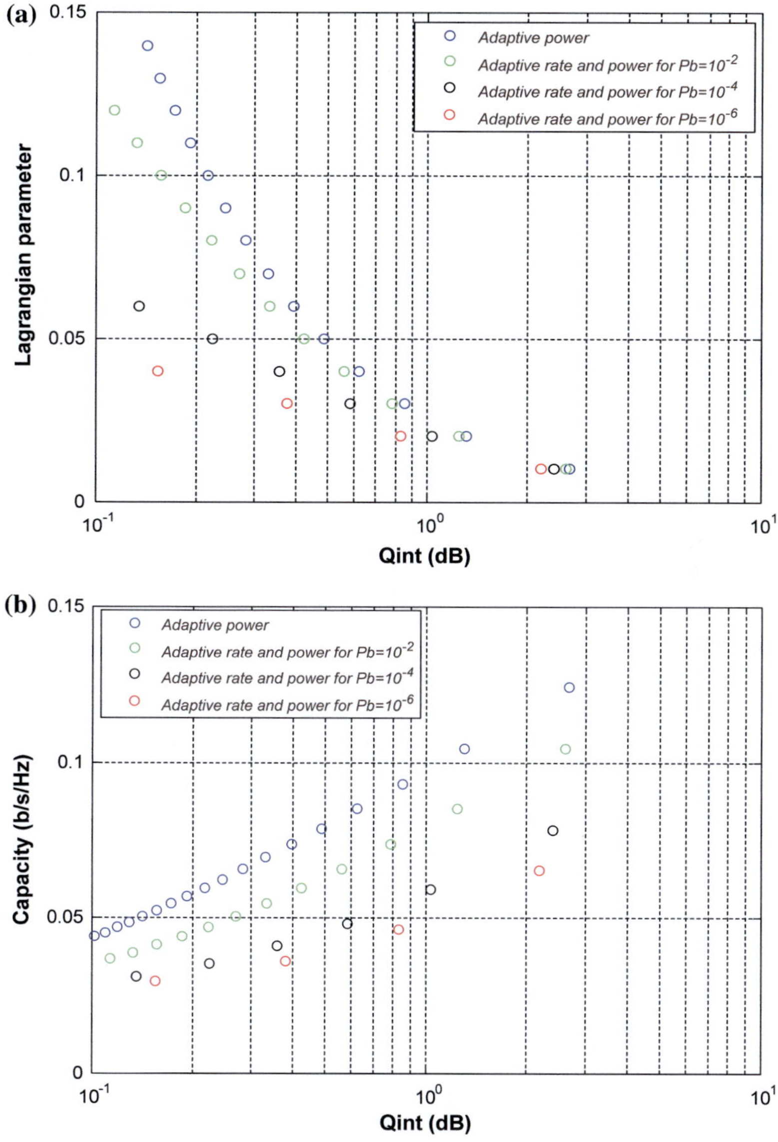

Fig. 9.6 The response of primary receiver interference power constraint for the adaptive power and adaptive rate and power transmission policies in the Rician fading channel for M-QAM modulation and $\gamma_{\mu}(\xi) = 0.8$ over **a** the Lagrangian parameter, and **b** Ergodic channel capacity

severe primary channel Rayleigh fading gives an advantage to the secondary transmitter to increase its transmission power while keeping the interference power constraint constant in comparison to the Rician fading channel with $m = 2$, which is less severe due to the presence of a line-of-sight (LOS) component. Moreover,

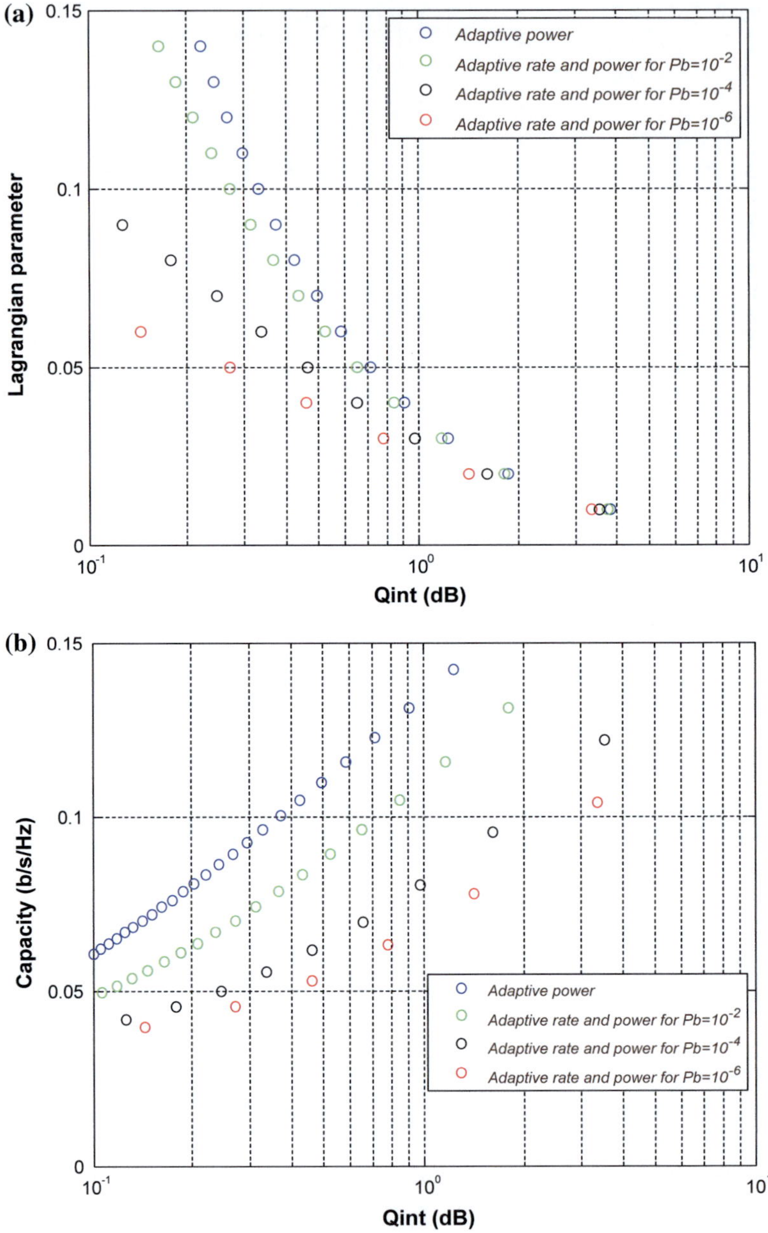

Fig. 9.7 The response of primary receiver interference power constraint for the adaptive power and adaptive rate and power transmission policies in the Rician fading channel for *M*-QAM modulation and $\gamma_{\mu}(\xi) = 1.2$ over **a** the Lagrangian parameter, and **b** Ergodic channel capacity

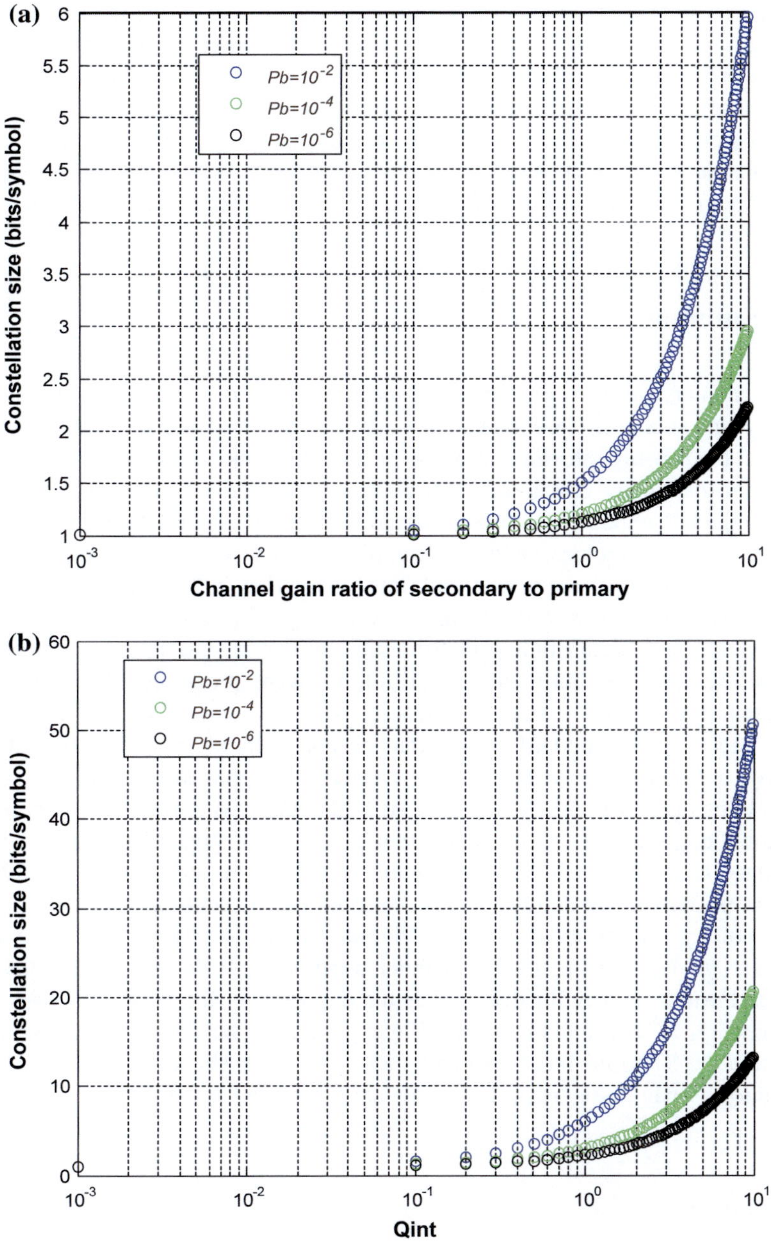

Fig. 9.8 The constellation size adaptation **a** according to the signal-to-noise power ratios of secondary-to-primary user for $Q_{int} = 0dB$, and **b** with the interference power constraint, for the given signal-to-noise power ratios (10 dB) of secondary-to-primary user for adaptive power and rate transmission policy with $P_b = 10^{-2}$, 10^{-4}, and 10^{-6}

Fig. 9.8a, b shows the adaptation in the constellation size according to the channel gain ratio of the secondary-to-primary user and average interference power for different BER, respectively. It is also clear from Fig. 9.8a, b that the number of bits per symbol or the constellation size of the modulation technique increases as the channel gain ratio of the ST to PR increases, or the average interference power limit at PR increases for the chosen BER. Thus significantly better channel conditions of the secondary link lead to the adaptation of higher modulation format.

9.6 Summary

In this chapter, we have considered a spectrum sharing concept for the cognitive radio system where the secondary user's transmit power and rate can be adjusted based on the sensing information of the primary user and secondary user, as well as secondary-to-primary user's fading environment. In addition, the spectrum sharing system operates under the average interference power constraints of the PR. In this context, we have demonstrated the Ergodic capacity of the cognitive radio communication system with power and rate adaptation policy in different fading environments for a chosen BER. Since the Nakagami-*m* distribution is fit for both the Rayleigh and Rician fading distributions by varying the fading parameter, the Ergodic capacity for both these distributions were presented. The numerically simulated results for the Ergodic capacity were presented for both the adaptive power and adaptive rate and power transmission policies, which revealed that the adaptive power transmission has more capacity than that of the adaptive rate and power transmission policy at the cost of BER. Moreover, we have demonstrated that knowledge of the sensing parameter provides an opportunity to control the secondary user's transmission parameters, such as rate and power, according to different primary users activity levels observed by the sensing detector. However, the secondary transmitter can adapt different modulation by varying the value of M in M-QAM according to the channel conditions, BER and interference constraints. Further, it was illustrated that the capacity, in the case of Rician fading environment, is lower than that of Rayleigh fading because a LOS component present in the ST to PR has provided a more prominent effect on the capacity of the secondary user in comparison to that present in the ST-to-SR link.

References

1. E. Biglieri, J. Proakis, S. Shamai, Fading channels: Information theoretic and communications aspects. IEEE Trans. Inf. Theory **44**(6), 2619–2692 (1998)
2. A.-J. Goldsmith, P. Varaiya, Capacity of fading channels with channel side information. IEEE Trans. Inf. Theory **43**(6), 1986–1992 (1997)

3. M.-A. Khojastepour, B. Aazhang, The capacity of average and peak power constrained fading channels with channel side information, *Proceedings of IEEE Wireless Communication and Networking Conference* (WCNC'04) (Atlanta, CA, USA, Mar. 2004), pp. 77–82
4. A. Ghasemi, E.S. Sousa, Capacity of fading channels under spectrum-sharing constraints, *Proceedings of IEEE International Conference on Communication* (ICC'06) (Istanbul, Turkey, June 2006), pp. 4373–4378
5. L. Musavian, S. Aissa, Ergodic and outage capacities of spectrum sharing systems in fading channels, *Proceedings of IEEE Global Telecommunication Conference* (GLOBECOM'07) (Washington, DC, USA, Nov. 2007), pp. 3327–3331
6. A. Ghasemand, E.-S. Sousa, Fundamental limits of spectrum-sharing in fading environments. IEEE Trans. Wirel. Commun. **6**(2), 649–658 (2007)
7. B. Wang, K.-J.-R. Liu, Advances in cognitive radio networks: a survey. IEEE J. Sel. Top. Sign. Proces. **5**(1), 5–23 (2011)
8. V. Asghari, S. Aissa, Resource sharing in cognitive radio systems: outage capacity and power allocation under soft sensing. *Proceedinds of IEEE Global Telecommunication Conference* (GLOBECOM'08) (New Orleans, LA, USA, Nov. 2008), pp. 1–5
9. I.-F. Akyildiz, W.-Y. Lee, M.-C. Vuran, S. Mohanty, NeXt generation/dynamic spectrum access/cognitive radio wireless networks: a survey. Comput. Netw. **50**(13), 2127–2159 (2006)
10. S. Pandit, G. Singh, Throughput maximization with reduced data loss rate in cognitive radio network. Telecommun. Syst. **57**(2), 209–215 (2014)
11. T. Yucek, H. Arslan, A survey of spectrum sensing algorithms for cognitive radio applications. IEEE Commun. Surv. Tutorials **11**(1), 116–130 (2009)
12. H. Urkowitz, Energy detection of unknown deterministic signals. Proc. IEEE **55**(4), 523–531 (1967)
13. S. Srinivasa, S.A. Jafar, Soft sensing and optimal power control for cognitive radio. IEEE Trans. Wirel. Commun. **9**(12), 3638–3649 (2010)
14. Z. Rezki, M.-S. Alouini, Ergodic capacity of cognitive radio under imperfect channel state information. IEEE Trans. Veh. Technol. **61**(5), 2108–2119 (2012)
15. S. Pandit, Medium access control protocol for spectrum sharing in cognitive radio communication system. *Ph D Thesis, Jaypee University of Information Technology*, Waknaghat, Solan-173234, India
16. S. Parsaeefard, A.-R. Sharafat, Robust distributed power control in cognitive radio networks. IEEE Trans. Mob. Comput. **12**(4), 609–620 (2013)
17. G. Zhao, C. Yang, G.-Y. Li, D. Li, A.-C.-K. Soong, Power and channel allocation for cooperative relay in cognitive radio networks, IEEE J. Sel. Top. Sign. Process. **5**(1), 151–159 (2011)
18. X. Kang, H.-K. Garg, Y.-C. Liang, R. Zhiang, Optimum power allocation for OFDM-based cognitive radio with new primary transmission protection criteria. IEEE Trans. Wirel. Commun. **9**(6), 2066–2075 (2010)
19. Q. Li, K.-H. Li, K.-C. Teh, Diversity-multiplexing tradeoff of wireless communication systems with user cooperation. IEEE Trans. Inf. Theory **57**(9), 5794–5819 (2011)
20. A. Sendonaris, E. Erkip, B. Aazhang, User cooperation diversity—part I: system description. IEEE Trans. Commun. **51**(11), 1927–1938 (2003)
21. V. Asghari, S. Aissa, Adaptive rate and power transmission in spectrum-sharing systems. IEEE Trans. Wirel. Commun. **9**(10), 3272–3280 (2010)
22. A.-J. Goldsmith, S.-G. Chua, Variable-rate variable-power MQAM for fading channels. IEEE Trans. Commun. **45**(10), 1218–1230 (1997)
23. M. Abramowitz, I.A. Stegun, *Handbook of Mathematical Functions: with Formulas, Graphs, and Mathematical Tables*, 9th edn. (Dover, New York, 1972)
24. S.-S. Tzeng, C.-W. Huang, Effective throughput maximization for in-band sensing and transmission in cognitive radio networks. Wireless Netw. **17**(4), 1025–1029 (2011)
25. S. Haykin, Cognitive radio: brain-empowered wireless communications. IEEE J. Sel. Areas Commun. **23**(2), 201–220 (2005)

26. P. Potier, C. Sorrells, Y. Wang, L. Qian, H. Li, Spectrum inpainting: a new framework for spectrum status determination in large cognitive radio networks. Wireless Netw. **20**(3), 423–439 (2014)
27. J.-K. Cavers, Variable rate transmission for Rayleigh fading channels. IEEE Trans. Commun. **20**(1), 15–22 (1972)
28. W.-T. Webb, R. Steele, Variable rate QAM for mobile radio. IEEE Trans. Commun. **43**(7), 2223–2230 (1995)
29. M.-S. Alouini, A.-J. Goldsmith, Adaptive modulation over Nakagami fading channels. Wireless Pers. Commun. **13**(1–2), 119–143 (2000)
30. X. Liu, S. Shankar, Sensing-based opportunistic channel access. Mob. Netw. Appl. **11**(4), 577–591 (2006)
31. S. Boyd, L. Vandenberghe, *Convex Optimization* (Cambridge University Press, New York, 2004)
32. D.P. Bertsekas, *Nonlinear Programming* (Athena Scientific, Belmont, MA, 1995)
33. H. Suzuki, A statistical model for urban multipath propagation. IEEE Trans. Commun. **25**(7), 673–680 (1977)
34. A.-U. Sheikh, M. Abdi, M. Handforth, Indoor mobile radio channel at 946 MHz: measurements and modeling. *Proceedings of IEEE Vehicular Technology Conference* (VTC '93) (Secaucus, NJ, May 1993), pp. 73–76

Chapter 10
Framework for Cross-Layer Optimization in Cognitive Radio Network

10.1 Introduction

A layered architecture, such as that of the seven-layer Open Systems Interconnection (OSI) and five-layer Transmission Control Protocol (TCP) models, divides the overall networking task into several layers and defines a hierarchy of services to be provided by the individual layers, as shown in Fig. 10.1. The services at the layers are realized by designing protocols for them. This architecture forbids direct communication between non-adjacent layers, and communication between adjacent layers is also limited such that higher-layer protocol merely makes use of services at the lower layers and is not concerned with the details of how the service is being provided. As next-generation wireless communication networks are increasingly occupying center stage in research and development in communication networks, the layered protocol architecture for these networks is coming under close scrutiny from the research community. It is repeatedly argued that although layered architectures have served as an effective model for wired networks, they are not suitable for wireless networks, as there are ample opportunities for introducing new services in wireless and mobile communication, and layered architecture becomes a bottleneck to these opportunities. Supporting multimedia applications and services over wireless networks in a layered architecture becomes challenging due to constraints and heterogeneities such as limited battery power, limited bandwidth, random time-varying fading effects, different protocols and standards, and stringent quality-of-service (QoS) requirements [1]. Hence, protocols can be designed for wireless networks by violating the reference layered architecture, for example, by allowing direct communication between the protocols at non-adjacent layers. The cross-layer design is one such violation of a layered architecture with respect to the reference layered architecture, and represents one way of addressing the challenges and providing reliable and high-quality end-to-end performance for next-generation wireless multimedia communications [1]. Cross-layer design thus includes creating new interfaces between the layers, redefining the layer boundaries,

S. Pandit and G. Singh, *Spectrum Sharing in Cognitive Radio Networks*,
DOI 10.1007/978-3-319-53147-2_10

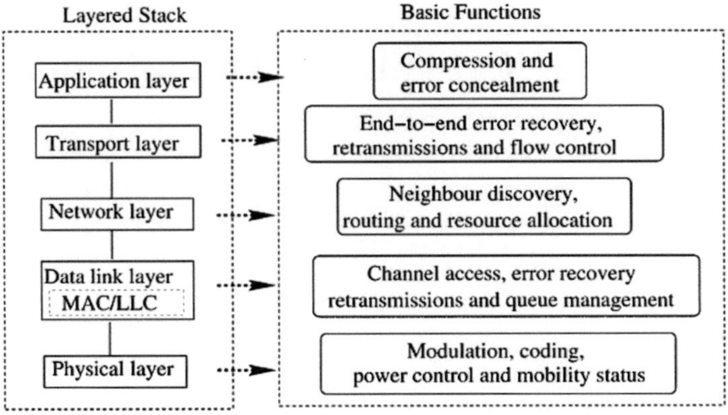

Fig. 10.1 The layered architecture, with the main functions in each layer [3]

designing protocol at a layer based on the details of how another layer is designed, and joint tuning of parameters across layers. Cross-layer design solutions allow optimized operation for mobile devices in the modern heterogeneous wireless environment [2]. Furthermore, protocols for different layers are no longer designed independently, and in this context, there have been a multitude of cross-layer design proposals in the literature. Generally, cross-layer design refers to protocol design performed by actively exploiting the dependency between protocol layers of OSI and TCP models to yield performance gains. This is unlike layering, where the protocols at the different layers are designed independently.

In recent years, there has been an avalanche of cross-layer design proposals for wireless networks. Various researchers have looked at specific aspects of network performance and, approaching the cross-layer design via their interpretation, have presented several design proposals. These involve different layers of the protocol stack, and address both centralized and ad hoc networks. Studies have also investigated the implementation of cross-layer interactions. In [4], the authors present a survey of the literature in the area of cross-layer design and highlight ongoing work in this area. In previous chapters, we have optimized only single MAC layer parameters and have not discussed its impact on the other layer parameters of the cognitive radio network. For example, maximized throughput and energy efficiency of the MAC layer may not reduce the probability of detection errors at the physical (PHY) layer and the expected delays. In order to achieve a more reliable system, an exchange of information among layers must take place. This chapter highlights various research efforts in cross-layer communication in wireless networks. For example, Maharshi et al. [5] proposed a joint PHY and MAC layer design to improve performance. One of the difficulties of such a joint design in general is the lack of analytical expressions that relate the MAC performance to the PHY layer parameters. In this regard, the authors in [5] developed analytical expressions and modeled the MAC layer scheme as a finite-state Markov chain. The networks considered were the ad hoc network and cellular network comprising a base station. In addition, a novel distributed adaptive opportunistic distance sensing multiple access (DAO-DSMA) scheme is proposed in [6] for cross-layer interaction for spectrum sharing in the cooperative preceded cyclic prefix-orthogonal frequency-division multiplexing (PCP-OFDM) system [7, 8]. With joint adaptive protocol design and utility maximization, the proposed design in [6] focuses on an adaptation of the opportunistic router within the framework of a heterogeneous [9] cognitive network (HCN). The authors in [6] also introduced a protocol integrating functionalities of all the layers, from the PHY to the application layer, into the cross-layer protocol. The proposed spectrum sharing scheme in [10] among femtocells can increase spatial reuse in heterogeneous cognitive radio networks. In the link layer, the cognitive networks have priority-based packet transmission [11, 12] for spectrum access. In the network layer, the increased number of mobile routers can improve the performance of the cognitive network by reducing the mean delay and drop probability [6]. In the transport layer, cross-layer TCP throughput optimization [13] can be employed to augment the performance of the cognitive system. The proposed application layer in [6] can automatically select the radio bands and operating modes among interweave, underlay, and overlay paradigms [14], according to the cognitive user data rate, transmission distance, active femto base station density [15], transmission capacity [16], and TCP throughput [6, 17].

The cross-layer-designed cognitive radio is highly interdisciplinary, being concerned with distinct engineering and computer science disciplines including signal processing, communication protocols, and machine learning [18]. Therefore, cognitive radio issues can span all layers of the communication protocol stack, but its basics are largely limited to the PHY and MAC layers. While spectrum sensing is restricted to the PHY and MAC layers, spectrum management (e.g., spectrum handover, decision making, and scheduling) can be related to all the upper layers, which necessitates interaction and coordination between the layers of the protocol stack of the cognitive radio [19]. The exchange of information between the MAC layer and other layers of the protocol stack in the cross-layer design of cognitive radio networks is detailed below, along with related works [19]:

(i) **MAC layer and PHY layer interaction**
The channel state information (CSI) obtained by the cognitive radio terminal's PHY layer can be utilized by the MAC layer to improve the efficiency of the communication system—for example, through beamforming, interference decoding methods, and throughput. In addition to the CSI obtained by the terminal itself, the information obtained by other terminals in the network is also useful. The interference power and activity of interfering devices measured by the cognitive radio terminal at the PHY- layer can also be useful for MAC layer design. The PHY layer has direct access to the hardware resources of the network and is well aware of the real-time operation information; therefore, the information on static (e.g., due to the architecture of the device) and dynamic (e.g., due to the available battery) hardware characteristics could be relevant for MAC protocol design [20]. In

addition, the adaptive modulation and coding information available at the PHY layer is useful to improve the automatic repeat request at the data link layer [20]. In [21], a joint PHY-MAC layer optimization policy for multi-channel ALOHA random access in wireless networks is presented in which users are necessarily within the transmission range of one another, and each user may have packets to send to receive from other users. This joint PHY-MAC policy exploits decentralized CSI, and achieves multi-user diversity through cross-layer design. A decentralized optimization for multichannel random access (DOMRA) is proposed that consists of three steps: (1) neighborhood information collection, (2) transmission control of the MAC layer based on instantaneous CSI, and (3) power allocation for each traffic flow on each sub-channel [21]. System performance is optimized while proportional fairness is obtained with the consideration of the inhomogeneous characteristics of the traffic spatial distribution. Simulation results show that the proposed scheme significantly outperforms existing channel-aware ALOHA schemes due to its exploitation of both multiuser diversity through cross-layer design and the inhomogeneous characteristics of traffic spatial distribution in the network. The methodology can be easily adapted to improve the performance of various wireless networks. For example, the backoff technique proposed for the cognitive radio MAC protocol in previous chapters could utilize the transmission of request-to-send (RTS) to compete for channel access and could also be designed according to the proposed DOMRA to further reduce collision risk and increase the success probability of users with better channel power gain. DOMRA can also be applied to other types of wireless networks, such as wireless sensor networks (WSNs) and mobile ad hoc networks, to improve QoS.

(ii) **MAC layer and network layer interaction**
In the proposal by Jia and Zhang [22], routing is computed by a cross-layer entity, and the results are given only to the network layer that constructs the routing table. The cognitive radio-based routing protocols must also take into account the activity of and consequences for primary users (e.g., service interruption losses) to determine the best routes. This is the main difference with traditional routing protocols, where a particular layer optimizes its entity without considering its effect on other layers. In [23], a cross-layer opportunistic spectrum access and dynamic routing algorithm for cognitive radio networks—the routing and dynamic spectrum allocation (ROSA) algorithm—is proposed. Akyildiz et al. [24] classified existing works in cognitive radio routing according to three distinct approaches: (1) routing with spectrum decision (i.e., joint selection of the spectrum and next hop), (2) routing with joint spectrum decision and primary user awareness (i.e., establishing routes that avoid locations with primary user activity), and (3) routing with joint spectrum decision and re-configurability (i.e., establishing routes that recover from primary user appearance). These approaches thus provide joint

optimization of various parameters of different layers. Wang et al. [25] also examined the correlation between the dynamic frequency assignment, routing in the network layer, and scheduling of access in MAC layers in wireless networks. These components should be treated jointly, and hence support the cross-layer design [19].

(iii) **MAC layer and transport layer interaction**
A cognitive user in the network is unable to forward packets during sensing. Therefore, sensing periods must also be considered at the transport layer in order to avoid excessive retransmissions and packet losses on the paths with any cognitive radio node in sensing state, especially for the multi-hop distributed networks. This makes interaction between MAC and transport layer entities necessary. There are two possible approaches: (a) stopping transmission at the transport layer, or (b) reducing its rate towards an optimal value, which avoids buffer overflow at intermediate nodes while maintaining transmission [19]. However, transport protocols in CR scenarios must be spectrum-aware, and therefore require new algorithms (e.g., for congestion window scaling in TCP) [24]. Akyildiz et al. [24] described a TCP-based protocol for cognitive radio ad hoc networks (CRAHNs). Theirs is the first study aiming to address transport layer challenges in CRAHNs. Finally, Issariyakul et al. [26] also address performance issues of the transport layer in cognitive radio networks.

(iv) **MAC layer and application layer interaction**
According to Yu et al. [27], the perceived reduction in QoS at the application layer by the presence of cognitive users limits the success of cognitive radio technologies. As reported in [19], various previous works on CR have dealt only with the maximization of throughput in time-slotted cognitive networks, and have largely ignored other QoS measures (e.g., distortion for multimedia applications). Recent work in cross-layer design shows that maximizing throughput does not necessarily promote QoS at the application layer for applications such as video communication. Therefore, the authors [19] considered the QoS at the application layer in cognitive radio proposals and described a design approach that is an extension of partially observable Markov decision process (POMDP)-based MAC [28], which optimizes the application layer QoS for multimedia transmission together with spectrum access and spectrum sensing in CR networks.

This chapter is organized as follows: Sect. 10.2 presents the cross-layer optimization framework presented by various researchers for cognitive radio and wireless network. Section 10.3 discusses an important parameter of wireless network design: energy efficiency, and cross-layer design issues related to it. Various challenges in cross-layer design are discussed in Sect. 10.4, and a summary is provided in Sect. 10.5.

10.2 Cross-Layer Optimization

Cross-layer-based opportunistic multi-channel MAC protocols for wireless ad hoc networks are proposed in [29]. The proposed schemes integrated the spectrum sensing policy at the PHY layer with packet scheduling at the MAC-layer. Under these schemes, each cognitive user consists of a control transceiver working on a dedicated control channel and a software-defined radio (SDR)-based transceiver that can be dynamically tuned to any one of the licensed channels to sense for spare spectrum, and then to receive/transmit the secondary users' packets. For the detection of unutilized channel availability, two channel sensing policies are proposed: a simple but efficient random sensing policy, and a performance-enhanced negotiation-based sensing policy. In addition, in the non-saturation network scenario, the authors identified a trade-off between the aggregate traffic throughput and the packet transmission delay, which provides insightful guidelines for improving the delay-QoS condition over cognitive radio wireless networks with the help of cross-layer design. It is well known that the intelligent capability of cognitive radio network provides for better throughput, even in congested spectrum, along with better propagation characteristics. The routing in the cognitive radio environment is a challenging task, as channel availability is constrained by the presence of the primary user, as discussed in [30, 31]. The problem of routing in CRAHNs targets the creation and maintenance of wireless multi-hop paths among the cognitive nodes by deciding both the spectrum to be used and the relay nodes of the path. In [30], the authors propose a cognitive cross-layer multipath probabilistic routing for cognitive radio-based networks. In addition, knowledge of the topology of the cognitive network is necessary for establishing optimal routes for information. Therefore, for optimal performance, the updated information on the availability of links and their capacities and latencies will have to be exchanged between the lower protocol stack. This exchange is particularly important for wireless links where this information changes over time [20]. The position information gathered, for example, by dedicated hardware for angle or time of arrival estimation may be integrated into the routing protocol by adopting a position-aware routing algorithm. The positions of licensed users can be used as constraints in the selection of the end-to-end path by the network layer [20]. The optimization of parameters during interaction between the PHY and network layers is based mainly on the spectrum access model used by the cognitive user, as follows [20]:

(i) **Spectrum underlay cognitive radio network**
In the underlay cognitive radio network, the perceived power from nearby primary transmitters can be taken as input from the PHY layer and can be integrated into the cost function used by routing protocols in the network layer to evaluate the cost of the link in order to select the most cost-efficient link for routing.

(ii) **Spectrum overlay cognitive radio network**
In this case, a network operating on multiple channels is considered, which tries to select on each hop a channel unused by the licensed user. The PHY

layer input may consist in an indication of which channels are considered free/busy, and in joint channel allocation/routing solutions, this information may lead to the selection of paths by the network layer that characterized by stability or by the lowest number of channel switches among the path.

In addition, the CSI available at the PHY layer is useful for operation of the application layer that can adapt its performance parameters towards improving QoS [20]. The TCP throughput of the transport layer is maximized by employing a cross-layer scheme that adapts the modulation and coding scheme at the PHY layer. Adaptive modulation and coding also play an important role in improving the achievable data rate and symbol error rate parameters of the application layer. This adaptation is possible when sensing information at the PHY layer is exchanged with the application layer [20]. A cross-layer optimization framework, presented in Fig. 10.2, is also developed to investigate the performance of the adaptive opportunistic distance sensing multiple access (AODSMA) scheme, including spatial diversity, opportunistic transmission, distributed scheduling, and successive interference cancellation (SIC) [8] in pre-coded cyclic prefix-OFDM (PCP-OFDM)-based cognitive radio networks [7].

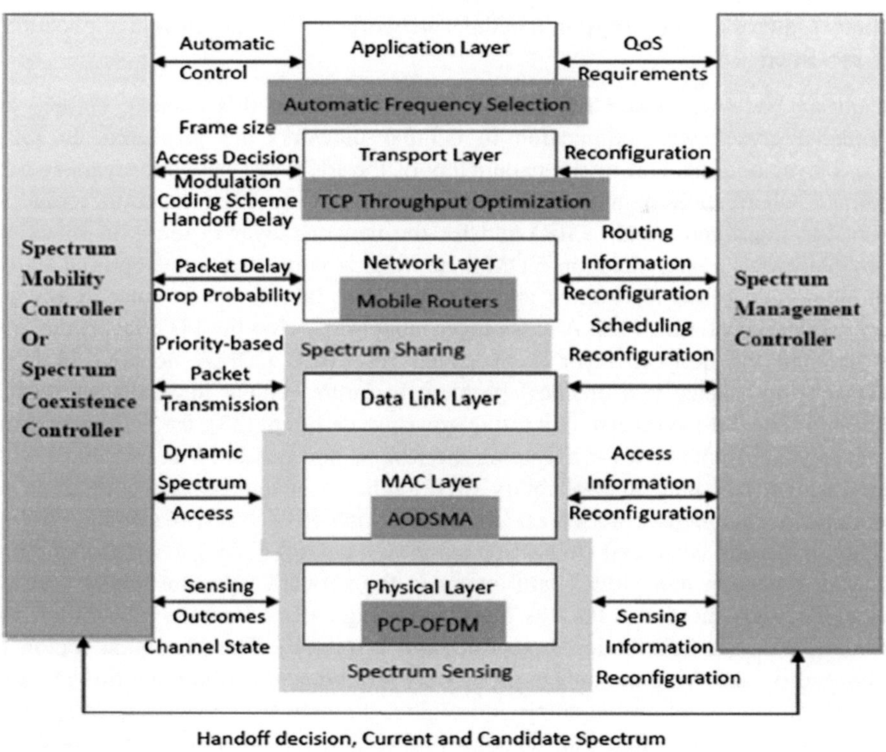

Fig. 10.2 Cross-layer optimization for a CR network [7]

Recent research on potential context-aware communications has led to the introduction of features and algorithms that rely on the presence of accurate context-rich information, which require cross-layer information exchanges. Cognitive radio, in particular, is expected to benefit from context awareness [20]. The classification and description of the information exchanged in a cognitive radio network between the layers of a generic protocol stack and between each layer and the cognitive radio is presented in [20]. For each layer, the key services provided are presented, followed by a catalog of exchanged parameters. Since most of the research activities have focused on the lower layers, however, significant performance improvements are expected by inclusion of the higher layers in the information exchange loop, which is discussed in [20]. The impact of cross-layer information exchanges in a cognitive radio framework is the basis for a discussion of implementation challenges and identification of the most promising partitioning of functions and tasks between layers and cognitive radio. The implementation of a CR network in a real-world scenario will most likely require the distribution of decision-making processes between the cognitive radio terminal and protocol layers [20].

Several current research efforts are seeking ways to integrate cross-layer design solutions into next-generation wireless communication standards for the purpose of allocating resources to cellular users, scheduling access to shared resources with higher throughput, and achieving better QoS for multimedia applications. In this context, a cross-layer design in a cellular network and for multimedia applications is presented.

Cellular Network and Cross-Layer Optimization In this context, studies on potential cross-layer optimization in cellular networks are presented. In [32], cross-layer optimization of the parameters of the PHY and network layers is performed where QoS requirements are achieved for the PHY layer in terms of signal-to-interference ratio (SIR) and for the network layer in terms of blocking probability. A next-generation CDMA network is considered, and optimal linear programming-based algorithms are presented that take into account SIR outage probability constraints [32]. A cross-layer model involving the PHY layer, the link layer, and the network layer is presented for future CDMA network in [33]. Capacity imbalance is a practical problem in future cellular networks where the forward link and the reverse link traffic are asymmetric. In [33], the authors present an analytical framework for balancing the reverse and forward link capacities with an adaptive soft handoff probability (SHP) scheme in multiservice code-division multiple-access cellular networks. The SHP in the PHY layer, the outage probability in the link layer, and connection admission control (CAC) schemes including complete sharing and virtual partitioning in the network layer are jointly considered. The QoS metrics in the link layer, including SIR and outage probability, are derived with the information from the PHY layer [33]. The admission region is obtained by satisfying the outage probability requirements in both the forward and reverse links. Based on the admission region, the network layer grade of service, including the new connection blocking probability, the handoff connection dropping probability, and the throughput, is formulated as the performance metrics. The optimal SHP is selected by looking for the lowest penalty of connection blocking, which encompasses both new connection blocking and handoff connection dropping probabilities. The cross-influences between the selection of the optimal SHP and the CAC schemes have also been investigated [33]. Cross-layer approaches are being increasingly studied as a design approach for confronting unintended performance degradations as well as supporting new processes for beyond 3G mobile communication networks. In [2], the authors predicted an evolution towards 4G networks based on these kinds of functional entities, protocols, and processes that provide optimization and reconfiguration at both the user terminal and the network. In [34], the interaction between the link and network layers for IEEE 802.16e beyond 3G wireless metropolitan area network (WMAN) is presented in order to improve handover performance, and two handover protocols are proposed for the IP layer and the MAC layer used in future mobile networks. The proposed cross-layer design [34] reduces the handover latency of the users in the network, and hence is an efficient solution for broadband networks.

Multimedia Traffic and Cross-Layer Optimization Multicast video streaming over multi-rate wireless LANs imposes strong demands on video codecs (encoder and decoder) and the underlying network. It is not sufficient that only the video codec or only the underlying protocols adapt to changes in wireless link quality [35, 36]. Cross-layer design is a new paradigm that addresses this challenge by optimizing communication network architectures across traditional layer boundaries. In this context, the authors in [35] jointly consider the PHY, data link, and application layers in order to provide adaptive video multicast streaming over multi-rate wireless LANs, where layer-specific information is passed in both directions, top-down and bottom-up. The outcomes in [35] show that the real-time video quality of the overall system can be greatly improved by cross-layer signaling, and the authors have provided optimized parameters. In addition, when cognitive radio technology is used in real-time video applications, the user-perceived video quality experienced by cognitive/secondary users is a very important performance metric for evaluating the effectiveness of the cognitive radio network. Therefore, in [37], the authors have proposed a cross-layer scheme for improving the user experience for secondary users of wireless video services over cognitive radio networks. In the proposed cross-layer design, the authors have improved the system performance of the upper layers and optimized the system to maximize the expected user-perceived video quality at the receiver end under the constraint of a packet delay bound [37]. They have also jointly formulated network functions including encoder behavior, cognitive MAC scheduling, and transmission, as well as modulation and coding, into a distortion-delay optimization framework. Important system parameters residing in different network layers are jointly optimized in a systematic way to achieve the best user-perceived video quality for secondary users in cognitive radio networks [37].

A cross-layer design for low-latency media streaming over ad hoc networks is shown in Fig. 10.3 [38]. In order to extend the achievable capacity region of the network at the data link layer, adaptive modulation is used to maximize the link

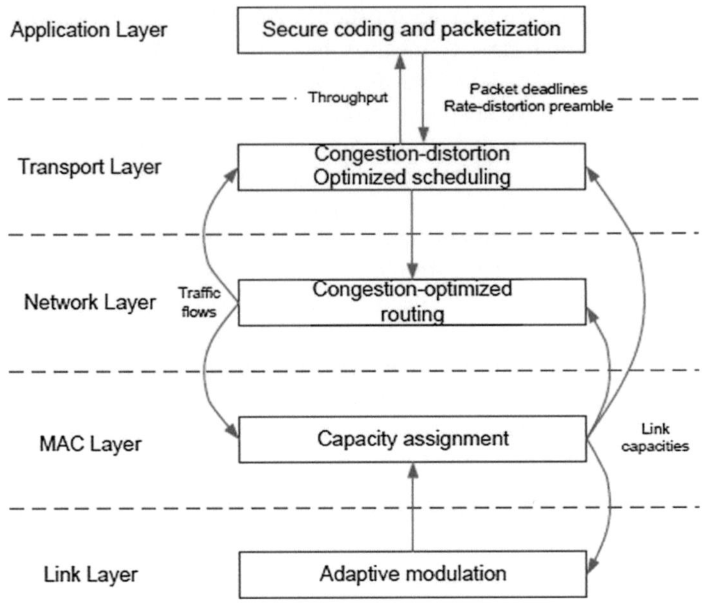

Fig. 10.3 A cross-layer design for low-latency media streaming over ad hoc networks [38]

rates under varying channel conditions [38]. Based on link state information shared by the adaptive modulation, the MAC layer selects one point to assign time slots, codes, or frequency bands to each of the links [38]. The MAC layer operates jointly with the network layer to be aware of traffic flows to determine the set of network flows that minimize congestion (shown as link capacities in Fig. 10.3) [38]. Successive suboptimal solutions are exchanged iteratively between the MAC layer and network layer in order to yield an optimal solution for the capacity assignment and network flows [38]. The transport layer uses congestion distortion-optimized scheduling to control the transmission and retransmission of video packets. The application layer determines the encoding rate to achieve the most efficient streaming [38, 39]. This process is executed at a node in the ad hoc networks, and there is no centralized node to control this process. Therefore, this scheme is considered a distributed cross-layer design method.

For wireless networks operating in an unlicensed band or cognitive radio networks utilizing a licensed band, evaluating overall performance of the network largely involves addressing coexistence issues which are associated with the contemporaneous presence of true and interfering signals at the PHY layer. This task is difficult to fulfill on the basis of only single-layer measurements, as merely a partial perspective of network behavior would be gained. With this concern in mind, a cross-layer approach is presented in [40] that provides for several measurements to be carried out concurrently at different layers through a proper automatic station, aiming to correlate the values of the major PHY layer quantities (e.g., channel power and SIR) exhibited by those characterizing the key higher-layer parameters (e.g., packet-loss ratio and one-way delay) in the presence of interference [40]. Cross-layer design thus represents a first step towards a full characterization of how the effect of a problem experienced at the PHY layer propagates along the whole protocol stack [40].

Cross-layer scheduling [41] is a promising solution for improving the efficiency of emerging broadband wireless systems by adapting to the dynamic nature of the wireless environment and the evolving nature of traffic. Demand (traffic) and provision (air interface) as the two major elements of existing and possible future impacts on radio resource management have been explored in [41]. This is illustrated with the aid of a customer-provider model from the field of economics. Additionally, the authors in [41] have evaluated some classical scheduling algorithms within the context of the OFDM, which is a dominant candidate platform for next-generation broadband wireless systems. Various basic practical limitations, such as interference, affect the efficiency of scheduling algorithms as an external factor, and this should be alleviated based on interference management methods. The imposed signaling overhead and complexity are other important factors requiring further consideration in the design of cross-layer scheduling algorithms. The possible evolution of scheduling techniques is also described based on the characteristics of traffic and air interface in emerging broadband wireless systems [41]. Finally, potential challenges are identified, and careful consideration is given to the evolution of traffic as well as air interface within individual scheduling algorithms [41]. In [42], the use of distributed random access for the performance of central schedulers is presented.

In [39], the authors discuss three goals of cross-layer design, namely, security, QoS, and mobility, as shown in Fig. 10.4, and briefly described in Table 10.1. For achieving these goals, cross-layer designs are classified in two ways. The first is how information is shared among the five layers of the TCP/IP model, and the cross-layer design is classified into two categories, non-manager and manager, as described in Table 10.2 [39]. The second involves the organization of the network, and cross-layer design is classified into two categories, centralized and distributed, also detailed in Table 10.2 [39].

Layerless technology has provided a new approach (Fig. 10.5) [45] for researchers and engineers in designing next-generation cognitive radio communication systems and enhancing the efficiency of precious spectrum resources. However, the exact means of implementing a layerless design concept within a layered structure is still an open research issue. We believe that cross-layer design is the most feasible among proposed solutions at the current stage. In [45], analyses of the random distance between node pairs were used to study the effects of shadow fading channels on link connectivity and node degree in wireless ad hoc networks, and a cross-layer concept was used to construct a critical node first (CNF)-based clustering algorithm to design a cluster-based network.

For WSN-based smart grid applications, electromagnetic interference, equipment noise, and multi-path effects make QoS a challenging task, and to address these challenges, a cognitive communication-based cross-layer framework is proposed by authors in [46]. The proposed framework exploits emerging cognitive radio

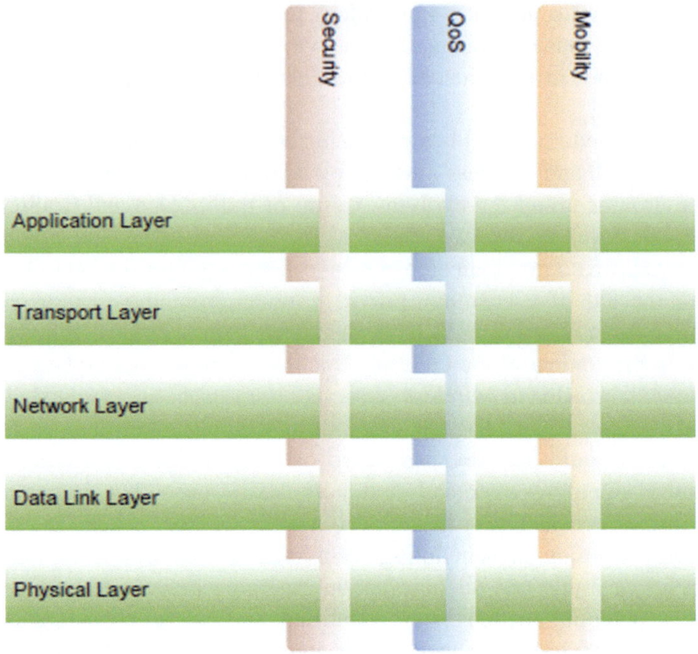

Fig. 10.4 The goals of cross-layer designs: security, QoS, and mobility [2, 4, 43, 44]. A cross-layer design scheme normally aims to achieve at least one of these three goals

Table 10.1 Goals of cross-layer design [39]

Objective	Explanation
Security	Security issues across the five TCP/IP layers are considered in some cross-layer designs. Encryption methods such as Secure SHell (SSH) or Wi-Fi protected access might be deployed in a cross-layer design aimed at secure communication
QoS	To improve the QoS in wireless communication across the five layers, some cross-layer designs enable cross-layer communication between the upper layers (application and transport layers) and the lower layers (physical layer and the data link layer) [2]
Mobility	Several cross-layer designs aim to guarantee uninterrupted communication in a wireless network, since node movement, which would cause channel switching, route change, and other problems, is common in wireless networks

technology to mitigate the noisy and congested spectrum bands, yielding reliable and high-capacity links for wireless communication in smart grids. In [13], the authors present a cross-layer design approach to jointly consider spectrum sensing, access decision, physical layer modulation and coding scheme, and data link layer frame size in CR networks to maximize the TCP throughput in CR networks. This cross-layer optimization architecture is shown in Fig. 10.6. Based on the history of observations and action decisions, the cognitive user will determine whether to sense

Table 10.2 Cross-layer design classification [39]

Classification	Method	Explanation
First classification: by how to share information among layers in one node	Non-manager method	The data exchange takes place directly between any two layers
	Manager method	There is vertical data exchange management between layers
Second classification: by network organization for cross-layer information sharing	Centralized method	Uses a centralized node or tier in a hierarchical manner to achieve communication between nodes. Typically used in cellular networks
	Distributed method	No centralized node or tier is used. Typically used in ad hoc networks

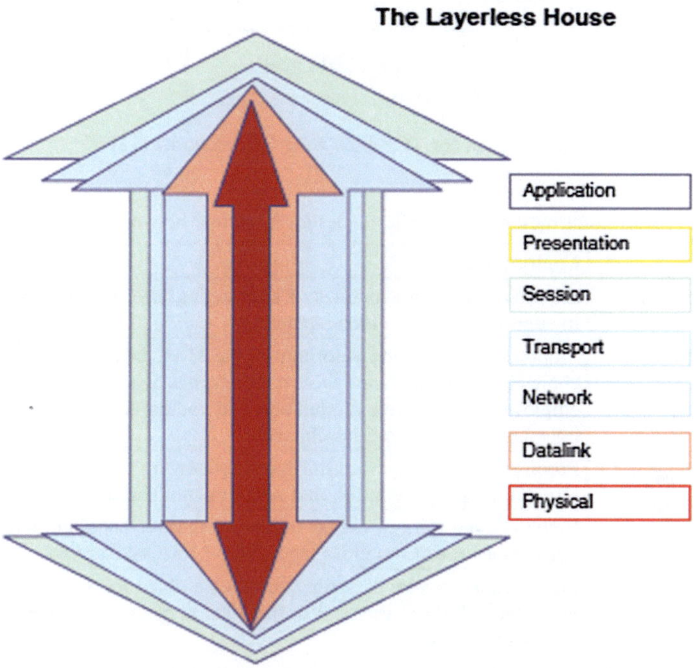

Fig. 10.5 A layerless house concept [1]

the channel. If a sensing action is selected, the cognitive sensor will observe the channel and obtain the sensing outcomes, which are sent directly to the TCP layer. To maximize TCP throughput, the cognitive user will determine whether to access the channel and the corresponding modulation and coding scheme at the PHY layer and the frame size at the data link layer. Consequently, it will feed back the three decisions to the MAC layer, the PHY layer, and the data link layer, respectively [13].

Finally, Table 10.3 summarizes major contributions of cross-layer design optimization in ad hoc networks.

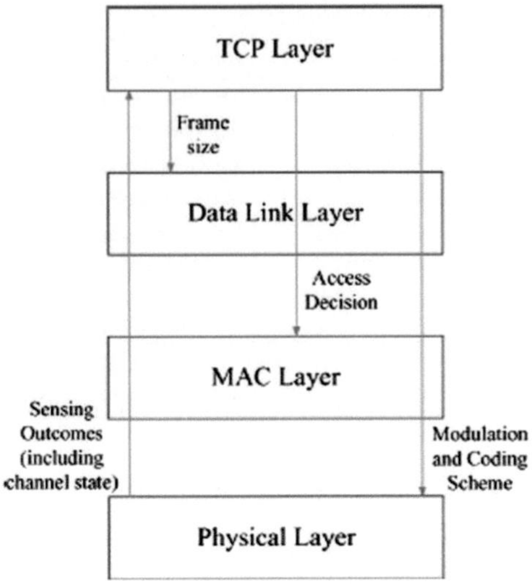

Fig. 10.6 Cross-layer optimization for TCP flow over CR networks [13]

Table 10.3 Major contributions of cross-layer design in ad hoc networks [3]

Authors	Contribution
Setton et al. [38]	Explored potential synergies of exchanging information between layers to support real-time video streaming
Liu et al. [47]	Proposed a scheduling algorithm at the MAC layer for multiple connections under varied QoS requirements, where each connection employs both adaptive modulation and coding at the PHY•layer for transmission over wireless channels
Huang and Letaief [48]	Proposed a cross-layer optimization framework to jointly design the scheduling, power control, and adaptive modulation
Zhang and Zhang [49]	Reviewed the state of the art on the cross-layer paradigm and discussed open issues related to cross-layer design for QoS support
Oh and Chen [50]	Presented a cross-layer design for reliable video transmission based on a multichannel MAC protocol in the context of time division multiple access (TDMA)
Chu and Wang [51]	Presented cross-layer centralized and distributed scheduling algorithms which exploit the PHY layer channel information to opportunistically schedule cooperative spatial multiplexed transmissions between multiple-input/multiple-output (MIMO)-based nodes
Ghosh and Hamouda [52]	Proposed a cross-layer antenna selection algorithm for improving the transmission efficiency in cognitive MIMO-based ad hoc networks
Mardani et al. [53]	Jointly considered flow control multipath routing and random access control based on network utility maximization
Uddin et al. [54]	Studied cross-layer design in random access-based fixed wireless multi-hop networks under a physical interference model
Tang et al. [55]	Proposed a cross-layer distributed approach for maximizing network throughput by jointly selecting stable routes and assigning channels based on mobility prediction

10.3 Energy Efficiency and Cross-Layer Design

The major concern in cognitive radio networks, and especially in ad hoc networks, is the energy conservation, although reliability is also very critical. Therefore, effective and efficient mechanisms should be provided to achieve reliability with low energy expenditure. However, different applications have diverse reliability requirements. Industrial control or military applications might require nearly 100% reliability [56], while other monitoring applications might tolerate data loss, leading to a trade-off between energy conservation and reliability [56]. Most PHY layer solutions focus on reducing energy consumption for transmission by using transmit power control, rate selection, or both [57]. However, these solutions either completely ignore the energy cost in idle time or assume fixed power consumption [57–59]. Liu and Zhong in [60] have shown that wireless interfaces spend a very high percentage of time and energy in short idle periods, even during active transmission time. However, such periods, tens of milliseconds or less, are out of the reach of conventional power-saving mechanisms provided by higher layers of the protocol stack, such as IEEE 802.11 MAC power-saving mode (PSM), other proposed power-saving protocols [61], and application-specific solutions [62, 63]. Therefore, to achieve energy efficiency, the protocol stack of cognitive radio must be tuned according to the actual requirements. Moreover, the traffic and network conditions of the licensed user are often very dynamic; thus, energy-aware and reliable data collection mechanisms should be able to adapt to the actual operating conditions. In this context, maximizing the energy efficiency of the cognitive radio network becomes challenging. Various researchers have proposed cross-layer designs to deal with energy efficiency, and some of the recent work in this direction is highlighted in this section.

The cross-layer framework in [56] involves an energy-aware adaptation module that captures the application's reliability requirements in terms of the target delivery ratio, and autonomously configures the MAC layer based on the network topology and the traffic conditions in order to minimize power consumption. The authors proposed a low-complexity distributed algorithm, the ADaptive Access Parameters Tuning (ADAPT), which has high energy efficiency and can effectively meet application-specific reliability requirements under a wide range of operating conditions for both single-hop and multi-hop networking scenarios. In [3], a cross-layer operation across the PHY layer, the data link layer, and the network layer is illustrated for improving the energy efficiency of the ad hoc network. There are two basic methods of information sharing in the cross-layer design [64]. One, referred to as a layer-centric solution, makes the variables of a specific layer visible to the other layers. The other, called a centralized solution, relies on shared middleware [64, 65], which provides the service of storage/retrieval of information to all the layers. Figure 10.7 illustrates the operation of these two cross-layer solutions. The basic principles of this pair of cross-layer solutions are described in the following section

The Layer-Centric Solution Certain layers are allowed to be a central layer, which controls the cross-layer adaptation by accessing the internal protocol parameters and

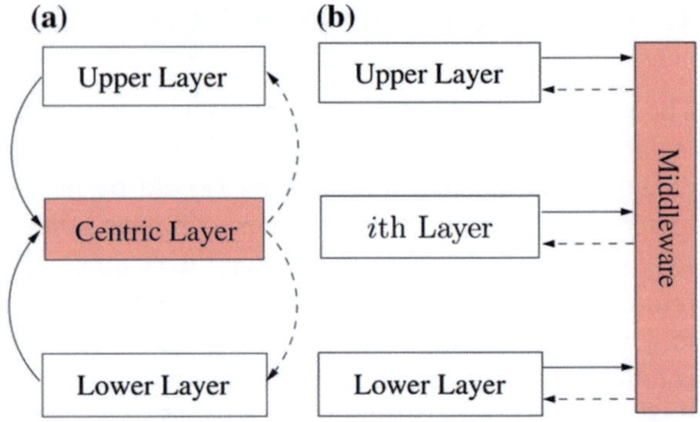

Fig. 10.7 Conceptual illustration of cross-layer design methods: **a** layer-centric solution and **b** centralized solution [3]

algorithms of the other layers, as shown in Fig. 10.7a. Although this approach improves the attainable system performance significantly, it violates the layered architecture, since it requires access to the internal variables of other layers [3].

The Centralized Solution A middleware or a system-level monitor (centralized optimizer) is employed for estimating both the availability of resources and the environmental dynamics in order to coordinate the allocation of resources across diverse applications as well as nodes, and for adapting the protocol's parameters within each layer based on the dynamics experienced, as shown in Fig. 10.7b [3]. This approach requires each layer to forward complete information characterizing its protocol parameters and algorithms to the middleware or system monitor. It also requires each layer to carry out the actions requested by the central optimizer. This approach also violates the layered architecture. The MobileMan [66] and CrossTalk [67] protocols are examples of centralized cross-layer solutions.

In [3], a method for reducing energy consumption by exploiting the benefits of the coordination between the PHY and network layers is presented. Throughput and energy consumption constitute a pair of important specifications in analyzing a network's performance, and critically depend on the parameters of the different OSI layers. Hence, combining the functions of multiple layers with the aid of cross-layer operation is useful in terms of improving attainable performance using several cross-layer-aided routing algorithms designed for ad hoc networks [3]. In addition, the authors have considered the factors influencing the design of both the PHY and network layers, and have characterized the influence of the data link layer in the cross-layer-aided routing design [3]. The objective is to achieve throughput improvement and energy reduction. To this end, they present the number of maximum MAC retransmissions as a representative factor in the data link layer [3]: the greater the maximum number of MAC retransmissions, the more energy will be consumed and the delay increased. As an advantage, successful packet reception probability is improved. Hence, we must find the most appropriate maximum number of MAC retransmissions for the sake of striking an attractive compromise between throughput and energy efficiency [3]. If the transmit power of each node is assumed to be the same, a certain amount of extra energy will be dissipated, since

Table 10.4 Reported studies related to data link layer, network layer, and cross-layer energy-efficient designs

	Author	Remarks
Link layer	Zhang et al. [70]	The study examines the effects of fading channel for intra-cluster data transmission in cluster-based WSNs that employ TDMA-based channel access protocols. An efficient MAC layer algorithm is proposed. The packet error rate and energy efficiency can be greatly improved by restraining a node from transmitting data in its assigned time slot when its channel is in deep fading
	Soni and Chockalingam [71]	The authors analyze the throughput and energy efficiency of the user datagram protocol (UDP) using linear, binary exponential, and geometric backoff algorithms at the link layer on point-to-point wireless fading links. The multipath fading channel is modeled as a first-order Markov chain
	Cavalcanti et al. [72]	The authors analyze the energy efficiency and QoS performance of 802.11e for low-rate applications, compared to 802.15.4, under varying interference and traffic conditions. In certain scenarios, 802.11e can achieve higher energy efficiency and QoS
	Chan et al. [73]	The energy efficiency, throughput, and packet delay for both non-persistent and p-persistent carrier-sensed multiple access (CSMA) are investigated. The authors show that non-persistent CSMA has markedly higher energy efficiency than p-persistent CSMA. When the non-persistent CSMA is optimized for energy efficiency, the throughput and delay are negatively impacted, whereas p-persistent CSMA can effectively optimize these three parameters
	Choi and Park [74]	The study analyzes the energy efficiency of a block acknowledgement mechanism and MIMO transceiver by assuming the knowledge of the power information of receiver side and channel path loss value. The results show that higher energy efficiency is achieved with a lower number of antennas, higher modulation, larger data payload size, and burst transmission
	Ci et al. [75]	An optimal frame size predictor based on a Kalman filter is proposed, which can greatly reduce the average number of transmissions, thus improving energy efficiency while retaining good throughput performance
	Mouzehkesh et al. [76]	An enhancement in mobility-aware MAC for sensor networks (MS-MAC) is proposed, and its dynamic approach is used to increase energy efficiency by preventing the nodes from getting unnecessarily involved inside the active zones, which is based on computing the distance of the border node from the border region
	Wang et al. [77]	The authors present an analysis of energy efficiency in 802.11 distributed coordinated function (DCF), and compare the impact of various contention windows and packet sizes. They show that in error-prone environments, optimal packet size can produce a greater improvement in energy efficiency than optimal contention window, and a combination of the two can achieve maximum optimization

(continued)

Table 10.4 (continued)

	Author	Remarks
Network layer	Alippi and Vanini [78]	With a static/semi-static medium-sized network, the nodes periodically acquire and process sensory data, and outputs are conveyed to the central unit. The proposed routing algorithm uses a global power-aware strategy and utilizes application- and environment-based optimization
	Bernardos et al. [79]	A hybrid (proactive-reactive) algorithm pertaining to Zone Routing Protocol is presented, the objective of which is to maintain the smallest possible number of transmissions and avoid redundant message sending, while locally minimizing the transaction delay
	Jung et al. [80]	By employing an applied adaptive load balancing technique to the MANET routing protocols, better performance and energy efficiency were achieved. New energy efficiency metrics were also proposed for the MANET routing protocol
	Rajan [81]	The author proposes a framework for studying the delay allocation problem in the ad hoc wireless network. A closed-form expression for the total required power in a network is derived as a function of the delay allocation, and approximation is exploited to find near-optimal schedulers
	Varaprasad [82]	The author discusses the need for power-aware routing and proposes a power-aware routing algorithm for MANET using gateway node, in order to minimize the number of control message packets and energy consumption and to increase the throughput without considering the packet loss
	Zhang and Soong [83]	A channel aware geographic-informed forwarding (CAGIF) routing algorithm is proposed that chooses the next hop relay node by considering the underlying channel conditions and analyzes the achievable energy efficiency. The results show that CAGIF significantly outperforms pure geographic-informed forwarding (PGIF)
Cross-layer	Betz and Poor [84]	The cross-layer design issue of energy-efficient communication using a distributed non-cooperative model is presented. In this game, users are allowed to choose their transmit power and uplink receivers to maximize their utility
	Ghasemi and Faez [85]	A design methodology is presented for a power-aware MAC for multi-hop wireless networks that uses CDMA at its PHY layer. The results emphasize that in multi-hop wireless networks, the MAC should be designed by considering both time and space contentions between links, which in turn are provided by adjusting the links' attempt rate and power
	Kuo [86]	A model of energy consumption is constructed by jointly considering the interactions between IEEE 802.11a PHY and MAC layers, and the effects of different PHY and MAC layer parameters on the energy efficiency of IEEE 802.11a are investigated

(continued)

Table 10.4 (continued)

	Author	Remarks
	Masurkar et al. [87]	The authors present the computation of transmission powers, rates, and link schedule for an energy-constrained wireless network to jointly maximize the network lifetime
	Wang et al. [88]	The authors propose a novel multi-rate-oriented approach with MAC and PHY cross-layer scheme to support distributed source coding (DSC)-based signal processing applications, in order to achieve high energy efficiency in WSNs. The scheme controls the optimization of the transmission power based on the desirable BER and the application's required data rate
	Xianling et al. [89]	The authors address the issue of joint design of power control and connected dominating set formation for power energy saving in ad hoc wireless networks. An energy-efficient cross-layer broadcast (CLBA) algorithm is proposed, which performs well in terms of reachability, average broadcast latency, and energy efficiency compared with flooding and TOP algorithms
	Yu [90]	Since MAC protocols are critical to energy efficiency, the author proposes an approach for MAC protocols to reduce the processing time spent on data moving, and hence energy consumption, with a cross-layer design. The end-to-end delay can be reduced, with a dramatic reduction in nodal processing time
	Zhong et al. [91]	A novel cross-layer power control game algorithm based on neural fuzzy connection admission controller (NFCAC) is proposed to effectively utilize location marking information and address the performance issues

the distance between each pair of nodes is different, thus necessitating a different amount of transmit energy. An energy-efficient algorithm has thus been designed in [3] using hop-length-dependent power allocation, which also jointly considers the frame error ratio (FER) in the PHY layer, the maximum number of retransmissions in the data link layer, and the number of hops in the network layer in order to achieve improved network throughput and reduced energy consumption.

The authors in [68] propose cross-layer fixed-power allocation schemes and adaptive power allocation algorithms that depend on the delay QoS constraint and instantaneous channel conditions. Table 10.4 [69] presents some of the literature on energy efficiency related to the data link layer, network layer, and cross-layer designs.

10.4 Potential Challenges in Cross-Layer Design

In this chapter, we have discussed the ongoing work in the area of cross-layer design and looked at different cross-layer proposals. We have also examined different ideas on how cross-layer interactions can be implemented. However, in discussing the different cross-layer proposals for wireless networks, various challenges must also be considered. Since there are a number of cross-layer design proposals in the literature today, it is unclear which of these are the most relevant or how these proposals coexist with one another. Therefore, a thorough cost-benefit analysis of the design proposals in terms of implementation complexity versus performance improvement is needed. Furthermore, while standardization of cross-layer solutions will provide universal cross-layer design, the investigation, specification, development, and ultimately the standardization of cross-layer entities, interfaces, and algorithms to meet the need for cross-layer optimization and dynamic interaction patterns between the protocol layers remain an open technical challenge [2]. In the proposed cognitive radio cross-layer design in [20], the protocol layers and cognitive engine form a full mesh network, posing an optimization problem significantly more difficult than that of the traditional layered model, where the lack of cross-layer interfaces significantly reduces the space for potential solutions, thus demanding a means of achieving a trade-off between complexity and efficiency. Other challenges in cross-layer design, the reasons, and their solutions are presented in Table 10.5 [20, 39].

For the optimization of each group of strategies, i.e., which strategies of different layers should be optimized jointly, one can use derivative and non-derivative methods (e.g., linear and nonlinear programming) [92]. Since this is a complex multivariate optimization with inherent dependencies across layers and among strategies, an important challenge is determining the best procedure for obtaining the optimal strategy. This involves determining the initialization, grouping of strategies at different stages, a suitable order for strategy optimization, and even which parameters, strategies, and layers should be considered based on their impact on quality, delay, or power of a particular service [92]. Further, the associated complexity in the layerless approach [1] will be immense, involving the exhaustive evaluation of all possible strategies and their parameters for choosing a composite strategy leading to the best quality of performance. Completely new system-level implementations may be required as the ways that protocols are organized presently are subjected to changes.

Since the current networks are not layerless, the strong legacy environment will pose market challenges for this design [1]. Telecommunication operators comprise a vast enterprise, and implementing such change will require an enormous new market or great disruption to their business model, and new products will need to be compatible with their existing business and technical approaches [93]. The introduction of layerless communications will also require new control planes for interoperation [1]. Given the scale of communications technologies and the

Table 10.5 Challenges and related solutions in cross-layer design [20, 39]

Challenge	Reason	Solution
Coexistence of different cross-layer designs	Each cross-layer design has its own standard in the communication between layers	Standardization of cross-layer communication, which can provide a unique vehicle for smoothly deploying various cross-layer design solutions in next-generation mobile communication networks
Cross-layer signaling	There is no uniform format or manner of exchange of cross-layer information in the network	Using packet headers, Internet Control Message Protocol (ICMP) messages, or network services for cross-layer data exchange
Overhead caused by cross-layer signaling	To exchange cross-layer information between the nodes, cross-layer signaling results in extra overhead	Although overhead must exist, a better design and implementation of cross-layer signaling may reduce overhead
Lack of a universal cross-layer design	Different applications have different requirements for cross-layer design	A universal cross-layer design is not likely feasible
Timing constraints	Varying delay for collection, storage, and exchange of context information	Highly dynamic information, such as real-time path loss or instantaneous power distribution, should be stored as close as possible to the point of decision. Conversely, information of a less dynamic nature, such as maps and characteristics of a building, the position of access points, propagation, and traffic pattern models, could be stored in more distant databases
Spatial constraints	Due to random deployment of wireless nodes	Data that are computationally complex to process should be stored close to network elements with high computational power
The vulnerability of the signal to the effects of fading and interference	Time-varying characteristics of the wireless channel leading to potential performance degradation within the higher layers	Careful adaptation of the protocol stack should be used at each layer to compensate for the variations at a particular layer, depending on the specific time scale of these variations

diversity of legacy infrastructure involved, the move towards full layerless design in one step would be immensely difficult [1]. Therefore, to begin the process, two adjacent layers could be merged into one layer, with the ultimate goal of complete layerless-ness.

10.5 Summary

In this chapter, we have discussed numerous studies of cross-layer solutions for improving the performance of wireless communication systems and protocol stacks in selected application areas. These works have shown that cross-layer designs allow for information sharing through the layer boundaries to improve network performance and reliability—increasing throughput, reducing latency, and minimizing bit error rate—by controlling the input to another layer [2]. Cross-layer designs are able to make the hidden information (e.g., CSI) in each layer visible to the other layers [2, 44]. Works related to the implementation of cross-layer interactions by various authors has been presented in this chapter, and it is high time that these individual efforts are put into perspective, and a more holistic view is taken [4]. Cross-layer design has substantial benefits, but it has its unique disadvantages as well. For example, the cross-layer interactions create dependencies among layers which affect not only the layer concerned, but the other layers as well [3]. A complete redesign of the operational networks and protocols will lead to high implementation costs [64]. Therefore, the cross-layer design must be carefully crafted, because once the seven-layer OSI structure is violated, the benefits of independent, layer-specific protocol design will disappear [4, 94, 95]. Furthermore, any protocol chosen in any single layer must be considered from the perspective of the effects on the overall system [3].

References

1. R. Prasad, A perspective of layerless architecture. Wirel. Pers. Commun. **44**(1), 95–100 (2008)
2. F. Foukalas, V. Gazis, N. Alonistioti, Cross-layer design proposals for wireless mobile networks: a survey and taxonomy. IEEE Commun. Surv. Tutor. **10**(1), 70–85 (2008)
3. J. Zuo, C. Dong, S.X. Ng, L.-L. Yang, L. Hanzo, Cross layer aided energy efficient routing design for ad hoc networks. IEEE Commun. Surv. Tutor. **17**(3), 1214–1238 (2015)
4. V. Srivastava, M. Motani, Cross-layer design: a survey and the road ahead. IEEE Commun. Mag. **43**(12), 112–119 (2005)
5. A. Maharshi, L. Tong, A. Swami, Cross-layer designs of multichannel reservation MAC under Rayleigh fading. IEEE Trans. Signal Process. **51**(8), 2054–2067 (2003)
6. D. Ye, Cross-layer optimization with MIPv6-based multiple mobile routers for cognitive networks. Wirel. Netw. **22**(1), 193–209 (2016)
7. X. Wang, H. Li, H. Lin, A new adaptive OFDM system with precoded cyclic prefix for dynamic cognitive radio communications. IEEE J. Sel. Areas Commun. **29**(2), 431–442 (2011)
8. D. Lopez-Perez, X. Chu, A.-V. Vasilakos, H. Claussen, Power minimization based resource allocation for interference mitigation in OFDMA femtocell networks. IEEE J. Sel. Areas Commun. **32**(2), 333–344 (2014)
9. M.A. Khan, H. Tembine, A.-V. Vasilakos, Game dynamics and cost of learning in heterogeneous 4G networks. IEEE J. Sel. Areas Commun. **30**(1), 198–213 (2012)

10. M. Andrews, V. Capdevielle, A. Feki, P. Gupta, in *Autonomous Spectrum Sharing for Mixed LTE Femto and Macro Cells Deployments*. Proceedings of IEEE INFOCOM Conference on Computer Communications Workshops (San Diego, CA, 2010), pp. 1–5
11. Y.-J. Chang, F.-T. Chien, C.-C. Jay Kuo, Cross-layer QoS analysis of opportunistic OFDM-TDMA and OFDMA networks. IEEE J. Sel. Areas Commun. **25**(4), 657–666 (2007)
12. T. Jiang, H. Wang, A.V. Vasilakos, QoE-driven channel allocation schemes for multimedia transmission of priority-based secondary users over cognitive radio networks. IEEE J. Sel. Areas Commun. **30**(7), 1215–1224 (2012)
13. C. Luo, F.R. Yu, H. Ji, V.C.M. Leung, Cross-layer design for TCP performance improvement in cognitive radio networks. IEEE Trans. Veh. Technol. **59**(5), 2485–2495 (2010)
14. V. Chakravarthy, X. Li, Z. Wu, M.A. Temple, F. Garber, R. Kannan, A. Vasilakos, Novel overlay/underlay cognitive radio waveforms using SD-SMSE framework to enhance spectrum efficiency-part i: theoretical framework and analysis in AWGN channel. IEEE Trans. Commun. **57**(12), 3794–3804 (2009)
15. S.M. Cheng, W.C. Ao, F.M. Tseng, K.C. Chen, Design and analysis of downlink spectrum sharing in two-tier cognitive femto networks. IEEE Trans. Veh. Technol. **61**(5), 2194–2207 (2012)
16. A.W. Min, K.-H. Kim, J.P. Singh, K.G. Shin, *Opportunistic spectrum access in mobile cognitive radios* (Shanghai, Proc. IEEE INFOCOM, 2011), pp. 2993–3001
17. S. Huang, X. Liu, Z. Ding, Optimal transmission strategies for dynamic spectrum access in cognitive radio networks. IEEE Trans. Mob. Comput. **8**(12), 1636–1648 (2009)
18. K. Chowdhury, T. Melodia, Platforms and test beds for experimental evaluation of cognitive ad hoc networks. IEEE Commun. Mag. **48**(9), 96–104 (2010)
19. J. Marinho, E. Monteiro, Cognitive radio: survey on communication protocols, spectrum decision issues, and future research directions. Wirel. Netw. **18**(2), 147–164 (2012)
20. A. Kliks, D. Triantafyllopoulou, L. De Nardis, O. Holland, L. Gavrilovska, A. Bantouna, Cross-layer analysis in cognitive radio—context identification and decision making aspects. IEEE Trans. Cogn. Commun. Netw. **1**(4), 450–463 (2015)
21. G. Miao, G.Y. Li, A. Swami, Decentralized optimization for multichannel random access. IEEE Trans. Commun. **57**(10), 3012–3023 (2009)
22. J. Jia, Q. Zhang, in *A Testbed Development Framework for Cognitive Radio Networks*. Proceedings of IEEE International conference on Communications (2009), pp. 4606–4610
23. L. Ding, T. Melodia, S.N. Batalama, J.D. Matyjas, M.J. Medley, Cross-layer routing and dynamic spectrum allocation in cognitive radio ad hoc networks. IEEE Trans. Veh. Technol. **59**(4), 1969–1979 (2010)
24. I. Akyildiz, W. Lee, K. Chowdhury, CRAHNs: cognitive radio ad hoc networks. Ad Hoc Netw. **7**(5), 810–836 (2009)
25. X. Wang, J. Garcia-Luna-Aceves, Collaborative routing, scheduling and frequency assignment for wireless ad hoc networks using spectrum-agile radios. Wirel. Netw. **17**(1), 167–181 (2011)
26. T. Issariyakul, L. Pillutla, V. Krishnamurthy, Tuning radio resource in an overlay cognitive radio network for TCP: greed isn't good. IEEE Commun. Mag. **47**(7), 57–63 (2009)
27. F. Yu, B. Sun, V. Krishnamurthy, S. Ali, Application layer QoS optimization for multimedia transmission over cognitive radio networks. Wirel. Netw. **17**(2), (2011)
28. Q. Zhao, L. Tong, A. Swami, Y. Chen, Decentralized cognitive MAC for opportunistic spectrum access in ad hoc networks: a POMDP framework. IEEE J. Sel. Areas Commun. **25** (3), 589–600 (2007)
29. H. Su, X. Zhang, Cross-layer based opportunistic MAC protocols for QoS provisionings over cognitive radio wireless networks. IEEE J. Sel. Areas Commun. **26**(1), 118–129 (2008)
30. D. Singhal, R.M. Garimella, Cognitive cross-layer multipath probabilistic routing for cognitive Networks, multi-hop routing with cooperative transmission: a cross-layer approach. Wirel. Netw. **21**(4), 1181–1192 (2015)
31. S. Abdulhadi, M. Jaseemuddin, A. Anpalagan, Multi-hop routing with cooperative transmission: a cross-layer approach. Wirel. Netw. **20**(7), 1685–1697 (2014)

32. F. Yu, V. Krishnamurthy, V.C.M. Leung, Cross-layer optimal connection admission control for variable bit rate multimedia traffic in packet wireless CDMA networks. IEEE Trans. Signal Process. **54**(2), 542–555 (2006)
33. J. Yao, D.T.C. Wong, Y.H. Chew, Capacity balancing between the reverse and forward links in multiservice CDMA cellular networks with cross-Layer design. IEEE Trans. Veh. Technol. **55**(4), 1397–1411 (2006)
34. Y.H. Han, H. Jang, J. Choi, B. Park, J. Mcnair, A cross-layering design for IPv6 fast handover support in an IEEE 802.16e wireless MAN. IEEE Netw. **21**(6), 54–62 (2007)
35. J. Villalon, P. Cuenca, L. Orozco-Barbosa, Y. Seok, T. Turletti, Cross-layer architecture for adaptive video multicast streaming over multirate wireless LANs. IEEE J. Sel. Areas Commun. **25**(4), 699–711 (2007)
36. D. Triantafyllopoulou, N. Passas, E. Zervas, L. Merakos, Analysis and optimization of a cross-layer adaptation mechanism for real-time applications in wireless networks. Wirel. Netw. **17**(5), 1339–1354 (2011)
37. H. Luo, S. Ci, D. Wu, A cross-layer design for the performance improvement of real-time video transmission of secondary users over cognitive radio networks. IEEE Trans. Circ. Syst. Video Technol. **21**(8), 1040–1048 (2011)
38. E. Setton, T. Yoo, X. Zhu, A. Goldsmith, B. Girod, Cross-layer design of ad hoc networks for real-time video streaming. IEEE Wirel. Commun. **12**(4), 59–65 (2005)
39. B. Fu, Y. Xiao, H. Deng, H. Zeng, A survey of cross-layer designs in wireless networks. IEEE Commun. Surv. Tutor. **16**(1), 110–126 (2014)
40. L. Angrisani, M. Vadursi, Cross-layer measurements for a comprehensive characterization of wireless networks in the presence of interference. IEEE Trans. Instrum. Meas. **56**(4), 1148–1156 (2007)
41. M. Shariat, A.U. Quddus, S.A. Ghorashi, R. Tafazolli, Scheduling as an important cross-layer operation for emerging broadband wireless systems. IEEE Commun. Surv. Tutor. **11**(2), 74–86 (2009)
42. G. Miao, Y. Li, A. Swami, Channel-aware distributed medium access control. IEEE/ACM Trans. Netw. **20**(4), 1290–1303 (2012)
43. G. Carneiro, J. Ruela, M. Ricardo, Cross-layer design in 4G wireless terminals. IEEE Wirel. Commun. **11**(2), 7–13 (2004)
44. H. Zeng, K. Kwak, J. Deng, B. Fu, Y. Xiao, J. Jeski, in *Proactive and Adaptive Reconfiguration for Reliable Communication in Tactical Networks*. Proceedings of SPIE 8405, Defense Transformation and Net-Centric Systems (Baltimore, Maryland, USA, 2012), 84050D
45. C.-C. Tseng, K.-C. Chen, Layerless design of a power-efficient clustering algorithm for wireless ad hoc networks under fading. Wirel. Pers. Commun. **44**(1), 3–26 (2008)
46. G.A. Shah, V.C. Gungor, O.B. Akan, A cross-layer QoS-aware communication framework in cognitive radio sensor networks for smart grid applications. IEEE Trans. Ind. Inf. **9**(3), 1477–1485 (2013)
47. Q. Liu, X. Wang, G.B. Giannakis, A cross-layer scheduling algorithm with QoS support in wireless networks. IEEE Trans. Veh. Technol. **55**(3), 839–847 (2006)
48. W.L. Huang, K.B. Letaief, Cross-layer scheduling and power control combined with adaptive modulation for wireless ad hoc networks. IEEE Trans. Commun. **55**(4), 728–739 (2007)
49. Q. Zhang, Y.-Q. Zhang, Cross-layer design for QoS support in multihop wireless networks. Proc. IEEE **96**(1), 64–76 (2008)
50. J. Oh, C.W. Chen, A cross-layer approach to multichannel MAC protocol design for video streaming over wireless Ad Hoc networks. IEEE Trans. Multimedia **11**(6), 1052–1061 (2009)
51. S. Chu, X. Wang, Opportunistic and cooperative spatial multiplexing in MIMO ad hoc networks. IEEE/ACM Trans. Netw. **18**(5), 1610–1623 (2010)
52. A. Ghosh, W. Hamouda, Cross-layer antenna selection and channel allocation for MIMO cognitive radios. IEEE Trans. Wirel. Commun. **10**(11), 3666–3674 (2011)
53. M. Mardani, S.-J. Kim, G.B. Giannakis, Cross-layer design of wireless multihop random access networks. IEEE Trans. Signal Process. **60**(5), 2562–2574 (2012)

54. M. Uddin, C. Rosenberg, W. Zhuang, P. Mitran, A. Girard, Joint routing and medium access control in fixed random access wireless multihop networks. IEEE/ACM Trans. Netw. **22**(1), 80–93 (2014)
55. F. Tang, L. Barolli, J. Li, A joint design for distributed stable routing and channel assignment over multihop and multiflow mobile ad hoc cognitive networks. IEEE Trans. Ind. Informat. **10** (2), 1606–1615 (2014)
56. M.D. Francesco, G. Anastasi, M. Conti, S.K. Das, V. Neri, Reliability and energy-efficiency in IEEE 802.15.4/ZigBee sensor networks: an adaptive and cross-layer approach. IEEE J. Sel. Areas Commun. **29**(8), 1508–1524 (2011)
57. D. Qiao, S. Choi, A. Jain, K.G. Shin, in *MiSer: An Optimal Low-Energy Transmission Strategy for IEEE 802.11a/h*. Proceedings of International Conference on Mobile Computing and Networking (2003), pp. 161–175
58. E. Uysal-Biyikoglu, A. El Gamal, On adaptive transmission for energy efficiency in wireless data networks. IEEE Trans. Inf. Theory **50**(12), 3081–3094 (2004)
59. S. Cui, A. Goldsmith, A. Bahai, Energy-efficiency of MIMO and cooperative MIMO techniques in sensor networks. IEEE J. Sel. Areas Commun. **22**(6), 1089–1098 (2004)
60. J. Liu, L. Zhong, in *Micro Power Management of Active 802.11 Interfaces*. Proceedings of ACM/USENIX International Conference on Mobile Systems, Applications and Services (2008), pp. 146–159
61. R. Kravets, P. Krishnan, in *Power Management Techniques for Mobile Communication*. Proceedings of International Conference on Mobile Computing and Networking (1998), pp. 157–168
62. T. Simunic, L. Benini, P. Glynn, G.D. Micheli, in *Dynamic Power Management for Portable Systems*. Proceedings of International Conference on Mobile Computing and Networking (2000), pp. 11–19
63. S. Chandra, A. Vahdat, in *Application-Specific Network Management for Energy-Aware Streaming of Popular Multimedia Formats*. Proceedings of USENIX Annual Technical Conference (2002), pp. 329–342
64. F. Fu, M.V.D. Schaar, A new systematic framework for autonomous cross-layer optimization. IEEE Trans. Veh. Technol. **58**(4), 1887–1903 (2009)
65. R. Jurdak, *Wireless Ad Hoc and Sensor Networks: A Cross-layer Design Perspective* (Springer, New York, 2010)
66. M. Conti, G. Maselli, G. Turi, S. Giordano, Cross-layering in mobile ad hoc network design. Computer **37**(2), 48–51 (2004)
67. R. Winter, J.H. Schiller, N. Nikaein, C. Bonnet, Crosstalk: cross-layer decision support based on global knowledge. IEEE Commun. Mag. **44**(1), 93–99 (2006)
68. S. Efazati, P. Azmi, Cross-layer power allocation for selection relaying and incremental relaying protocols over single relay networks. IEEE Trans. Wirel. Commun. **15**(7), 4598–4610 (2016)
69. K. Zheng, Optimization and Learning in Energy Efficient Cognitive Radio System, Ph.D. Thesis, University of Tennessee, 2012
70. X. Zhang, H. Wang, A. Khokhar, in *An Energy-Efficient MAC Layer Transmission Algorithm Considering Fading Channels for Cluster-Based Sensor Networks*. Proceedings of IEEE Global Telecommunications Conference (GLOBECOM) (2008), pp. 1–5
71. P. Soni, A. Chockalingam, Performance analysis of UDP with energy efficient link layer on Markov fading channels. IEEE Trans. Wirel. Commun. **1**(4), 769–780 (2002)
72. D. Cavalcanti, R. Schmitt, A. Soomro, in *Achieving Energy Efficiency and QoS for Low-Rate Applications with 802.11e*. Proceedings of IEEE Wireless Communications and Networking Conference (2007), pp. 2143–2148
73. D. Chan, T. Berger, R. Bridgelall, in *Energy Efficiency of CSMA Protocols for Wireless Packet Switched Networks*. Proceedings of Wireless Communications and Networking Conference (2004), pp. 447–452

74. H. Choi, S. Park, in *Analysis of Energy Efficiency for Low Power Transmission in WLAN with MIMO-OFDM and Block Ack Mechanism*. Proceedings of Advanced Communication Technology (Phoenix Park, 2006), pp. 3, 1464
75. S. Ci, H. Sharif, D. Peng, in *An Effective Scheme for Energy Efficiency in Mobile Wireless Sensor Networks*. Proceedings of IEEE International Conference in Communications, vol. 6 (2004), pp. 3486–3490
76. N. Mouzehkesh, N. Noordin, M. Rasid, in *Maximizing the Energy Efficiency in MS-MAC Using a Dynamic Approach*. Proceedings 2nd Malaysia Conference on Photonics Telecommunication Technologies (2008), pp. 284–288
77. X. Wang, J. Yin, D. Agrawal, in *Effects of Contention Window and Packet Size on the Energy Efficiency of Wireless Local Area Network*. Proceedings of IEEE Wireless Communications and Networking Conference (2005), pp. 94–99
78. C. Alippi, G. Vanini, in *Application-Based Routing Optimization in Static/Semi-Static Wireless Sensor Networks*. Proceedings of IEEE International Conference on Pervasive Computing and Communications (Pisa, 2006), p. 51
79. A. Bernardos, P. Tarrio, J. Casar, in *An Energy Aware Routing Algorithm for Ad Hoc and Sensor Networks: Concept and Performance*. Proceedings of IEEE International Symposium on Personal, Indoor and Mobile Radio Communications (2006), pp. 1–5
80. S. Jung, N. Hundewale, A. Zelikovsky, in *Energy Efficiency of Load Balancing in MANET Routing Protocols*. Software Engineering, Artificial Intelligence, Networking and Parallel/Distributed Computing, 2005 and First ACIS International Workshop on Self-Assembling Wireless Networks (2005), pp. 476–483
81. D. Rajan, Power efficient delay allocation in multihop wireless networks. IEEE Trans. Veh. Technol. **56**(4), 1813–1825 (2007)
82. G. Varaprasad, in *New power aware routing algorithm for MANETs using gateway node*. Proceedings of International Conference on Parallel and Distributed Systems, vol. 2 (2007), pp. 1–4
83. L. Zhang, B.-H. Soong, Energy efficiency analysis of channel aware geographic-informed forwarding (CAGIF) for wireless sensor networks. IEEE Trans. Wirel. Commun. **7**(6), 2033–2038 (2008)
84. S.M. Betz, H.V. Poor, Energy efficient communications in CDMA networks: a game theoretic analysis considering operating costs. IEEE Trans. Signal Process. **56**(10), 5181–5190 (2008)
85. A. Ghasemi, K. Faez, Power-aware MAC for multihop wireless networks: a cross layer approach. IEEE Trans. Wirel. Commun. **7**(10), 3917–3929 (2008)
86. W.-K. Kuo, Energy efficiency modelling for IEEE 802.11a distribution coordination function system without finite retry limits. IET Commun. **1**(2), 165–172 (2007)
87. M.S. Masurkar, G.M. Asutkar, K.D. Kulat, in *Lifetime Maximization in Wireless Sensor Network Using Cross Layer Design: A Design Review*. Proceedings of International Conference on Emerging Trends in Engineering and Technology (Nagpur, Maharashtra, 2008), pp. 234–237
88. W. Wang, D. Peng, H. Wang, H. Sharif, H.-H. Chen, in *Taming Underlying Design for Energy Efficient Distributed Source Coding in Multirate Wireless Sensor Network*. Proceedings of IEEE Vehicular Technology Conference (2007), pp. 124–129
89. X. Lu, H. Wang, Y. Sun, in *Energy Efficient Cross Layer Design for Broadcast in Ad Hoc Wireless Networks*. Proceedings of International Conference on Wireless Communications, Networking and Mobile Computing (Shanghai, 2007), pp. 1604–1607
90. Y. Yu, in *A Cross Layer Approach for Energy Efficient MAC Layer in Wireless Sensor Networks*. Proceedings of International Conference on Wireless Communications, Networking and Mobile Computing (2008), pp. 1–3
91. Y. Zhong, Q. Yang, Z. Bai, K.S. Kwak, D. Yuan, H.H. Park, in *A Novel Cross Layer NFCAC Game Algorithm in IEEE 802.16-based Multi-Radio Heterogeneous Ad Hoc Networks*. Proceedings of International Conference on Advanced Communication Technology (Gangwon-Do, 2008), pp. 2100–2103

92. M. van Der Schaar, N. Sai Shankar, Cross-layer wireless multimedia transmission: challenges, principles, and new paradigms. IEEE Wirel. Commun. **12**(4), 50–58 (2005)
93. J. Saarnio, R. Aguiar, I.V. Kumar, Layereless communications: from dream to reality? Wirel. Pers. Commun. **44**(1), 51–55 (2008)
94. V. Kawadia, P.R. Kumar, A cautionary perspective on cross-layer design. IEEE Wirel. Commun. **12**(1), 3–11 (2005)
95. S. Ohteru, K. Kishine, H. Ichino, MAC protocol based on cross-layer design methodology for fast link in wireless communication systems. IEICE Electron. Express **4**(19), 593–599 (2007)

Index

S. Pandit and G. Singh, *Spectrum Sharing in Cognitive Radio Networks*,
DOI 10.1007/978-3-319-53147-2

FSC
www.fsc.org
MIX
Papier aus verantwortungsvollen Quellen
Paper from responsible sources
FSC® C105338

If you have any concerns about our products, you can contact us on
ProductSafety@springernature.com

In case Publisher is established outside the EU, the EU authorized representative is:
Springer Nature Customer Service Center GmbH
Europaplatz 3, 69115 Heidelberg, Germany

Printed by Libri Plureos GmbH
in Hamburg, Germany